BIO-INSPIRED COMPUTING FOR IMAGE AND VIDEO PROCESSING

BIO-INSPIRED COMPUTING FOR IMAGE AND VIDEO PROCESSING

D. P. ACHARJYA
V. SANTHI

CRC Press
Taylor & Francis Group
Boca Raton London New York

CRC Press is an imprint of the
Taylor & Francis Group, an **informa** business

A CHAPMAN & HALL BOOK

CRC Press
Taylor & Francis Group
6000 Broken Sound Parkway NW, Suite 300
Boca Raton, FL 33487-2742

© 2018 by Taylor & Francis Group, LLC
CRC Press is an imprint of Taylor & Francis Group, an Informa business

No claim to original U.S. Government works

Printed on acid-free paper
Version Date: 20171120

International Standard Book Number-13: 978-1-4987-6592-3 (Hardback)

This book contains information obtained from authentic and highly regarded sources. Reasonable efforts have been made to publish reliable data and information, but the author and publisher cannot assume responsibility for the validity of all materials or the consequences of their use. The authors and publishers have attempted to trace the copyright holders of all material reproduced in this publication and apologize to copyright holders if permission to publish in this form has not been obtained. If any copyright material has not been acknowledged please write and let us know so we may rectify in any future reprint.

Except as permitted under U.S. Copyright Law, no part of this book may be reprinted, reproduced, transmitted, or utilized in any form by any electronic, mechanical, or other means, now known or hereafter invented, including photocopying, microfilming, and recording, or in any information storage or retrieval system, without written permission from the publishers.

For permission to photocopy or use material electronically from this work, please access www.copyright.com (http://www.copyright.com/) or contact the Copyright Clearance Center, Inc. (CCC), 222 Rosewood Drive, Danvers, MA 01923, 978-750-8400. CCC is a not-for-profit organization that provides licenses and registration for a variety of users. For organizations that have been granted a photocopy license by the CCC, a separate system of payment has been arranged.

Trademark Notice: Product or corporate names may be trademarks or registered trademarks, and are used only for identification and explanation without intent to infringe.

Library of Congress Cataloging-in-Publication Data

Names: Acharjya, D. P., 1969- editor. | Santhi, V., 1971- editor.
Title: Bio-inspired computing for image and video processing / [edited by] D.P. Acharjya and V. Santhi.
Description: Boca Raton : CRC Press, [2017] | Includes bibliographical references and index.
Identifiers: LCCN 2017022417| ISBN 9781498765923 (hardback : acid-free paper) | ISBN 9781315153797 (ebook)
Subjects: LCSH: Natural computation. | Image processing--Mathematical models. | Image analysis--Mathematical models.
Classification: LCC QA76.9.N37 B556 2017 | DDC 006.3/8--dc23
LC record available at https://lccn.loc.gov/2017022417

Visit the Taylor & Francis Web site at
http://www.taylorandfrancis.com

and the CRC Press Web site at
http://www.crcpress.com

Dedicated to my beloved mother, Pramodabala Acharjya
D.P. Acharjya

Dedicated to my beloved parents
V. Santhi

Contents

List of Figures   ix

List of Tables   xvii

Preface   xxi

Acknowledgments   xxvii

Editors   xxix

Contributors   xxxi

I   Bio-Inspired Computing Models and Algorithms   1

1   Genetic Algorithm and BFOA-Based Iris and Palmprint Multimodal Biometric Digital Watermarking Models   3
S. Anu H. Nair and P. Aruna

2   Multilevel Thresholding for Image Segmentation Using Cricket Chirping Algorithm   31
S. Siva Sathya and Jonti Deuri

3   Algorithms for Drawing Graphics Primitives on a Honeycomb Model-Inspired Grid   59
M. Prabukumar

4   Electrical Impedance Tomography Using Evolutionary Computing: A Review   93
Wellington Pinheiro dos Santos, Ricardo Emmanuel de Souza, Reiga Ramalho Ribeiro, Allan Rivalles Souza Feitosa, Valter Augusto de Freitas Barbosa, Victor Luiz Bezerra Arajo da Silva, David Edson Ribeiro, and Rafaela Covello de Freitas

II   Bio-Inspired Optimization Techniques   129

5 An Optimized False Positive Free Video Watermarking System in Dual Transform Domain 131
   L. Agilandeeswari and K. Ganesan

6 Bone Tissue Segmentation Using Spiral Optimization and Gaussian Thresholding 161
   Hugo Aguirre-Ramos, Juan-Gabriel Avina-Cervantes, and Ivan Cruz-Aceves

7 Digital Image Segmentation Using Computational Intelligence Approaches 205
   S. Vijayakumar and V. Santhi

8 Digital Color Image Watermarking Using DWT SVD Cuckoo Search Optimization 227
   S. Ganesh Babu and B. Sarojini Ilango

9 Digital Image Watermarking Scheme in Transform Domain Using the Particle Swarm Optimization Technique 245
   Sarthak Nandi and V. Santhi

III Bio-Inspired Computing Applications to Image and Video Processing 265

10 Evolutionary Algorithms for the Efficient Design of Multiplier-Less Image Filter 267
   Abhijit Chandra

11 Fusion of Texture and Shape-Based Statistical Features for MRI Image Retrieval System 297
   N. Kumaran and R. Bhavani

12 Singular Value Decomposition–Principal Component Analysis-Based Object Recognition Approach 323
   Chiranji Lal Chowdhary and D.P. Acharjya

13 The KD-ORS Tree: An Efficient Indexing Technique for Content-Based Image Retrieval 343
   N. Puviarasan and R. Bhavani

14 An Efficient Image Compression Algorithm Based on the Integration of a Histogram Indexed Dictionary and the Huffman Encoding for Medical Images 369
   D.J. Ashpin Pabi, P. Aruna, and N. Puviarasan

Index 395

List of Figures

1.1	Flow diagram of the GA and proposed BFOA watermarking system	7
1.2	Conversion of polar to rectangular form of iris	8
1.3	Iris modality extraction	9
1.4	Palmprint modality extraction	10
1.5	Sparse representation method	18
1.6	Output of different fusion methods, such as average, maximum, minimum, IHS, and PCA	19
1.7	Output of different fusion methods, such as Laplacian pyramid, gradient pyramid, DWT, SWT, and sparse representation	20
1.8	Sample output obtained by applying GA watermarking system	23
1.9	Sample output obtained by applying BFOA watermarking system	26
1.10	Performance of BFOA watermarking model vs. other watermarking models in the literature	28
2.1	Resultant images after applying the CCA to the set of benchmark images (Cameraman and Zebra) using Kapur's function	44
2.2	Resultant images after applying the CCA to the set of benchmark images (Sea Fish and Boat Man) using Kapur's function	44
2.3	Resultant images after applying the CCA to the set of benchmark images (Ostrich and Boat) using Kapur's function . .	45
2.4	Resultant images after applying the CCA to the set of benchmark images (Tree and Snake) using Kapur's function . . .	45
2.5	Resultant images after applying the CCA to the set of benchmark images (Cameraman and Zebra) using Otsu's function	48
2.6	Resultant images after applying the CCA to the set of benchmark images (Sea Star and Boat Man) using Otsu's function	49
2.7	Resultant images after applying the CCA to the set of benchmark images (Ostrich and Boat) using Otsu's function . . .	49
2.8	Resultant images after applying the CCA to the set of benchmark images (Tree and Snake) using Otsu's function	50
3.1	Line on hexagonal grid	64

List of Figures

3.2	Circle on hexagonal grid	65
3.3	Structure of a real hexagonal pixel	68
3.4	Different hexagonal structures presented using square grids	69
3.5	The selected hexagonal pixel structure	69
3.6	Hexagonal grids constructed by the selected hexagonal pixels	69
3.7	Modified hexagonal pixel structure	70
3.8	Six neighbor pixels using modified hexagonal pixel structure	70
3.9	The hexgonal grids after modification	71
3.10	Coordinate system in hexagonal grids	71
3.11	The actual coordinate system on the monitor (gray) and the applied coordinate system (dark gray)	72
3.12	Different initial positions in various columns	72
3.13	Distance between a pixel and its above pixel	73
3.14	Distance between a pixel and its right bottom pixel	73
3.15	The sampling of a curve in square grids	74
3.16	Approach to calculating the deviation in square sampling	74
3.17	Approach to calculating the deviation in hexagonal grids	75
3.18	Scan-converted lines of angle from 0 to 350 degrees	77
3.19	Scan-converted circles of radii R = 1, 5, 10, 15, 20, 25, and 30	80
3.20	Scan-converted ellipses of a = 10 to 25 incremented by 5, b = 30	81
3.21	Scan-converted character	84
3.22	Scan-converted character	84
3.23	Scan-converted character	84
3.24	Scan-converted character	84
3.25	Scan-converted character	85
3.26	Scan-converted character	85
3.27	Scan-converted character	85
3.28	Scan-converted character	85
3.29	Scan-converted character	86
3.30	Scan-converted character	86
3.31	Scan-converted character	86
3.32	Scan-converted character	86
3.33	Scan-converted character	87
4.1	EIT schematization	97
4.2	Mesh of triangular finite elements generated by EIDORS	98
4.3	Ground-truth images used as in the experiments	99
4.4	Genetic algorithm scheme	102
4.5	Results using GA for an object placed in the center (a1, b1, and a3), between the center and the border (b1, b2, and b3), and near the border (c1, c2, and c3) of the circular domain for 50, 300, and 500 iterations	104
4.6	Relative error along the iterations for GA	105
4.7	Temperature during the cooling time	106

4.8	Results of using SA for an object placed in the center, between the center and the border, and near the border of the circular domain	109
4.9	Relative error along the iterations for SA	109
4.10	Birds flock flight (adapted from: media.salon.com)	110
4.11	Flowchart of PSO algorithm	111
4.12	Results of using PSO for an object placed in the center, between the center and the border, and in the border of the circular domain for 50, 300, and 500 iterations	113
4.13	Relative error along the objective function calculation number for an object placed at the border, between the center and the border, and in the border	113
4.14	Differential evolution algorithm's scheme	114
4.15	Results using DE for an object placed in the center (a1, b1, and a3), between the center and the border (b1, b2, and b3), and in the border (c1, c2, and c3) of the circular domain for 50, 300, and 500 iterations	118
4.16	Relative error along iterations to an isolated object placed at the center, between the center and the border, and at the border, for DE, 5th version	118
4.17	Results using FSS for an object placed in the center (a1, b1, and a3), in the border (b1, b2, and b3), and between the center and the border (c1, c2, and c3) of the circular domain for 50, 300, and 500 iterations	122
4.18	Error decreasing according to the number of calculations of the fitness functions, considering an object placed in the center, in the border, and between the center and the border of the domain	122
5.1	Schematic diagram of contourlet transform	139
5.2	CT on mountain.avi	139
5.3	Proposed nature-inspired embedding algorithm	141
5.4	Proposed firefly-based extraction algorithm	144
5.5	Firefly-based optimal embedding strength selection	145
5.6	Video frame covers	148
5.7	Original watermark image	148
5.8	Watermarked video frames	150
5.9	Extracted watermark from watermarked video	150
5.10	Imperceptibility and robustness measure for various spatial attacks on watermarked videos and extracted watermark . .	152
5.11	Imperceptibility and robustness measure for various temporal attacks on watermarked videos and extracted watermark VITlogo	154
5.12	Spatial attacks vs. PSNR and NCC	155
5.13	Temporal attacks vs. NCC	156

List of Figures

6.1 Challenges in bone segmentation over CT images (top), its gold standard (center), and gray level histogram (bottom) for: (a) non-homogenous structure, (b) reduced size, (c) radiological similarity, (d) closeness of other bones or organs 163

6.2 Spirals in nature: (a) spiral galaxy, (b) aloe polyphylla, and (c) hurricane 165

6.3 Trajectory of $X^{(3)} = (10, 0, 0)$ after multiple rotations by $M^{(3)}(\theta_{12}, \theta_{13}, \theta_{23})$ matrix for: (a) $M^{(3)}(0, \pi/18, 0)$, (b) $M^{(3)}(\pi/18, \pi/18, 0)$, and (c) $M^{(3)}(\pi/18, \pi/18, \pi/18)$ 166

6.4 Trajectories for a spiral with $\theta = \pi/4$ and an initial point $X_1^{(2)}(0) = (75, 75)$ for: (a) $r = 0.5$, (b) $r = 0.92$, and (c) $r = 0.98$ 168

6.5 Trajectories for a search point with $\rho = 0.95$ from initial point $X^{(2)}(0) = (75, 75)$ for: (a) $\theta = \pi/6$, (b) $\theta = 2\pi/3$, and (c) $\theta = 5\pi/6$ 168

6.6 Sphere function assessment for $s = 5$ and $s = 10$ with $S(0.95, \pi/4)$ after 1000 iterations 169

6.7 Different SO search paths using two spirals $S(0.95, \pi/6)$ after 100 iterations 170

6.8 Sphere function $f_{Obj}^2(x) = x_1^2 + x_2^2$ for $x_i \in [-100, 100]$: (a) isometric view, (b) lateral view, and (c) level plot 172

6.9 Sphere function evaluation for $s = 6$, $r = 0.95$, and $\theta = \{\pi/6, \pi/4, \pi/2\}$, after 700 iterations 172

6.10 Sphere function assessment for $s = 6$, $\theta = \pi/4$, and $r = \{0.90, 0.95, 0.99\}$ after 1000 iterations 173

6.11 Sphere function evaluation for different spiral numbers $s = \{2, 3, 5, 6, 10\}$ with $S(0.95, \pi/6)$ after 1000 iterations 175

6.12 Spiral and optimal paths for sphere function evaluation with $s = 2$, $S_2(0.95, \pi/6)$, for two different random variables: (a) $\varphi_1 = U(0, 1)$ and (b) $\varphi_2 = N(0.5, 0.25)$, after 100 iterations 177

6.13 Sphere function evaluation after one run with an SO model with $S(0.95, \pi/6)$, for $\varphi = \{1, U(0, 1), N(0.5, 0.2)\}$ with $s = \{10, 2, 2\}$, respectively, after 1000 iterations 177

6.14 Details of a cervical vertebrae from a CT image 181

6.15 Real part of a Gabor filter with $\tau = 4$, $\lambda = 6$, and $\alpha = 0$, for different window size: (a) 21×21, 61×61, and (c) 121×121, and their respective effects over a medical image (d–f) ... 182

6.16 Magnitude of the (a) real, (b) imaginary, and (c) complex kernel for a Gabor filter, with $\tau = 50$, $\lambda = 1.5$, $\alpha = 0$, and $\rho = 20$, and their respective effects over a medical image .. 182

6.17 Real part of a Gabor filter with $\tau = 50$, $\alpha = \pi/4$, and $\rho = 25$ for three different aspect ratios: (a) $\lambda_1 = 1$, (b) $\lambda_2 = 4$, and (c) $\lambda_3 = 6$, and their respective effects over a medical image 184

List of Figures

6.18 Real part of a Gabor filter with $\lambda = 6$, $\alpha = \pi/4$, and $\rho = 50$, for three different sine periods: (a) $\tau_1 = 2$, (b) $\tau_2 = 4$, and (c) $\tau_3 = 8$, and their respective effects over a medical image . . 185

6.19 Real part of a Gabor filter with $\tau = 3$, $\lambda = 8$, and $\rho = 30$, for three different rotation angles: (a) $\alpha_1 = -\pi/4$, (b) $\alpha_2 = 0$, and (c) $\alpha_3 = \pi/4$, and its effects over a medical image . . . 186

6.20 The proposed method: A threshold U is iteratively set, based on the intersection of two Gaussian models generated from an SO-seeded grown region over the Gabor filter response of an input image, until the models converge 187

6.21 Example of the masks used on image preprocessing: (a) DR mask, and (b) PM mask 187

6.22 Real Gabor filter kernels used, in the GFB defined with $\tau = 8$ and $\lambda = 1.5$, and oriented at different angles, $\alpha_m \in (-\pi/2, \pi/2)$, with steps of $\alpha = \pi/30$ 188

6.23 (a) Single-point seed for different images in row (b) and their respective SBRG outputs: k_result (c) and $k\_enclosing$ . . . 190

6.24 Selection of threshold U as the intersection of the Gaussian models $N_p(\mu_p, \sigma_p)$ and $N_n(\mu_n, \sigma_n)$ 191

6.25 (a) Different stages of post-processing for images in column (b) result after threshold, (c) erode, (d) *in-filling*, and (e) border correction. 192

6.26 (a) Seed-based region growing results for the seed shown, (b) over an image without Gabor filtering, (c) and an image with Gabor filtering. 193

6.27 SOIGT results for image (a) considering: (b) random seeds, (c) seeds $= (x,y) > 0.5\max(F(x,y))$, (d) seeds $= (x,y) > 0.75\max(F(x,y))$, (e) seeds $= \max(F(x,y))$, and (f) seeds obtained by SO. 194

6.28 Bone pixels from image in column one. From column two to seven: gold standard, K_2, R_2, O_5 (best for O), KM_7 (best for KM), FCM_7 (best for FCM), and the $SOIGT$ 198

7.1 Shortest path identification by an ant to reach its destination 207
7.2 Representation of clusters using fuzzy c-mean algorithm . . 211
7.3 Representation of clusters using FCM algorithm 213
7.4 The most predominantly classified results among 30 trial runs 220
7.5 The most predominantly classified results among 30 trail runs 221
7.6 The most predominantly classified results among 30 trial runs 222
7.7 The most predominantly classified results among 30 trial runs 223
7.8 Error percentage comparison of various techniques over best, worst and average cases 224

8.1 The egg-laying radius (ELR) 232
8.2 Cuckoo optimization algorithm flowchart 233

List of Figures

8.3	Cuckoo's immigration	235
8.4	Watermarking system in DWT	237
8.5	DWT decomposition with two levels	238
8.6	Host image and watermark image before embedding	240
8.7	Host image after watermarking and the extracted watermark image	240
8.8	Histogram of host image and watermarked image R, G, and B channels	241
9.1	Watermark embedding process	251
9.2	Watermark extraction process	254
9.3	Original images	255
9.4	Original watermark	255
9.5	Watermarked images	255
10.1	Schematic diagram of differential evolution algorithm	272
10.2	Flow chart for the design of the SORIGAML2D filter	276
10.3	Flow chart for the design of the DEML2D filter	277
10.4	2D view of frequency response corresponding to (a) mean, (b) circular averaging, (c) Gaussian, (d) DEML2D, (e) SORIGAML2D filter of size 3×3	278
10.5	2D view of frequency response for the filters in (a) Sriranganathan, (b) Siohan, (c) Lee, (d) Chen, (e) DEML2D5X5, (f) DEML2D7X7, (g) SORIGAML2D5X5, (h) SORIGAML2D7X7	279
10.6	3D view of frequency response corresponding to (a) mean, (b) circular averaging, (c) Gaussian, (d) DEML2D, (e) SORIGAML2D filter of size 3×3	280
10.7	3D view of frequency response for the filters in (a) Sriranganathan, (b) Siohan, (c) Lee, (d) Chen, (e) $DEML2D_{5\times 5}$ (f) $DEML2D_{7\times 7}$	281
10.8	3D view of frequency response for the filters in (g) Sriranganathan and (h) Siohan	282
10.9	(a) Original, (b) noisy (Gaussian noise with $\sigma = 0.25$), (c) mean-filtered (d) circular-filtered Lena image	283
10.10	(e) Gaussian-filtered, (f) DEML2D-filtered, (g) SORIGAML2D-filtered Lena image	283
10.11	(a) Original (b) noisy (Gaussian noise with $\sigma = 0.25$) Moon image	284
10.12	(c) Mean-filtered, (d) circular-filtered, (e) Gaussian-filtered, (f) DEML2D-filtered, (g) SORIGAML2D-filtered Moon image	284
10.13	(a) Original, (b) noisy (Gaussian noise with $\sigma = 0.25$), (c) mean-filtered, (d) circular-filtered, (e) Gaussian-filtered, (f) DEML2D-filtered, (g) SORIGAML2D-filtered MRI image .	285

List of Figures

10.14	(a) Original, (b) noisy (Gaussian noise with $\sigma = 0.25$), (c) mean-filtered, (d) circular-filtered, (e) Gaussian-filtered, (f) DEML2D-filtered, (g) SORIGAML2D-filtered Cameraman image	286
11.1	Block diagram of the proposed work	303
11.2	(3×3) pixel matrix	304
11.3	Example of texture unit transformation	304
11.4	GWT of an input image	307
11.5	(3×3) convolution kernels for Prewitt operator	309
11.6	Results of Prewitt operator-based edge detection	310
11.7	Results of region selection for Prewitt operator-based edges	310
11.8	K-NN illustration (k = 5)	314
11.9	Accuracy of proposed system	318
11.10	F-measure of proposed system	319
12.1	Typical object recognition system	325
12.2	Columbia object image library $(COIL - 100)$	330
12.3	Training set of an object sampled at every 50^0	331
12.4	Training set of an object sampled at every 25^0	332
12.5	Eigenspace manifolds for an object in the database with the position angle manifold being sampled at 50^0	332
12.6	Eigenspace manifolds for an object in the database with the position angle manifold being sampled at 25^0	333
12.7	Comparision between PCA, ICA, SVD-PCA and percentage recognition with n^{th} nearest matching	334
12.8	(a) Comparision between PCA, ICA, and PCA-SVD for position angle sampling of 50^0 (b) Comparison between PCA, ICA, and PCA-SVD for position angle sampling of 25^0	335
12.9	Comparison between PCA, ICA, and PCA-SVD for position angle sampling of 10^0	335
12.10	Results with position estimation	338
13.1	Block diagram of the proposed CBIR system	346
13.2	Contrast detection filter showing inner and outer square region	348
13.3	Sample input images from the COIL-100 database	349
13.4	Sample input images from the Oxford flower-102 database	349
13.5	Sample segmented images of the COIL-100 DB shown in Figure 13.3	350
13.6	Sample segmented images of the Oxford flower-102 database shown in Figure 13.4	350
13.7	Orientation vector visualization of the COIL-100 database using HOG	354

List of Figures

13.8	Orientation vector visualization of the Oxford flower-102 database using HOG	355
13.9	Magnitude vector visualization of the COIL-100 database using HOG	356
13.10	Magnitude vector visualization of the Oxford flower-102 database using HOG	357
13.11	Example for 2d-tree	358
13.12	Insertion of a new node $(0.0553, 0.0621)$ in the 2d-tree . . .	358
13.13	Query images and retrieval of images according to various algorithms with respect to COIL-100 database	363
13.14	Query images and retrieval of images according to various algorithms with respect to Oxford flower-102 database . . .	363
13.15	Average F-score values and accuracy values for image categories in the COIL-100 database	364
13.16	Average F-score values and accuracy values for image categories in the Oxford flower-102 database	364
14.1	Block diagram of general image storage system	371
14.2	The basic flow of image compression coding	372
14.3	The stages of the proposed HID system	374
14.4	The block diagram for the encoder of the proposed HID image compression	375
14.5	The block diagram for the decoder of the proposed HID image compression	375
14.6	2D-DCT	377
14.7	DCT basis functions	377
14.8	Zig-zag scan	378
14.9	Huffman encoding	380
14.10	Tested medical images	382
14.11	Quality metrics of the proposed compression technique over some other filtering methods	386
14.12	Comparison of PSNR over different blocks of images at different threshold values	386
14.13	Reconstructed images when the threshold TH = 10	387
14.14	Reconstructed images when the threshold TH = 80	387
14.15	Comparison of PSNR value and the bpp of the proposed reconstructed image for various image blocks	389
14.16	Results of tested images for the proposed HID method and the existing methods SVD, WDR, and JPEG	390
14.17	Comparison of PSNR value and the CR of the proposed reconstructed image over other existing methods	391

List of Tables

1.1	Performance analysis for different watermarked images using GA watermarking system.	21
1.2	Performance analysis for different watermarked images using GA watermarking system.	27
1.3	Performance analysis for different watermarked images using BFOA watermarking system.	27
2.1	Control parameters of CCA.	42
2.2	Best results after applying the CCA using Kapur's function to the set of benchmark images.	46
2.3	Best results of applying the CCA using Otsu's function to the set of benchmark images.	47
2.4	p-values produced by Wilcoxon's test comparing Otsu vs. Kapur over PSNR.	51
2.5	Comparison of CCA and MTEMO using Kapur's method.	52
2.6	Comparison of CCA with MTEMO using Otsu's method.	53
2.7	ANOVA test over the CCA and MTEMO based on Kapur's method.	54
2.8	ANOVA test over the CCA and MTEMO based on Otsu's method.	55
3.1	Deviation of lines in different angles in both square and hexagonal grids.	78
3.2	Deviation of circles in different radii in both square and hexagonal grids.	80
3.3	Deviation of ellipse in both square and hexagonal grids.	82
4.1	Typical values of tissue's electrical conductivity.	96
4.2	EIDORS parameter configurations.	98
4.3	Parameters used for genetic algorithms.	103
4.4	Parameters used to simulate annealing.	107
4.5	Mutation equations.	115
5.1	No attack vs. imperceptibility and robustness values.	151
5.2	Spatial attacks and their attack index.	151
5.3	Spatial attacks and their attack index.	153

List of Tables

6.1	Statistics for the sphere evaluation after 50 runs, using $s = 6$, $r = 0.95$, for different rotation angles $\theta = \{\pi/6, \pi/4, \pi/2\}$ and maximum iteration number $k_{max} = \{50, 100, 400, 700, 1000\}$.	173
6.2	Statistical analysis of the sphere function evaluation after 50 runs, using $s = 6$, $\theta = \pi/4$, for different convergence rates $r = \{0.20, 0.50, 0.90, 0.95, 0.99\}$ and maximum iteration numbers $k_{max} = \{50, 100, 400, 700, 1000\}$.	174
6.3	Statistics for the $f_{sphere}^{(2)}(x)$ assessment after 50 runs for a different combination of k_{max} and s, with $S(0.95, \theta = \pi/4)$.	176
6.4	Sphere function evaluation for different k_{max} and s values using the ϕ-model with $S(0.95, \pi/4)$, for the two random variables $\varphi_1 = U(0,1)$ and $\varphi_2 = N(0.5, 0.25)$, after 50 runs.	179
6.5	Comparative analysis of the proposed and the comparative methods, using statistical measures.	196
6.6	Comparative analysis of the proposed and the comparative methods, using statistical measures.	197
9.1	Calculated scaling and embedding parameters from test images.	256
9.2	Similarity measure between original and extracted watermark.	256
9.3	Extracted watermark's calculated PSNR and NCC values for gray scale test image House.	258
9.4	Extracted watermark calculated PSNR and NCC after various attacks from color test image Lena.	260
10.1	Comparison in terms of PSNR (DB) resulting from various filters at different noise intensities for Moon image.	287
10.2	Comparison in terms of SSIM resulting from various filters at different noise intensities for Moon image.	288
10.3	Comparison in terms of IEF resulting from various filters at different noise intensities for Moon image.	289
10.4	Comparison in terms of IQI resulting from various filters at different noise intensities for Moon image.	290
10.5	Comparison among various filters in terms of SSIM at different noise intensities for Lena image.	290
10.6	Comparison among various filters in terms of SSIM at different noise intensities for Moon image.	291
10.7	Comparison among various filters in terms of SSIM at different noise intensities for Peppers image.	291
11.1	Total number of optimum features selected using FGFS.	314
11.2	Precision rate of proposed system.	318
11.3	Recall rate of proposed system.	319

List of Tables

12.1	Number of correct recognitions by using PCA, ICAs and SVD-PCA in considering position angle sampling 50^0.	336
12.2	Number of (%) recognitions by using PCA, ICA, and SVD-PCA when position angle sampling is 50^0.	336
12.3	Number of correct recognitions by using PCA, ICA, and SVD-PCA when position angle sampling is 25^0.	336
12.4	Number of (%) recognitions by using PCA, ICA, and SVD-PCA when position angle sampling is 25^0.	336
12.5	Number of correct recognition by using PCA, ICA, and SVD-PCA when position angle sampling is 10^0.	336
12.6	Number of (%) recognitions by using PCA, ICA, and SVD-PCA when position angle sampling is 10^0.	336
12.7	Average success rate for COIL-100 objects database.	337
13.1	Components of the proposed approach.	351
13.2	Average precision and average recall for all indexing methods.	362
13.3	Computation time of different indexing methods in seconds.	365
14.1	Indexed dictionary.	380
14.2	Influence of threshold over compression, bits per pixel, and PSNR for a medical image, Brain image 1.	384
14.3	Influence of threshold over compression, bits per pixel, and PSNR for a standard Cameraman image.	385
14.4	The quality of the compressed image for various image blocks when applying different filtering models.	385
14.5	Objective quality metrics of the compressed image for various image blocks when the threshold TH = 10.	388
14.6	Objective quality metrics of the compressed image for various image blocks when the threshold TH = 30.	388
14.7	Objective quality metrics of the compressed image for various image blocks when the threshold TH = 80.	389
14.8	Quality metrics of different images for the proposed method.	390

Preface

These days a huge amount of data is being generated across the world in the form of images and video sequences. Due to the rapid development of computational facilities, the area of signal, image, and video processing has undergone remarkable changes for more than a decade. A revolution has taken place in terms of memory, computational speed, and the complexity of algorithms. This has inspired researchers throughout the globe to inculcate numerous promising concepts in the arena of image and video processing. One such computational model has been developed to make a justified correlation with a number of social or genetic behaviors of animals and other creatures. As a matter of fact, digital image and video processing are considered to be essential research areas in the current scenario. We are required to analyze these images and video signals, to extract meaningful information from them in order to solve many engineering and scientific problems. But analyzing and processing these data is a tedious task using classical algorithms and techniques, because real world data contains uncertainties. In order to address these challenges, the concept of bio-inspired computing could be used. It is an exciting and relatively recent field in which concepts from biology are used to design and implement new and improved computing methods to handle uncertainty. Many of these bio-computing techniques have been used successfully to find solutions in a wide range of areas, such as combinatorial optimization, classification and decision making, pattern recognition, machine learning, computer security, biometrics, time series prediction, data mining, image processing, and much more. Bio-inspired approaches include genetic algorithms, biodegradability prediction, cellular automata, and emergent systems such as ants, termites, bees, wasps, neural networks, artificial life, artificial immune systems, and bacterial colonies. The way in which bio-inspired computing differs from traditional artificial intelligence is in how it takes a more evolutionary approach to learn, as opposed to what could be described as "creationist" methods used in the traditional artificial approach. In addition, bio-inspired computing takes a more bottom-up, de-centralized approach; bio-inspired techniques often involve the method of specifying a set of simple rules, a set of simple organisms that adhere to those rules, and a method of iteratively applying those rules.

The multimodal biometric watermarking system using multiple sources of information has been widely recognized. However, computational models for multimodal biometrics recognition have only recently received attention. Chapter 1 presents a novel biometric watermarking system that uses biometric traits such as fingerprints, iris prints and palm prints. At the modality extraction level, the modalities are extracted from different biometric traits and then blended to produce a joint template, which is the basis for the watermarking system. Fusion at modality extraction level generates a homogeneous template for fingerprint, iris print, and palm print features. The fused templates are analyzed by using difference metrics. The obtained results, using the discrete wavelet transformation fused template, provides better results than other fused templates. The best fused template is applied as input, along with the cover image, to the different watermarking models using a genetic algorithm and a bacterial foraging optimization algorithm. The quality of the cover images is measured using peak signal-to-noise ratio, while normalized absolute error and normalized cross correlation are used for analyzing the robustness of the algorithm with various cover images.

One of the most important tasks in image processing is segmentation, which deals with the classification of pixels into two or more groups. This forms the basis for other complex image recognition and processing systems. In Chapter 2, a multi-level thresholding method for image segmentation is formulated using the cricket chirping algorithm, which is a heuristics algorithm that is inspired by the chirping behavior of crickets. In this method, random samples from a feasible search space depending on the image histogram are encoded as candidate solutions, and Otsu's and Kapur's methods are used to evaluate the quality. Guided by these objective values, the set of candidate solutions are evolved through the cricket chirping algorithm operators until an optimal solution is found. The performance of the proposed algorithm is tested in terms of accuracy, convergence, and speed, and compared with the other popular optimization techniques using prevailing parameters such as fitness function, peak signal-to-noise ratio, structural similarity index, and execution time.

Chapter 3 investigates the relative benefits of the lattices for representing lines and curves. The hexagonal simulator for rasterizing the basic output primitives such as lines, circles, and ellipses on square and hexagonal grids is introduced. The pros and cons of the hexagonal grid-based rendering algorithm over conventional square grid algorithms are demonstrated. It is observed that the hexagonal grid is less jagged on regions of higher curvature than the square grid, and the regions with low curvature are found to on hexagonal grid be more jagged than on square grids. This feature of hexagonal grids is reflected in the figures as well.

Electrical impedance tomography is an image technique based on the application of alternating electrical currents on electrodes placed around the region to be imaged and on the measurement of the resulting electrical potentials in the remaining electrodes. Therefore, the EIT images represent a map of the estimation of the electrical proprieties of the medium. EIT has the advantage of being non-invasive, free of ionizing radiation, and low cost compared to other image techniques, having applications in geophysics and in industrial, biological and medical areas. However, the EIT image reconstruction process is an inverse ill-posed and ill-conditioned problem that results in high computational cost and images with low spatial resolution. One way to reconstruct an EIT image is by treating this problem as an optimization problem that can be solved by methods of evolutionary computing. The aim of this approach is to minimize the root mean square relative error between the electrical potentials measured and simulated. Chapter 4 addresses the basic principles of EIT and image reconstruction as an optimization problem using genetic algorithms, simulated annealing, particle swarm optimization, and fish school search.

In Chapter 5, a novel optimization-based video watermarking technique is proposed to increase robustness and imperceptibility. It uses dual transforms, namely, Hilbert and Contourlet transform with a sub-band selection method using principal component analysis (PCA) for determining the suitable location for embedding. The optimal scaling parameters are calculated using a modified firefly optimization algorithm for embedding the watermark within the video frame covers in a hybrid domain. Since the embedding process is done in Hilbert space of video frame covers, the imperceptibility of the watermark is improved and it is also free from the false positive detection problem. The experimental results prove that the proposed method is more robust against different kinds of spatial attacks, such as noise addition, filtering attack, contrast adjustment, histogram equalization, row-column deletion, geometrical attack, JPEG, and video attacks, namely, frame dropping, frame averaging, frame swapping or exchange attacks, and MPEG compression when compared to the similar existing algorithms.

A novel strategy for automatic bone tissue segmentation using the spiral optimization (SO) strategy and a Gaussian modeled thresholding is presented in Chapter 6. In the segmentation stage, the SO strategy is used to initialize region-growing operators according to its intensity over the Gabor filter response, and two Gaussian models are created from its result. Based on the intersection of these models, a thresholding strategy is introduced to classify bone and non-bone tissue pixels. The proposed method has been compared with different state-of-the-art segmentation methods and evaluated with the accuracy measure and the dice coefficient.

Chapter 7 gives an overview of image segmentation based on the swarm intelligence (SI) approaches. The segmentation process could be carried out after the application of image pre-processing techniques for extracting attributes from the images. Normally, the image segmentation process subdivides an image into its constituent objects or regions. Swarm intelligence is one of the modern and incipient digital image segmentation approaches inspired by nature. In this chapter, the study of various swarm intelligence-based approaches used for segmentation is carried out. Finally, the performance of various swarm intelligence approaches in image segmentation has been compared and presented.

Nowadays, in every field there is enormous use of digital contents. Information handled on the Internet and multimedia network systems is in digital form. The copying of digital content without quality loss is not so difficult. Due to this, there are more chances to copy such digital information. So, there is a great need for prohibiting such illegal copying of digital media. Digital watermarking is the powerful solution to this problem. In Chapter 8, a new metaheuristic algorithm called cuckoo search for digital watermarking is presented. Digital watermarking is nothing but the technology in which a piece of digital information is embedded in digital content to protect it from illegal copying, and later extracted. In digital watermarking, a watermark is embedded into a cover image in such a way that the resulting watermarked signal is robust to certain distortion caused by either standard data processing in a friendly environment or malicious attacks in an unfriendly environment.

Chapter 9 proposes a novel invisible watermarking scheme to embed a digital watermark in an image. The embedding and scaling factors are adaptively calculated using the particle swarm optimization technique, while embedding and extraction are carried out using discrete wavelet transform and singular value decomposition. The proposed technique can be used against copyright infringement of images by inserting a digital watermark in an image. The robustness of this technique has been verified by making the images watermarked using this technique subject to various attacks that degrade the quality of the image. The results obtained show that the proposed technique is robust to various attacks and can be used effectively to tackle copyright infringement.

Evolutionary computational algorithms are considered to be one of the most important constituents of this bio-inspired computing tool, which establishes different mathematical models corresponding to the evolution of species over generations. These algorithms have already confirmed their supremacy over the traditional optimization techniques in the domain of signal and image processing. Chapter 10 makes an attempt to throw sufficient light on the impact of a few such evolutionary models in some recent areas of image processing, which involves the design of hardware-efficient image filters. Reducing the hardware complexity of digital systems has received reasonable attention from the research community over the last few decades. The hardware cost of digital filters may have been minimized by encoding the individual coefficients as the sum of signed powers-of-two (SPT), which can substitute the operation of multiplication by means of simple addition/subtraction and shifting. Design of such multiplier-less low-pass image filters with the aid of evolutionary computational algorithms has been explicitly demonstrated in this chapter. Two such popularly employed recent evolutionary techniques, namely differential evolution (DE) and the self-organizing random immigrants genetic algorithm (SORIGA), are elaborately discussed in this regard. The objective function of the optimization technique has been formulated in such a way that it can approximate the ideal frequency response to a certain extent. Characteristics of the image filters have been analyzed from different perspectives and subsequently compared with some other standard two-dimensional filters. Finally, the designed filters have been employed to minimize the effect of Gaussian noise from a few benchmark test images. Performance of these image de-noising filters has been analyzed with respect to certain performance parameters, such as peak signal-to-noise ratio, structural similarity index measure, image enhancement factor, image quality index, and so on.

Content-based image retrieval (CBIR) is one of the largest and most powerful research fields. It is an area that has grown in the last two decades and it is extensively employed today, with applications in many fields. Content-based medical image retrieval (CBMIR) is one of the most essential applications of CBIR. In recent days, giant numbers of medical images are being produced in hospitals as a consequence of new technologies. The traditional concept-based image retrieval systems have more limitations for the retrieval of medical images from this immense database. This will direct various new systems for storage, organization, indexing, and retrieval of medical images using low-level contents. An efficient multi-features fusion-based hybrid framework for magnetic resonance imaging (MRI) medical image retrieval is discussed in Chapter 11.

A novel feature extraction scheme for object recognition tasks constructed on singular value decomposition–principal component analysis (SVD–PCA) is proposed in Chapter 12. The result of the proposed work is compared with additional existing transforms, such as principal component analysis (PCA) and independent component analysis (ICA). In this chapter, study of the appearance-based object recognition techniques for two-dimensional and three-dimensional images is carried out. In addition, it is observed that SVD–PCA performs significantly better than conventional and subsequent eigenvalue decompositions. Experimental results in appearance-based object recognition confirm that SVD–PCA offers better recognition rates over ICA and PCA. The excellent recognitions rates achieved in all experiments indicates that the proposed method is well suited for object recognition in applications like surveillance, robot vision, biometrics and security tasks, etc.

In content-based image retrieval (CBIR) applications, the idea of indexing is mapping the extracted descriptors from images into a high-dimensional space. In Chapter 13, visual features like color, texture, and shape are considered. After combining color, texture, and shape features, the feature vector is reduced using kernel principle component analysis (KPCA). Then, the knowledge discovery (KD)-tree is used for indexing the images. A new proposed optimised range search (ORS) algorithm is used to obtain the optimal range for retrieving the relevant images from the database. The proposed KD-ORS tree is compared with the other existing trees. It has been found experimentally that the proposed KD-ORS tree gives better performance than the other existing trees.

Image compression plays a vital role in medical diagnosis in terms of efficient storage and data transfer. Chapter 14 discusses an efficient image compression technique constructed by integrating the proposed histogram indexed dictionary (HID) into the standard Huffman encoding. The integration is evaluated over magnetic resonance imaging (MRI) of brain images. At the initialization stage, the given input medical image is decomposed into 8×8, or 16×16, or 32×32 blocks, and then the discrete cosine transform (DCT) is applied on each of these blocks with sliding neighborhood filtering. The second stage quantization truncates the high frequency content of the filtered DCT coefficients by applying the thresholding. The final stage is an encoding scheme. A novel HID is created and it is integrated into the standard Huffman encoding. The HID improves the performance of the Huffman encoding. It is quite useful and assists the radiologists in deciding and planning the treatment. The experimental results show that the proposed method provides greater improvement in PSNR values than the other existing techniques.

D.P. Acharjya
V. Santhi

Acknowledgments

It is with a great sense of satisfaction that we present our book, entitled *Bio-Inspired Computing for Image and Video Processing*, and we wish to express our views to all those who helped us in both direct and indirect ways to complete this work. First of all we would like to thank the authors who have contributed to this book. We acknowledge, with sincere gratitude, the kindness of the School of Computer Science and Engineering, Vellore Institute of Technology University, Vellore, India for providing an opportunity to carry out this research work. In addition, we are also thankful to VIT for providing facilities to complete this project.

We take this unique opportunity to express our deepest appreciation to Aastha Sharma, senior commissioning editor, CRC Press, for her effective suggestions, sincere instruction, and kind patience during this project. We are really very grateful to you for your inspiration and assistance. We are also thankful to Shikha Garg for her nice co-operation during this project. It will be highly regretted if we fail to express our sincere thanks to Shashi Kumar, for his whole-hearted assistance and timely help during manuscript preparation of this research work.

We would like to thank our friends and faculty colleagues at VIT, for the time they spared in helping us through difficult times. Special mention should be made of the timely help given by reviewers during this project, those whose names are not mentioned here. The suggestions they provided cannot be thrown into oblivion.

While writing, contributors have reference several books and journals; we take this opportunity to thank all those authors and publishers. We are extremely thankful to the reviewers for their support during the process of evaluation. Last but not least, we thank the production team of CRC Press for encouraging us and extending their full cooperation and help in the timely completion of this book.

D.P. Acharjya
V. Santhi

Editors

D.P. Acharjya received his Ph.D. in computer science from Berhampur University, India; his M.Tech. degree in computer science from Utkal University, India; and his M.Sc. from NIT, Rourkela, India. He is currently working as a Professor in the School of Computer Science and Engineering, VIT University, Vellore, India. He has authored more than 60 national and international journal articles, conference papers, and book chapters. He has also written four books: *Fundamental Approach to Discrete Mathematics, Computer Based on Mathematics, Theory of Computation; Rough Set in Knowledge Representation* and *Granular Computing*. In addition, he has edited seven books for IGI Global, Springer, and CRC Press. He has been awarded a Gold Medal from NIT, Rourkela; Eminent Academician Award from Khallikote Sanskrutika Parisad, Berhampur, Odisha; Outstanding Educator and Scholar Award from the National Foundation for Entrepreneurship Development, Coimbatore; and The Best Citizens of India Award from International Publishing House, New Delhi, India. He is an editorial board member of various journals and reviewer for *IEEE Fuzzy Sets and Systems, Applied Soft Computing* (Elsevier), and *Knowledge-Based Systems* (Elsevier). He is associated with many professional bodies, including CSI, ISTE, IMS, AMTI, ISIAM, OITS, IACSIT, CSTA, ACM, IEEE, and IAENG. He was founding secretary of the OITS Rourkela chapter. His current research interests include rough sets, formal concept analysis, knowledge representation, data mining, granular computing, and business intelligence.

V. Santhi received her Ph.D. in computer science and engineering from VIT University; her M.Tech. in computer science and engineering from Pondicherry University; and her B.Tech. in computer science and engineering from Bharathidasan University, India. She is currently working as an associate professor in the School of Computer Science and Engineering, VIT University. She has 18 years of experience in academia and 3 years of experience in industry. She had served as head of department for more than eight years in engineering colleges. She has received research awards for publishing research papers in refereed journals from VIT University 3 times consecutively. She has guided more than 80 graduate projects and more than 50 postgraduate level projects. Her publications are indexed by the IEEE Computer Society, SPIE Digital Library, NASA Astrophysics Data Systems, SCOPUS, and the ACM Digital Library. She is a reviewer for *IEEE Transaction on Multimedia Security, IEEE Transaction on Image Processing, IET Image Processing,* and the *International Journal of Information and Computer Security*. She is a member of the IEEE Signal Processing Society, the Computer Society of India, and a senior member of IACSIT. She has published many papers in refereed journals and has attended international conferences. She took part in various administrative activities. Currently, she is programme and division chair at the School of Computer Science and Engineering, VIT University. She has contributed many chapters and is currently in the process of editing books. Her current research included digital image processing, digital signal processing, multimedia security, soft computing, bio-Inspired computing, and remote sensing.

Contributors

D.P. Acharjya
School of CSE
VIT University
Vellore, India

L. Agilandeeswari
School of Information Technology
and Engineering
VIT University
Vellore, India

Hugo Aguirre-Ramos
CA de Telematica, DICIS
Universidad de Guanajuato
Salamanca, GTO, Mexico

P. Aruna
Department of CSE
Annamalai University
Chidamvaram, India

D.J. Ashpin Pabi
Department of CSE
Annamalai University
Chidamvaram, India

Juan-Gabriel Avina-Cervantes
CA de Telematica, DICIS
Universidad de Guanajuato
Salamanca, GTO, Mexico

S. Ganesh Babu
Robert Bosch Engineering and
Business Solutions Limited
Coimbatore, India

R. Bhavani
Department of CSE
Annamalai University
Chidamvaram, India

Abhijit Chandra
Department of Instrumentation and
Electronics Engineering
Jadavpur University, India

Chiranji Lal Chowdhary
School of Information Technology
and Engineering
VIT University
Vellore, India

Ivan Cruz-Aceves
CONACYT - Centro de Investigacion
en Matematicas A.C. (CIMAT)
Guanajuato, GTO, Mexico

Victor Luiz Bezerra Arajo da Silva
Escola Politécnica da Universidade
de Pernambuco
Recife, Brazil

Rafaela Covello de Freitas
Departamento de Engenharia
Biomédica
Universidade Federal de Pernambuco
Recife, Brazil

Valter Augusto de Freitas Barbosa
Departamento de Engenharia
Biomédica
Universidade Federal de Pernambuco
Recife, Brazil

Ricardo Emmanuel de Souza
Departamento de Engenharia
 Biomédica
Universidade Federal de
 Pernambuco
Recife, Brazil

Jonti Deuri
Department of CSE
Pondicherry University
Pondicherry, India

Allan Rivalles Souza Feitosa
Centro de Informática
Universidade Federal de
 Pernambuco
Recife, Brazil

K. Ganesan
School of Information Technology
 and Engineering
VIT University
Vellore, India

B. Sarojini Ilango
Department of Computing Science
Avinashilingam University
Coimbatore, India

N. Kumaran
Department of CSE
Annamalai University
Chidamvaram, India

S. Anu H. Nair
Department of CSE
Annamalai University
Chidamvaram, India

Sarthak Nandi
School of CSE
VIT University
Vellore, India

Wellington Pinheiro dos Santos
Departamento de Engenharia
 Biomédica
Universidade Federal de Pernambuco
Recife, Brazil

M. Prabukumar
School of Information Technology
 and Engineering
VIT University
Vellore, India

N. Puviarasan
Department of CSE
Annamalai University
Chidamvaram, India

David Edson Ribeiro
Departamento de Engenharia
 Biomédica
Universidade Federal de
 Pernambuco
Recife, Brazil

Reiga Ramalho Ribeiro
Departamento de Engenharia
 Biomédica
Universidade Federal de
 Pernambuco
Recife, Brazil

V. Santhi
School of CSE
VIT University
Vellore, India

S. Siva Sathya
Department of CSE
Pondicherry University
Pondicherry, India

S. Vijayakumar
School of CSE
VIT University
Vellore, India

Part I

Bio-Inspired Computing Models and Algorithms

Chapter 1

Genetic Algorithm and BFOA-Based Iris and Palmprint Multimodal Biometric Digital Watermarking Models

S. Anu H. Nair
Annamalai University, Chidamvaram, India

P. Aruna
Annamalai University, Chidamvaram, India

1.1	Introduction	4
1.2	Literature Review	5
1.3	Proposed Work	6
1.4	Biometric Modality Extraction	7
	1.4.1 Iris Modality Extraction	7
	1.4.2 Segmentation	8
	1.4.3 Normalization	9
	1.4.4 Palmprint Modality Extraction	9
	1.4.5 Preprocessing	10
	1.4.6 Binarization	10
	1.4.7 Cropping and Convolving	11
	1.4.8 Data Sets	11
1.5	Various Image Fusion Methods	11
1.6	Implementation of Image Fusion Algorithms	12
	1.6.1 Simple Average	12
	1.6.2 Select Maximum	12
	1.6.3 Select Minimum	12
	1.6.4 Intensity Hue Saturation Fusion	12
	1.6.5 Principal Component Analysis	13
	1.6.6 Pyramid Fusion Algorithm	14
	1.6.7 Discrete Wavelet Transform	16
	1.6.8 Stationary Wavelet Transform	16
	1.6.9 Sparse Representation Fusion	17
1.7	Performance Metrics Used	17
	1.7.1 Xydeas and Petrovic Metric	18
	1.7.2 Visual Information Fidelity	18

	1.7.3	Fusion Mutual Information	19
	1.7.4	Average Gradient	20
	1.7.5	Entropy	21
1.8	Watermarking	21
	1.8.1	The Process of Genetic Algorithm	22
	1.8.2	Bacterial Foraging Optimization Algorithm Watermarking Model	24
1.9	Performance Metrics	24
	1.9.1	Peak Signal-to-Noise Ratio	24
	1.9.2	Normalized Absolute Error	26
	1.9.3	Normalized Cross Correlation	26
1.10	Conclusion	27
	Bibliography	28

The multimodal biometric watermarking system using multiple sources of information has been widely recognized. However, computational models for multimodal biometrics recognition have only recently received attention. The biometric traits used in this research work are fingerprint, iris, and palmprint. At the modality extraction level, the modalities are extracted from different biometric traits. These modalities are then blended to produce a joint template, which is the basis for the watermarking system. Fusion at modality extraction level generates a homogeneous template for fingerprint, iris, and palmprint features. The multimodal biometric system is implemented using different fusion schemes. The fused templates are analyzed by using performance metrics like Qabf, VIF, MI, average gradient, entropy, etc. From the results obtained, the DWT fused template provided better results than other fused templates. The best fused template is applied as input along with the cover image to the different watermarking models, such as the genetic algorithm (GA) and bacterial foraging optimization algorithm (BFOA). Various metrics such as peak signal-to-noise ratio (PSNR), normalized absolute error (NAE), and normalized cross correlation (NCC) are used for analyzing the robustness of the algorithm with various cover images.

1.1 Introduction

Biometrics acts as a source for identifying a human being. This is used for authentication and identification purposes. A multimodal biometric system combines two or more biometric data recognition results, such as a combination of a subject's fingerprint, face, iris, and voice. This increases the reliability of personal identification systems that discriminate between an authorized person and a fraudulent person. Multimodal biometric systems have addressed some issues related to unimodal systems, such as non-universality or insufficient population coverage (reducing failure to enroll rate, which increases population coverage). It becomes more and more unmanageable for an impostor to imitate multiple biometric traits of a legitimately enrolled individual. Additionally, multimodal biometric systems effectively address the problem of noisy data (illness affecting voice, scar affecting fingerprint). For the implementation of a multimodal biometric system, the fusion of biometric features is proposed. Modality fusion is implemented by using several image fusion methods.

This chapter proposes a novel approach for creating a multimodal biometric system. The multimodal biometric system is implemented using different fusion schemes, such as average fusion, minimum fusion, maximum fusion, PCA fusion, DWT fusion, SWT fusion, IHS fusion, Laplacian gradient fusion, pyramid gradient fusion, and sparse representation fusion, to improve the performance of the system. At the feature extraction level, the information extracted from different modalities is stored in vectors on the basis of their modality. These modalities are then blended to produce a joint template, which is the basis for the watermarking system. Fusion at modality extraction level generates a homogeneous template for fingerprint, iris, and palmprint features.

Digital watermarking is the technology of embedding information into the multimedia data, such as image, audio, video, and text, which is the called a cover image. It is realized by embedding data that is invisible to the human visual system into a host image. Hence the term digital image watermarking is a procedure by which watermark data is covered inside a host image that imposes imperceptible changes to the picture. Watermarking techniques have been used in multimodal biometric systems for the purpose of protecting and authenticating biometric data and enhancing accuracy of identification. The fused image is applied as input, along with the cover image, to the GA and BFOA watermarking systems. Here the BFOA is a powerful optimization tool that inherits the foraging characteristics of bacteria for identifying optimized positions in an image. Until now, BFO has been applied successfully to some engineering problems, such as optimal control, harmonic estimation, transmission loss reduction, and machine learning. In this work, BFO is used for watermarking images. Various metrics, such as PSNR, NAE, and NCC, are used to measure the image quality.

1.2 Literature Review

Patil et al. [17] developed a novel approach to iris recognition using a lifting wavelet-based algorithm that enhances iris image, reduces the noise, and extracts important feature from the image. The similarity between the images is estimated using Euclidean distance. Liau et al. [13] implemented image segmentation, and creation of a noise mask. The iris region was segmented by locating the boundaries between the pupil and iris region and between the sclera and iris region using a method that is based on canny edge detection and circular Hough transform. Chowhan et al. [4] described iris recognition using a modified fuzzy hyper sphere neural network with a learning algorithm. The performance is evaluated using different distance measures. It is observed that the feasibility of a modified fuzzy hyper sphere neural network has been successfully appraised on the CASIA database. Kamlesh et al. [23] modeled a palmprint-based recognition system that uses texture and dominant orientation pixels as features. Xiumei Guo et al. [6] identified a palmprint recognition method that uses blanket dimension for extracting image texture information. Feng Yue et al. [24] presented a typical palmprint identification system that constructed a pattern from the orientation and response features. Nibouche et al. [16] designed a new palmprint matching system based on the extraction of feature points identified by the intersection of creases and lines. Jing Li et al. [12] proposed a competent representation method that can be used for classification. Canuto et al. [3] created a model that fused voice and iris biometric features. This model acted as a new representation of existing biometric data. Norman Poh et al. [18] proposed a user-specific and selective fusion strategy for an enrolled user. Asaari et al. [2] identified a new geometrical feature width centroid contour distance for finger geometry biometrics. Huang et al. [8] developed a face and ear biometric system that uses a feature weighing scheme called sparse coding error ratio. Meng Ding et al. proposed the fusion method based on a compressive sensing theory that contains a complete dictionary, an algorithm for sparse vector approximation, and a fusion rule. Aravinth et al. [1] identified the feature extraction techniques for three modalities: fingerprint, iris, and face. The extracted data is stored as a template that can be fused using density-based score level fusion. Sumit et al. [20] proposed a multimodal sparse representation method, which understands the test data by a sparse linear combination of training data, while restricting the observations from different modalities of the test subject to share their sparse representations. Khanduja et al. [11] demonstrated a novel method for watermarking relational databases for recognition and validation of ownership based on the secure embedding of blind and multi-bit watermarks, using the bacterial foraging optimization algorithm (BFOA). Riya Mary Thomas [22] provided an outline of a bacterial foraging optimization algorithm (BFOA) and its intermediated operations in grid scheduling. Sharma et al. [19] presented an application based review of variants of BFOA that have come up with faster convergence with higher accuracy and will be useful for new researchers exploring its use in their research problems.

1.3 Proposed Work

The proposed work describes the feature extraction of multimodal biometric images such as fingerprint, palmprint, and iris. The extracted information is

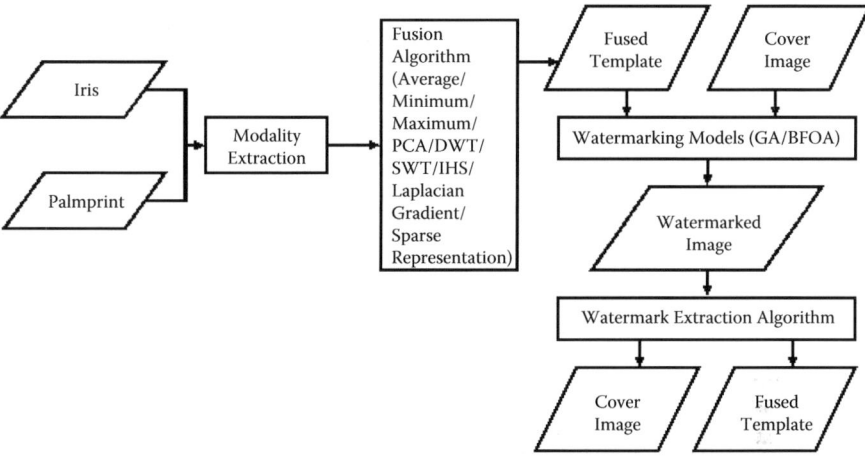

FIGURE 1.1: Flow diagram of the GA and proposed BFOA watermarking system

fused together using average fusion, minimum fusion, maximum fusion, PCA fusion, DWT fusion, SWT fusion, IHS fusion, Laplacian gradient fusion, pyramid gradient fusion, and sparse representation fusion. The best fused template is then watermarked into a host image using GA and BFOA algorithms and extracted further. Figure 1.1 depicts the flow diagram of the genetic algorithm and the proposed BFOA watermarking system.

1.4 Biometric Modality Extraction

This chapter explains individual extraction of iris and palmprint modalities. The processes involved are explained below.

1.4.1 Iris Modality Extraction

Iris recognition is an automated method of biometric identification that uses mathematical pattern-recognition techniques on video images of the irises of an individual's eyes, whose complex random patterns are unique and can be seen from some distance. Many millions of individuals in several nations about the globe have been enrolled in iris recognition systems, for convenience purposes, such as passport-free automated border-crossings, and some

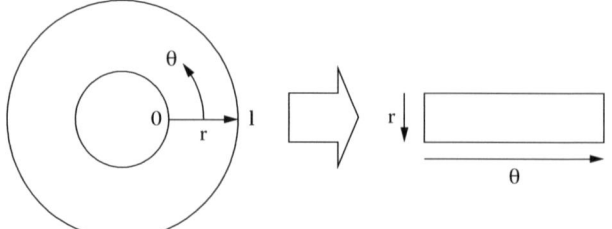

FIGURE 1.2: Conversion of polar to rectangular form of iris

national ID systems based on this technology are being deployed. A central advantage of iris recognition, besides its speed of matching and its extreme opposition to false matches, is the stableness of the iris as an inner, protected, yet externally visible organ of the optic system.

1.4.2 Segmentation

In this research work, segmentation of iris is the first phase of iris extraction. Here, the actual region is separated from the image. The region can be specified by 2 circles. The occlusion by eyelids and eyelashes must be eliminated to locate a circular region. Imaging quality plays a major role in the success of the extraction of the circular region. In the extraction method, biometric shapes like lines and circles are identified by the Hough transform. The circular Hough transform extracts the center and radius iris region. The first derivatives of intensities are calculated to produce an edge map. The definition of a circle is given below, where (x, y) represents the coordinate of a point on the circle with radius r and center $(0, 0)$.

$$x^2 + y^2 = r^2$$

The definition of Hough transform is given below. The linear Hough transform fits a line to the upper and lower eyelid. The second line, which is horizontal, is drawn. The first line intersects with the second at the edge of the iris. This allows maximum isolation of regions of eyelids. Canny edge detection is utilized to produce an edge map.

$$H(x_c, y_c, r) = \sum_{j=1}^{n} h(x_j, y_j, x_c, y_c, r) \tag{1.1}$$

where

$$h(x_j, y_j, x_c, y_c, r) = \begin{cases} 1 & if \quad g(x_j, y_j, x_c, y_c, r) = 0 \\ 0 & Otherwise \end{cases}$$

with $g(x_j, y_j, x_c, y_c, r) = (x_j - x_c)^2 + (y_j - y_c)^2 - r^2$

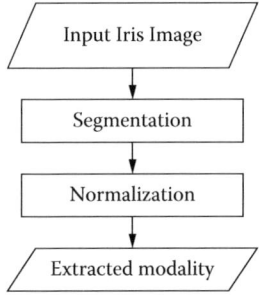

FIGURE 1.3: Iris modality extraction

1.4.3 Normalization

Generally, normalization is a linear procedure and it is implemented to negate the variable's impact on the information. Here, the output of the segmentation process provides a polar form of the iris modality. The iris modality is preferred to be a rectangular form for the fusion of images. Therefore, normalization appears to be a conversion process for this research work. Figure 1.2 displays the conversion of polar form to rectangular form for the iris. The rectangular coordinates are obtained using Equations (1.2) to (1.7), where $\theta = (2*\pi*X)/n$; (x_i, y_i) is the midpoint of the inner circle; (x_o, y_o) is the midpoint of the outer circle; and (x, y) is the midpoint of iris the image.

$$x_i = center(x) + innerradius * cos(\theta) \quad (1.2)$$

$$y_i = center(y) + innerradius * sin(\theta) \quad (1.3)$$

$$x_o = center(x) + outerradius * cos(\theta) \quad (1.4)$$

$$y_o = center(y) + outerradius * sin(\theta) \quad (1.5)$$

$$x = x_i + [(x_o - x_i) * y]/m \quad (1.6)$$

$$y = x_i + [(x_o - x_i) * y]/m \quad (1.7)$$

Equations (1.2) and (1.3) denote the calculation of the inner circle, whereas Equations (1.4) and (1.5) depict calculation of the outer circle. Equations (1.6) and (1.7) provide the fixed block 512 × 64 iris image. Figure 1.3 depicts the flow diagram of iris modality extraction.

1.4.4 Palmprint Modality Extraction

Palmprint has its own advantages compared with other methods of biometrics. Palmprint is hardly affected by age, whereas the problem of age is the main problem for face recognition. Palmprint contains more information and can use low resolution devices in comparison with finger printing. Palmprint cannot harm the health of people, and many people prefer palmprinting to

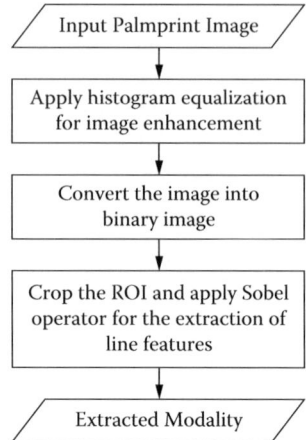

FIGURE 1.4: Palmprint modality extraction

iris recognition based on this very reason. Palmprints contain more information than fingerprints, so they are more distinctive. Additionally, palmprint-capture devices are much cheaper than iris devices. Palmprints also contain additional distinctive features, such as principal lines and wrinkles, which can be extracted from low-resolution images. A highly accurate biometrics system can be built by combining all features of palms, such as palm geometry, ridge and valley features, principal lines and wrinkles, etc. It is for these reasons that palmprint recognition has recently attracted an increasing amount of attention from researchers. Figure 1.4 depicts the modality extraction from a palmprint image.

1.4.5 Preprocessing

The global contrast is increased via the histogram equalization method, since the usable information is depicted by close contrast values. The distribution of the intensities on the histogram is implemented. Lower contrast values gain higher contrast by this method. Most frequency interest values are spread out effectively by the histogram equalization method. In this extraction method, an enhanced palmprint image is obtained after performing histogram equalization.

1.4.6 Binarization

This process converts an input gray level image into a binary one to improve the contrast between ridges in the biometric trait. This procedure analyses each pixel value and converts them to either 0 or 1, based on the threshold value. The output image is the enhanced palmprint image.

1.4.7 Cropping and Convolving

The image is subjected to the cropping process for the selection of the region of interest. The obtained image is convolved with the sobel operator. The Sobel operator is [12 1 0 0 0 − 1 − 2 − 1]. The resultant image would be the extracted modality of palmprint.

1.4.8 Data Sets

The proposed work is implemented using Matlab 2012 (Version 8.0) software. The hardware used was Intel(R) Atom (TM) CPU N570 @1.66GHz. In this work, the fingerprint images and iris images were obtained from Pondicherry Engineering College, Pondicherry. The palmprint images were obtained from the CASIA palmprint image database. A total of 100 images from each of the iris and palmprint biometric traits were taken and analyzed. The iris images of size 150 × 200 pixels and the palmprint images of size 640 × 480 pixels are taken as inputs.

1.5 Various Image Fusion Methods

There are various methods that have been developed to perform image fusion. Some well-known image fusion methods are listed below:

1. Simple image fusion
   - Simple average
   - Simple maximum or minimum
   - Intensity-hue-saturation (IHS) transform-based fusion
   - Principal component analysis (PCA)-based fusion

2. Multi-scale transform-based fusion
   - Pyramid method
     (a) Laplacian pyramid
     (b) Gradient pyramid
   - Wavelet transforms
     (a) Discrete wavelet transforms (DWT)
     (b) Stationary wavelet transforms

3. Sparse representation fusion

1.6 Implementation of Image Fusion Algorithms

To obtain a homogenous template, a fusion process is used. The modalities are extracted individually and combined together using fusion methods explained below.

1.6.1 Simple Average

Generally, areas in focus have greater intensity. This procedure acquires in-focus areas. In this research work, the pixel value $P(i,j)$ in each modality is chosen and summed. This total is then halved to get the average. The pixels of the output image are assigned the average values. Equation (1.8) gives the same. This is iterated for all pixel values, where $X(i,j)$ and $Y(i,j)$ are two input images.

$$K(i,j) = X(i,j) + Y(i,j)/2 \tag{1.8}$$

1.6.2 Select Maximum

An in-focus image has greater pixel values. Therefore, this procedure chooses the greatest value to obtain in-focus areas. The pixel values of input images are compared and the greatest value is assigned to the output image. In this research work, the input images are extracted biometric modalities, where $A(i,j)$ and $B(i,j)$ are different biometric modalities and $F(i,j)$ is the fused image.

$$F(i,j) = \sum_{i=1}^{m} \sum_{j=1}^{n} max(A(i,j), B(i,j)) \tag{1.9}$$

1.6.3 Select Minimum

This algorithm chooses the lowest pixel values among the two input images. The output image is assigned the lower values. In this research work, the input images are extracted biometric modalities.

$$F(i,j) = \sum_{i=1}^{m} \sum_{j=1}^{n} min(A(i,j), B(i,j)) \tag{1.10}$$

1.6.4 Intensity Hue Saturation Fusion

The intensity hue saturation (IHS) fusion technique is utilized in image fusion to make use of the complementary nature of MS images. First red, green, and blue space is changed into IHS space using an IHS fusion strategy, and then the intensity band is replaced by the image. This is converted into RGB space. Similarly the above process is applied for the other input image, which is added, resulting in the fused image. In this research work, the IHS fusion for every pixel can be modeled by the following process. The steps are as follows:

$$\begin{bmatrix} I \\ v_1 \\ v_2 \end{bmatrix} = \begin{bmatrix} 1/3 & 1/3 & 1/3 \\ -\sqrt{2}/6 & -\sqrt{2}/6 & 2\sqrt{2}/6 \\ 1/\sqrt{2} & -1/\sqrt{2} & 0 \end{bmatrix} \begin{bmatrix} R \\ G \\ B \end{bmatrix} \quad (1.11)$$

$$H = tan^{-1}(v_2/v_1) \quad (1.12)$$

$$S = \sqrt{v_1^2 + v_2^2} \quad (1.13)$$

The corresponding inverse transform is defined as

$$v_1 = SCos(H) \quad (1.14)$$

$$v_1 = SSin(H) \quad (1.15)$$

$$\begin{bmatrix} R' \\ G' \\ B' \end{bmatrix} = \begin{bmatrix} 1/3 & 1/3 & 1/3 \\ -\sqrt{2}/6 & -\sqrt{2}/6 & 2\sqrt{2}/6 \\ 1/\sqrt{2} & -1/\sqrt{2} & 0 \end{bmatrix} \begin{bmatrix} I \\ v_1 \\ v_2 \end{bmatrix} \quad (1.16)$$

Equations (1.11) to (1.16) state that the fused image can be obtained from the original image by adding the values of the IHS images. That is, the IHS method is efficiently implemented by this process. This algorithm is called the fast IHS fusion algorithm.

1.6.5 Principal Component Analysis

Principal component analysis (PCA) is a mathematical tool that changes various correlated variables into a number of uncorrelated variables. The applications of PCA are used in areas like image compression and image classification. PCA calculates a compact and optimal description of the data set. The first principal component represents as much of the variance in the data as possible, and each succeeding component represents as much of the remaining variance as possible. The first principal component is taken to be along the direction of the maximum variance. The second principal component is constrained to lie in the subspace perpendicular to the first. Within this subspace, this component points in the direction of maximum variance. The third principal component is taken to be in the maximum variance direction in the subspace perpendicular to the first two, and so on. Its basis vectors rely on the data set. In this research work, the biometric modalities are given as input images. The stepwise description of the PCA algorithm for fusion is explained below.

Algorithm

1. The column vectors are generated from the input image matrices.

2. The mean along each column is calculated, and this is subtracted from each column. The column vectors form a matrix X.

3. The covariance matrix of the column vectors found in step 1 is calculated by using the folowing equation:

$$C = XX^T \qquad (1.17)$$

4. The diagonal elements of the 2×2 covariance vector contain the variance of each column vector with itself, respectively.

5. The eigenvalues and the eigenvectors of the covariance matrix are computed.

6. Normalization of the column vector with the larger eigenvalue is implemented by dividing each element by the mean of the eigenvector. Suppose $(x, y)^T$ is a vector of the eigenvectors of the images A and B, the weight value of image A and image B are

$$\omega_A = x/x + y \qquad (1.18)$$

$$\omega_B = y/x + y \qquad (1.19)$$

7. The normalized eigenvectors act as the weights that are multiplied with every pixel value of the input images, respectively.

8. The total value of the two matrices calculated in step 7 is the fused image. Then, the fusion is accomplished using a weighted average as defined below, where where I_F is the fused image and I_A and I_B represent images A and B, respectively.

$$I_F = \omega_A I_A + \omega_B I_B \qquad (1.20)$$

1.6.6 Pyramid Fusion Algorithm

An image pyramid comprises a set of low pass or band pass copies of an image, each copy depicting pattern information on a different scale. The size of the pyramid is halved in the preceding level and the higher levels will focus on the lower spatial frequencies. The basic idea is to create the pyramid transform of the fused image from the pyramid transforms of the source images, and then the fused image is retrieved by applying inverse pyramid transform [21]. Every pyramid transform consists of three major stages. These are:

1. Decomposition: implement decomposition where a pyramid is produced subsequently at every level of the fusion. The 3 steps are:
   (a) The biometric modality images are subjected to low pass filtering using $W = [1/16, 4/16, 6/16, 4/16, 1/16]$.
   (b) The low pass filtered images obtained are subtracted and the pyramid is created.
   (c) The input images are decimated by dividing the number of rows and columns by 2.
2. Formation of the initial image for recomposition.
   (a) The input images are merged after the process of decomposition.
   (b) The output of this process is the input for the process of recomposition.
3. Recomposition
   (a) I/P image is undecimated.
   (b) The image is undecimated by copying every row and column and convolves the undecimated image with the filter vector's transpose.
   (c) The output of the step (2) is merged with the pyramid formed at the level of decomposition.
   (d) The output is the fused image.

A Laplacian pyramid is a set of band pass images, in which each image is a band pass filtered copy of its antecedent. The difference between low pass images at subsequent levels of a Gaussian pyramid is calculated. In the Laplacian fusion approach, the Laplacian pyramids for input images are utilized. In this research work, the input images are extracted biometric modalities [9].

A gradient pyramid is acquired by implementing a set of 4 directional gradient filters (horizontal, vertical, and 2 diagonal) to the Gaussian pyramid at every level. Merging of 4 directional gradient pyramids results in a combined pyramid at each level. The combined gradient is used instead of the Laplacian pyramid. In this research work, the inputs are extracted biometric modalities [9].

1.6.7 Discrete Wavelet Transform

The wavelets-based methodology is suitable for performing fusion tasks for three reasons. These are appropriate for different image resolutions; preserves image formation even during image decomposition; and data is preserved even though IDWT is applied. The symbol x is the DWT signal, g is the low pass filter, h is the high pass filter.

$$y[n] = (x * g)[n] = \sum_{k=-\alpha}^{x} x[k]g[n-k] \tag{1.21}$$

$$y_{low}[n] = \sum_{k=-\alpha}^{x} x[k]g[2n-k] \tag{1.22}$$

$$y_{high}[n] = \sum_{k=-\alpha}^{x} x[k]h[2n-k] \tag{1.23}$$

The 2 × 2 Haar matrix associated with the Haar wavelet is

$$\begin{bmatrix} 1 & 1 \\ 1 & -1 \end{bmatrix}$$

The wavelet transform transforms the image into low–high, high–low, and high–high spatial frequency bands at different scales, and the low–low band at the coarsest scale, which is demonstrated. The average image data is present in the LL band. Directional data is available in other bands because spatial orientation causes directional data. Edges or lines are depicted in higher bands of the image. In this research work, the basic steps performed in image fusion are as follows:

1. Perform independent wavelet decomposition of the two images until level 2 fusion of two input images' DWT coefficients takes place, by calculating the appropriate coefficient's average.

2. The fused image is procured by applying inverse DWT

1.6.8 Stationary Wavelet Transform

This method fuses two biometric modality images by means of stationary wavelet transform (DSWT or SWT). The stationary wavelet transform is a wavelet transform procedure that is modeled to overcome the absence of translation invariance of the discrete wavelet transform. Translation invariance is accomplished by eliminating the down samplers and the up samplers in the DWT, and up-sampling the filter coefficients by a component of $2(j-1)$ at the jth level of the procedure. In summary, the SWT method can be described as follows:

1. Decompose the two source images utilizing SWT at one level, resulting in three details sub-bands, and one approximation sub-band (HL, LH, HH and LL bands). Further, the approximate parts are averaged.

2. Choose the absolute values of horizontal details of the image and subtract the second part of the image from first. Compute

$$D = |H1L2| - |H2L2| \geq 0$$

3. For the fused horizontal part perform elementwise multiplication of D and horizontal detail of the first image, and then subtract another horizontal detail of the second image multiplied by logical not of D.

4. Find D for vertical and diagonal parts and acquire the fused vertical and details of the image. Further fused image is retrieved by applying ISWT.

1.6.9 Sparse Representation Fusion

The extracted features are fed as the input of the sparse fusion methodology. The input features were converted into matrix format to perform sparse representation [15]. The algorithm is as follows:

Algorithm

1. The images are fed as input.

2. The values of input image 1 and input image 2 are convolved.

3. The resultant matrix is converted into sparse representation by squeezing out the zero elements.

4. The sparse representation is further transformed into an image using orthogonal matching pursuit. The fused template is obtained.

1.7 Performance Metrics Used

To analyze and find the best fused image, some performance metrics are used. This identifies the best fused template, which is fed as input to the watermarking model.

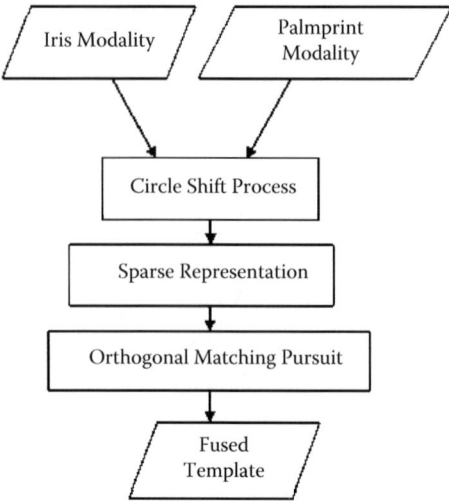

FIGURE 1.5: Sparse representation method

1.7.1 Xydeas and Petrovic Metric

A normalized weighted performance metric of a given process p that fuses A and B into F is given below [5], where A, B, and F represent the input and fused images, respectively.

$$Q^{AB/F} = \frac{\sum_{m=1}^{M}\sum_{n=1}^{N}[Q^{AF}(m,n)w^A(m,n) + Q^{BF}(m,n)w^B(m,n)]}{\sum_{m=1}^{M}\sum_{n=1}^{N}[w^A(m,n)w^B(m,n)]} \quad (1.24)$$

The definition of Q^{AF} and Q^{BF} are given below, where $Q^{*F}g$, $Q^{*F}\alpha$ are the edge strength and orientation values at location (m,n) for images A and B. The dynamic range for $Q^{AB/F}$ is $[0,1]$ and it should be close to one for better fusion.

$$Q^{AF}(m,n) = Q_g^{AF}(m,n)Q_\alpha^{AF}(m,n) \quad (1.25)$$

1.7.2 Visual Information Fidelity

Visual information fidelity (VIF) first decomposes the natural image into several sub-bands and parses each sub-band into blocks [7]. Further, it measures the visual information by computing mutual information in the different models in each block and each sub-band. Finally, the image quality value is measured by integrating visual information for all the blocks and all the sub-bands. This relies on modeling of the statistical image source, the image distortion channel, and the human visual distortion channel. Image quality assessment is done based on information fidelity, where the channel imposes fundamental

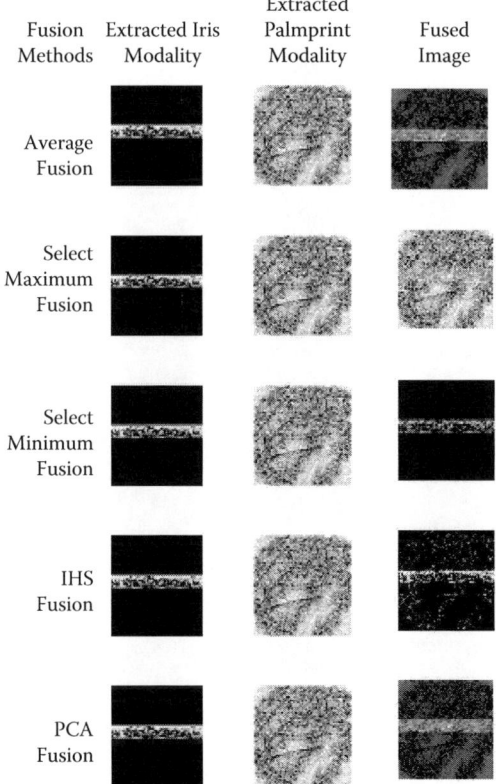

FIGURE 1.6: Output of different fusion methods, such as average, maximum, minimum, IHS, and PCA

limits on how much information could flow from the source (reference image), through the channel (image distortion process) to the receiver (human observer) as shown in Figure 1.6.

$$VIF = \frac{Distorted\ image\ information}{Reference\ image\ information} \quad (1.26)$$

1.7.3 Fusion Mutual Information

Mutual information calculates the degree of dependence of two images. If the joint histogram between $I_1(x,y)$ and $I_f(x,y)$ is defined as $h_{I_1 I_f}(i,j)$, and $I_2(x,y)$ and $I_f(x,y)$ is defined as $h_{I_2 I_f}(i,j)$, then fusion mutual information (FMI) is given as below:

$$FMI = MI_{I_1 I_f} + MI_{I_2 I_f} \quad (1.27)$$

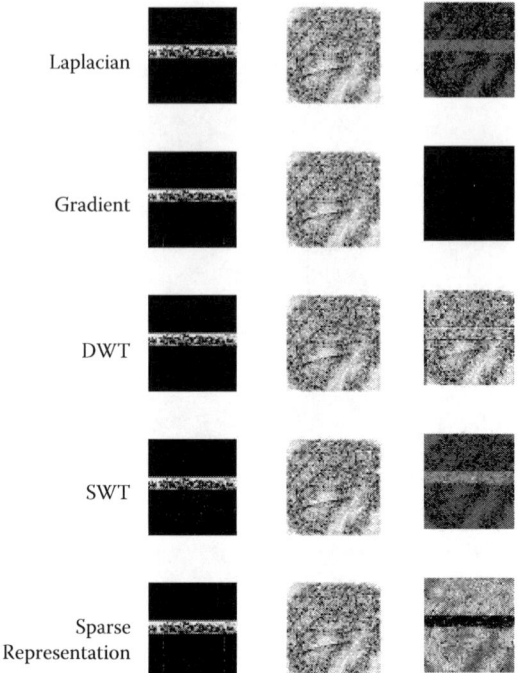

FIGURE 1.7: Output of different fusion methods, such as Laplacian pyramid, gradient pyramid, DWT, SWT, and sparse representation

where

$$MI_{I_{1}I_{f}} = \sum_{i=1}^{M}\sum_{j=1}^{N} h_{I_{1}I_{f}}(i,j) \log_2 \left(\frac{h_{I_{1}I_{f}}(i,j)}{h_{I_{1}}(i,j)h_{I_{f}}(i,j)} \right) \qquad (1.28)$$

$$MI_{I_{2}I_{f}} = \sum_{i=1}^{M}\sum_{j=1}^{N} h_{I_{2}I_{f}}(i,j) \log_2 \left(\frac{h_{I_{2}I_{f}}(i,j)}{h_{I_{1}}(i,j)h_{I_{f}}(i,j)} \right) \qquad (1.29)$$

The output of different fusion methods is depicted in Figure 1.7.

1.7.4 Average Gradient

The average gradient is applied to measure the detailed information in the images, where M and N are the rows and colums of images. The higher the average gradient, the better it describes the fused template.

$$g = 1/(M-1)(N-1) \sum_{x=1}^{N-1}\sum_{y=1}^{N-1} \sqrt{(x,y)^2 + (x,y)^2} \qquad (1.30)$$

TABLE 1.1: Performance analysis for different watermarked images using GA watermarking system.

Metrics fusion methods	Qabf	VIF	MI	Average gradient	Entropy
Gradient pyramid fusion	0.02	0.01	0.46	4.32	0.61
Sparse representation fusion	0.13	0.04	0.56	6.51	2.40
SWT fusion	0.14	0.05	1.71	8.63	3.48
Minimum fusion	0.15	0.13	2.00	10.63	3.97
Laplacian pyramid fusion	0.18	0.16	2.07	13.77	6.60
IHS fusion	0.19	0.19	2.28	13.78	7.03
PCA fusion	0.42	0.34	2.95	21.05	7.05
Average fusion	0.70	0.34	3.43	21.97	7.18
Maximum fusion	0.78	0.37	3.44	26.21	7.52

1.7.5 Entropy

Entropy is defined as the amount of information contained in a signal. The entropy of the image can be evaluated as shown below, where $P(d_i)$ is probability of occurrence of a particular gray level d_i. If the entropy of a fused image is higher than that of the parent image, then it indicates that the fused image contains more information. Table 1.1 depicts the performance analysis of various fusion methods.

$$H = -\sum P(i) * \log_2(P(d_i)) \quad (1.31)$$

The above results indicate that the DWT fusion method gave better results than other fusion algorithms. This fused template is fed into the watermarking model along with the cover image.

1.8 Watermarking

Digital watermarking is the technology of embedding information (watermark or host signal) into multimedia data (such as image, audio, video, and text), called cover signal, in order to produce a watermarked signal for use in certain applications, such as copyright protection, content authentication, copy protection, broadcast monitoring, and so on. It is realized by embedding data that is invisible to the human visual system into a host image. Thus, digital image watermarking is the process by which watermark data is hidden within a host image imposing imperceptible changes to the image. The root of watermarking as an information-hiding technique can be traced from ancient Greece as steganography he application of watermarking ranges from copyright protection to file tracking and monitoring. Watermarking techniques have been used in multimodal biometric systems for the purposes of protecting and authenticating biometric data, and enhancing accuracy of recognition. A multimodal biometric system combines two or more biometric data, which increases the reliability of personal recognition systems that discriminate between an authorized person and a fraudulent person. The applications of multimodal biometric systems include law enforcement, e-commerce, smart cards, passports and visas, etc.

In optimization of a design, the design objective could be easy to minimize the cost of production or to maximize the efficiency of production. An optimization algorithm is a procedure that is executed iteratively by comparing various solutions until an optimum or a satisfactory solution is found. With the advent of computers, optimization has become a part of computer-aided design activities.

1.8.1 The Process of Genetic Algorithm

A genetic algorithm (GA) is a heuristic search procedure for deciding the global maximum/minimum solutions for issues in evolutionary computation. Any optimization problem is modeled in GA by characterizing the chromosomal representation, fitness function, and application of the GA operators. It is an optimization search method based on the principles of genetics and natural selection. GA is made up of five components: arbitrary number generator, fitness for assessment unit, and genetic operators for reproduction, crossover, and mutation operation [10]. The arbitrary number generator issues a set of strings to initiate the algorithm. Every string is a fitness value processed by the assessment unit. The reproduction operator implements a natural selection function known as "seeded selection." The crossover operator selects pairs of strings at random and delivers new pairs. A mutation operator arbitrarily mutates or inverses the values of bits in a string. Here, the objective function used is PSNR. In this research work, the cover image is divided into 8×8 blocks, where GA is implemented to find the best block. The best block is embedded with the watermark. The reverse process gives the extracted watermark image. The algorithm used in this work is given below.

Here, the crossover probability is 0.8 and mutation probability is 0.1. Figure 1.8 provides the sample outputs obtained by applying the GA watermarking model, for various standard cover images and a DWT fused template of iris and palmprint. The watermark extracted and the cover images are also shown.

Algorithm [Genetic Algorithm]

1. Initialize the population size p
2. for $i = 1$ to p where i is the index for population
   (a) Selection is done by evaluating the fitness function. For each value, calculate the fitness value.
   (b) New population is obtained by performing crossover and mutation. Crossover is performed using tournament selection method. Here the best individual with crossover probability is chosen. The next best individual is crossover probability $\times$ (1- crossover probability). Mutation is performed by implementing the flip-bit mutation operator.
   (c) Calculate the fitness of the new offspring.
   (d) Replace the existing value with the obtained value if the current value is the best fit.
3. end for

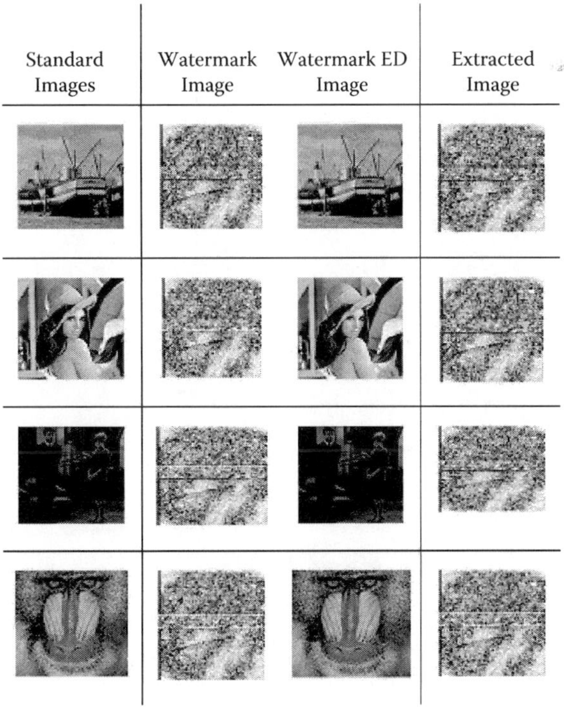

Standard Images	Watermark Image	Watermark ED Image	Extracted Image

FIGURE 1.8: Sample output obtained by applying GA watermarking system

1.8.2 Bacterial Foraging Optimization Algorithm Watermarking Model

The bacterial foraging optimization algorithm (BFOA) is a newcomer to the group of nature-inspired optimization procedures. In the last five decades, optimization procedures such as genetic algorithms, evolutionary programming (EP), and evolutionary strategies (ES), which draw their motivation from evolution and natural genetics, have been overwhelming the realm of optimization algorithms. Recently, natural swarm-inspired algorithms like particle swarm optimization (PSO) and ant colony optimization (ACO) have found their way into this domain and have demonstrated their effectiveness. Application of group foraging method of a swarm of E.coli bacteria in multi-optimal function optimization is the key thought for this new procedure. Bacteria search for nutrients in a manner such as to increase energy obtained per unit of time. An individual bacterium also converses with others by sending signals. A bacterium makes foraging choices in the wake of two previous factors. The procedure, in which a bacterium moves by taking small steps while searching for nutrients, is defined as chemotaxis, and the key thought of BFOA is emulating chemotactic development of virtual bacteria in the problem search space [19].

Based on the value of the fitness function, in reproduction the number of healthier values S is split in two. These are retained in the same location as their parents. Reproduction is the calculation of cumulative health of each value [22]. Elimination and dispersal is used to eliminate the weak values when healthy ones are added. Here, the objective function used is PSNR [14]. The algorithm used in this work is given below. Figure 1.9 provides the sample outputs obtained by applying the proposed BFOA watermarking model, for various standard cover images and a DWT-fused template of iris and palmprint. The watermark extracted and the cover images are also shown.

1.9 Performance Metrics

This section discusses various performance measures taken into consideration for the research work.

1.9.1 Peak Signal-to-Noise Ratio

Peak signal-to-noise ratio (PSNR) represents the measure of peak error. In order to compute PSNR, the mean squared error (MSE) is first calculated, where A, B are watermarked and cover images, respectively; and M, N are rows and columns of images. An increase in PSNR implies a high-quality

Algorithm [Bacterial Foraging Optimization Algorithm]

1. Initialize the values p: Dimension of the search space. S: The number of bacteria in the population. N_c: Chemotactic steps. N_s: Swimming length. N_{re}: The number of reproduction steps. N_{ed}: The number of elimination-dispersal events. P_{ed}: Elimination-dispersal with probability. $C(i)(i = 1, 2, \cdots S)$: The size of the step taken in the random direction specified by the tumble. $P(j, k, l) : P(j, k, l) = \theta(i, j, k, l)|i = 1, 2, \cdots S$.

2. For $l = 1$ to N_{ed} (Elimination dispersal loop)

3. For $k = 1$ to N_{re} (Reproduction loop)

4. For $j = 1$ to N_c (Chemotaxis loop)

    (a) Adapt a chemotactic step for every bacterium (i).

    (b) Calculate fitness function: $J(i, j, k, l)$ and assign $Jlast = J(i, j, k, l)$.

    (c) For $i = 1$ to S, decide the tumbling or swimming choice, where tumbling produces a random number θ with each element (i) where $m = 1, 2, \cdots, p$ and the random number lies at $[0, 1]$.

    (d) Swimming loop: Let $m = 0$ (counter for swim length) while $m < Ns, m = m+1$. If $J(i, j, k, l) < Jlast$ Let $Jlast = J(i, j, k, l), \theta^i(j+1, k, l) = \theta^i(j, k, l) + C(i)\Delta i/\sqrt{\Delta(i)\Delta(i)^T}$. Calculate fitness function $J(i, j, k, l)$ else let $m = N_s$.

    (e) Go to next bacterium

5. End for (Chemotaxis loop)

6. Reproduction: Compute the health of the bacterium i: Arrange the bacteria and chemotactic parameters $C(i)$ in order of ascending cost $Jhealth$. Bacteria with the highest $Jhealth$ values split into two.

7. End for (Reproduction loop)

8. Elimination-dispersal: Eliminate and disperse bacteria with probability P_{ed}.

9. End for (Elimination dispersal loop).

---

image.

$$MSE = 1/MN \sum_{i=0}^{M-1} \sum_{J=0}^{N-1} [A(i,j) - B(i,j)]^2 \qquad (1.32)$$

$$PSNR = 10\log_{10}[255^2/MSE] \qquad (1.33)$$

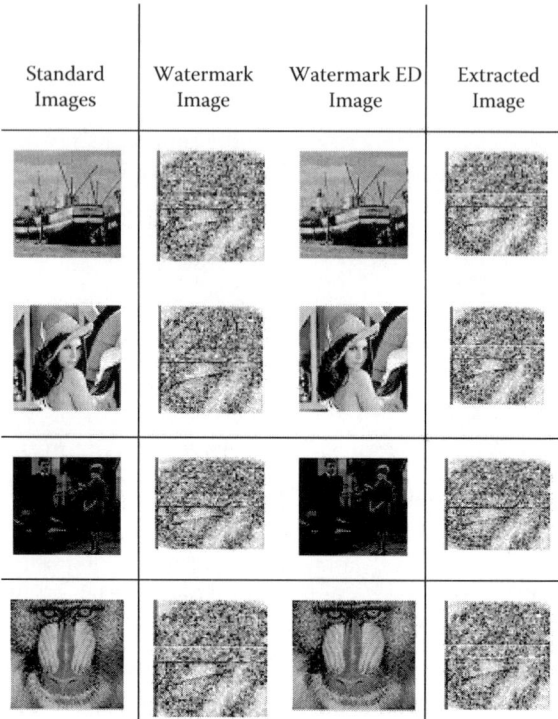

FIGURE 1.9: Sample output obtained by applying BFOA watermarking system

1.9.2 Normalized Absolute Error

Normalized absolute error (NAE) is the total normalized error values between the cover image and the watermarked images. The large value of normalized absolute error (NAE) means that the image is of poor quality. NAE is defined as follows:

$$NAE = \sum_{j=1}^{M}\sum_{k=1}^{N}|x_{jk} - x_{jk'}| / \sum_{j=1}^{M}\sum_{k=1}^{N}|x_{jk}| \qquad (1.34)$$

1.9.3 Normalized Cross Correlation

The normalized cross correlation (NCC) metric is calculated as the ratio between the net sum of the multiplication of the corresponding pixel densities of the watermarked image and the cover image, and the net sum of the squared values of the pixel densities of the cover image. The symbols A and B are images; M and N are rows and columns of images. The normalized cross

TABLE 1.2: Performance analysis for different watermarked images using GA watermarking system.

Standard Images	PSNR(dB)	NCC	NAE
BOAT	45.98	0.883	0.049
BABOON	45.23	0.88	0.045
LENA	46.32	0.89	0.047
COUPLE	47.89	0.896	0.043

correlation value would ideally be 1 if both images are identical.

$$NCC = \sum_{i=1}^{M}\sum_{j=1}^{N}(A_{i,j} * B_{i,j}) / \sum_{i=1}^{M}\sum_{j=1}^{N}(A_{i,j})^2 \qquad (1.35)$$

Tables 1.2 and 1.3 show the various performance analyses for different watermarked images using a GA watermarking system and the proposed BFOA watermarking system. Here, the standard cover images are taken into consideration.

From the above Tables 1.2 and 1.3, it is clear that the BFOA watermarking model performed better than the GA watermarking model. Figure 1.10 shows that BFOA performed better than other watermarking models. The standard images were taken and compared with the results in literature.

1.10 Conclusion

In this chapter, novel modality-level fusion algorithms for multimodal biometric features and genetic algorithm and bacterial foraging optimization algorithm watermarking models were implemented. Each biometric modality was individually extracted and the obtained modalities were fused together. As a result, the fusion methods successfully produced the fused templates and the

TABLE 1.3: Performance analysis for different watermarked images using BFOA watermarking system.

Standard Images	PSNR(dB)	NCC	NAE
BOAT	53.5	0.945	0.037
BABOON	52.78	0.94	0.034
LENA	54.8	0.937	0.036
COUPLE	55.7	0.952	0.031

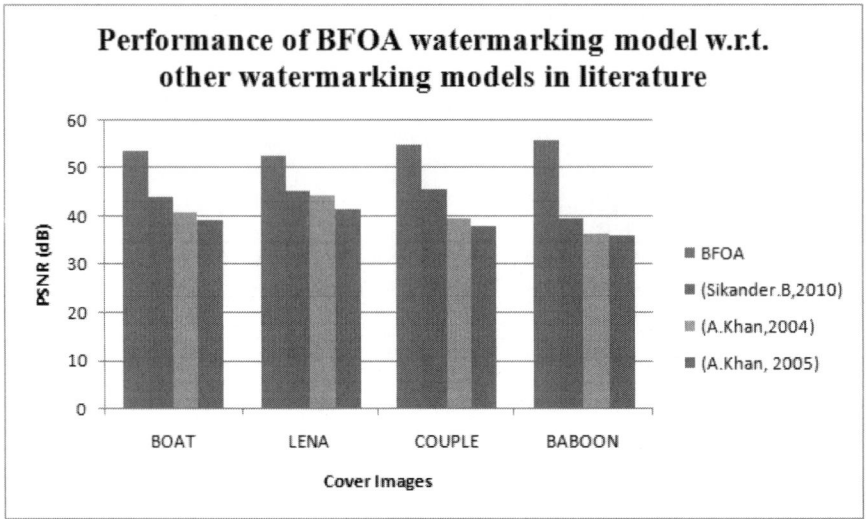

FIGURE 1.10: Performance of BFOA watermarking model vs. other watermarking models in the literature

best fused template was identified. The best fused template had been watermarked in different cover images using genetic algorithm and bacterial foraging optimization algorithm watermarking models. Various metrics such as PSNR, NAE, and NCC are used to measure the image quality. CASIA database is chosen for the biometric images. All the images are 8 bit gray-level JPEG images with a resolution of 320×280. The experimental results show that the bacterial foraging optimization algorithm for watermarking model provided better results than other watermarking models.

Bibliography

[1] J. Aravinth and S. Valarmathy. A novel feature extraction techniques for multimodal score fusion using density based Gaussian mixture model approach. *International Journal of Emerging Technology and Advanced Engineering*, 2(1):189–197, 2012.

[2] Mohd Shahrimie Asaari, Shahrel A. Suandi, and Bakhtiar Affendi Rosdi. Fusion of band limited phase only correlation and width centroid contour distance for finger based biometrics. *Expert Systems with Applications*, 41(7):3367–3382, 2014.

[3] Anne M.P. Canuto, Fernando Pintro, and João C. Xavier-Junior. Investigating fusion approaches in multi-biometric cancellable recognition. *Expert Systems with Applications*, 40(6):1971–1980, 2013.

[4] S. Chowhan, U.V. Kulkarni, and G.N. Shinde. Iris recognition using modified fuzzy hypersphere neural network with different distance measures. *International Journal of Advanced Computer Science and Applications*, 2(6):130–134, 2011.

[5] Nedeljko Cvejic, Artur Loza, David Bull, and Nishan Canagarajah. A similarity metric for assessment of image fusion algorithms. *International Journal of Signal Processing*, 2(3):178–182, 2005.

[6] Xiumei Guo, Weidong Zhou, and Yu Wang. Palmprint recognition algorithm with horizontally expanded blanket dimension. *Neurocomputing*, 127:152–160, 2014.

[7] Yu Han, Yunze Cai, Yin Cao, and Xiaoming Xu. A new image fusion performance metric based on visual information fidelity. *Information fusion*, 14(2):127–135, 2013.

[8] Zengxi Huang, Yiguang Liu, Chunguang Li, Menglong Yang, and Liping Chen. A robust face and ear based multimodal biometric system using sparse representation. *Pattern Recognition*, 46(8):2156–2168, 2013.

[9] N. Indhumadhi and G. Padmavathi. Enhanced image fusion algorithm using laplacian pyramid and spatial frequency based wavelet algorithm. *International Journal of Soft Computing and Engineering*, 1(5):298–303, 2011.

[10] Asifullah Khan, Anwar M. Mirza, and Abdul Majid. Optimizing perceptual shaping of a digital watermark using genetic programming. *Iranian Journal of Electrical and Computer Engineering*, 3(2):144–150, 2004.

[11] Vidhi Khanduja, Om Prakash Verma, and Shampa Chakraverty. Watermarking relational databases using bacterial foraging algorithm. *Multimedia Tools and Applications*, 74(3):813–839, 2015.

[12] Jing Li, Jian Cao, and Kaixuan Lu. Improve the two-phase test samples representation method for palmprint recognition. *Optik-International Journal for Light and Electron Optics*, 124(24):6651–6656, 2013.

[13] Heng Fui Liau and Dino Isa. Feature selection for support vector machine-based face-iris multimodal biometric system. *Expert Systems with Applications*, 38(9):11105–11111, 2011.

[14] S. Anu H. Nair and P. Aruna. Comparison of dct, svd and bfoa based multimodal biometric watermarking systems. *Alexandria Engineering Journal*, 54(4):1161–1174, 2015.

[15] S. Anu H. Nair, P. Aruna, and K. Sakthivel. Sparse representation fusion of fingerprint, iris and palmprint biometric features. *International Journal of Advanced Computer Research*, 4(1):46–53, 2014.

[16] Omar Nibouche and Jianmin Jiang. Palmprint matching using feature points and svd factorisation. *Digital Signal Processing*, 23(4):1154–1162, 2013.

[17] Chandrashekar M. Patil and Sudarshan Patilkulkarani. Iris feature extraction for personal identification using lifting wavelet transform. *International Journal of Computer Applications*, 1(14):68–72, 2010.

[18] Norman Poh, Arun Ross, Weifeng Lee, and Josef Kittler. A user-specific and selective multimodal biometric fusion strategy by ranking subjects. *Pattern Recognition*, 46(12):3341–3357, 2013.

[19] Vipul Sharma, S.S. Pattnaik, and Tanuj Garg. A review of bacterial foraging optimization and its applications. In *National Conference on Future Aspects of Artificial Intelligence in Industrial Automation, NCFAAIIA*, pages 9–12, 2012.

[20] Sumit Shekhar, Vishal M. Patel, Nasser M. Nasrabadi, and Rama Chellappa. Joint sparse representation for robust multimodal biometrics recognition. *IEEE Transactions on Pattern Analysis and Machine Intelligence*, 36(1):113–126, 2014.

[21] Sukhpreet Singh and Rachna Rajput. A comparative study of classification of image fusion techniques. *International Journal of Engineering And Computer Science*, 3(7):7350–7353, 2014.

[22] Riya Mary Thomas. Survey of bacterial foraging optimization algorithm. *International Journal of Science and Modern Engineering (IJISME)*, 1(4):11–12, 2013.

[23] Kamlesh Tiwari, Devendra K. Arya, G.S. Badrinath, and Phalguni Gupta. *Neurocomputing*, 116:222–230, 2013.

[24] Feng Yue and Wangmeng Zuo. Consistency analysis on orientation features for fast and accurate palmprint identification. *Information Sciences*, 268:78–90, 2014.

Chapter 2

Multilevel Thresholding for Image Segmentation Using Cricket Chirping Algorithm

S. Siva Sathya
Pondicherry University, Pondicherry, India

Jonti Deuri
Pondicherry University, Pondicherry, India

2.1	Introduction	32
2.2	Cricket Chirping Algorithm	34
2.3	Multilevel Thresholding (MT)	37
	2.3.1 Kapur's Method (Entropy Criterion Method)	38
	2.3.2 Otsu's Method (Between-Class Variance)	39
2.4	Multi-Level Thresholding Using Cricket Chirping Algorithm ...	41
2.5	Experiment Results	43
	2.5.1 Image Segmentation Using CCA and Kapur's Method .	43
	2.5.2 Image Segmentation Using CCA and Otsu's Method ...	48
2.6	Comparisons and Statistical Analysis	48
	2.6.1 Comparison between Otsu and Kapur in CCA	50
	2.6.2 Comparison of CCA with MTEMO	51
	2.6.3 Statistical Analysis	52
2.7	Conclusion	55
	Bibliography	55

One of the most important tasks in image processing is segmentation, which deals with the classification of pixels into two or more groups. This forms the basis for other complex image recognition and processing systems. In this work, a multi-level thresholding method for image segmentation is formulated using the cricket chirping algorithm (CCA), which is a metaheuristics algorithm inspired by the chirping behavior of cricket. In this method, random samples from a feasible search space depending on the image histogram are encoded as candidate solutions, and Otsu's and Kapur's methods are used to evaluate the quality. Guided by these objective values, the set of candidate solutions are evolved through the CCA operators until an optimal solution is found. The performance of the proposed algorithm is experimented in terms of accuracy, convergence, and speed, and compared with the other popular optimization techniques using prevailing parameters such as fitness function, peak-signal-to-noise ratio (PSNR), structural similarity (SSIM) index, and execution time.

2.1 Introduction

Image processing plays a crucial role in different fields, such as the medical discipline, agriculture, navigation, environment modeling, automatic event detection, surveillance, texture and pattern recognition, damage detection, etc. It is motivated by three major applications: the improvement of pictorial information for human perception, image processing for autonomous machine applications, and efficient storage and transmission. One of the major and primary tasks in image processing is image segmentation. Image segmentation is the partitioning of an image into multiple sets of pixels, segments, or regions that share some common characteristics, such as color, intensity, similarity, discontinuity, etc.

There are several image segmentation methods available, such as edge based, fuzzy theory based, partial differential equation (PDE) based, artificial neural network (ANN) based, threshold based, region based, etc. [17]. The threshold-based segmentation method is the most desirable and easy method among the existing methods. Image segmentation using thresholding can be categorized as bi-level (two classes having only one threshold value) and multilevel (more than two classes with different threshold values) threshold. Segmentation can also be grouped into local and global levels. Local segmentation segments sub-images that are small windows on a whole image. Global segmentation deals with the segmentation of the whole image. The global method is classified into two approaches: parametric and non-parametric approaches, which usually select thresholds by optimizing (maximization or minimization) some criterion functions defined from images [13].

There exist several classical thresholding methods: Otsu's [25] class variance method that maximizes the variance between classes; Kapur's [16] entropy criterion method that uses the maximization of the entropy to measure the homogeneity among classes; non-extensive or Tsallis entropy method [31, 7], etc. Since the classical methods search exhaustively for the best values to optimize the objective function for multilevel thresholding, they are computationally expensive. The use of evolutionary approaches for optimization has proved to be efficient. Nowadays, different heuristic and metaheuristic algorithms such as genetic algorithms, particle swarm optimization (PSO), bacterial foraging optimization (BFO), differential evaluation (DE), artificial bee colony (ABC), cuckoo search (CS), galaxy-based search algorithm (GSA), harmony search optimization (HS), the bat algorithm (BA), electro-magnetism optimization (EMO), the firefly algorithm (FA), hybrid method, etc. are widely used for solving the optimal multilevel image segmentation problem. The classical and optimization algorithm-based thresholding methods are employed to find the best possible threshold in the segmented histogram by satisfying some guiding parameters.

A general scheme to segment images by a genetic algorithm using an evaluation criterion was developed by S. Chabrier; it quantifies the quality of the image segmentation result [5]. This method can integrate a local ground truth when it is available, in order to set the desired level of precision of the final result. A genetic algorithm is then used in order to determine the best combination of information extracted by the selected criterion. A new framework for image segmentation based on multi-agent system (MAS) theory and a hybrid GA was proposed [22]. According to this approach, each segmentation agent performs the iterated conditional modes (ICM) algorithm starting from its own sub-optimal image and returns the resulting segmented image to the coordinator agent. This latter diversifies these initial sub-optimal images by applying the hybrid genetic operators, in order to produce promising new starting solutions, which are refined once again by the segmentation agents. Hammouche et al. proposed a method for image segmentation by combining a genetic algorithm with a wavelet transform [12]. First, the length of the original histogram is reduced by using the wavelet transform. Based on this lower resolution version of the histogram, the number of thresholds and the threshold values are determined by using a genetic algorithm. Similarly using PSO, Akhilesh Chander et al. presented a self-iterative method (Otsu's method) to find the suitable number of thresholds that should be used to segment an image [6]. The thresholds resulting from the iterative scheme are taken as initial thresholds and the particles are created randomly around these thresholds, for the PSO variant. This algorithm made a new contribution in adapting social and momentum components of the velocity equation for particle movement updates. A hybrid cooperative–comprehensive learning-based PSO algorithm for image segmentation using multilevel thresholding was developed by M. Maitra and A. Chatterjee [20], where an improved variant of PSO employs cloning of fitter particles at the expense of worst particles, in an attempt to further enhance the capability of the optimization strategy. PSO is also modified and hybridized with other algorithms for multilevel thresholding image segmentation [18, 33], while P. D. Sathya and R. Kayalvizhi [29, 28, 27] used a bacterial foraging algorithm to find the optimal threshold values for maximizing the Tsallis, Kapur's and Otsu's objective functions. Using the ABC algorithm, Ming-Huwi Horn proposed the maximum entropy-based artificial bee colony thresholding (MEABCT) method [15] for image segmentation, and ABC and improved ABC algorithms have been used for SAR image segmentation in [19, 14]. Diego Oliva et al. used the harmony search algorithm for multilevel thresholding image segmentation that encoded random samples from a feasible search space inside the image histogram as candidate solutions [24], whereas their quality was evaluated considering the objective functions that were employed by the Otsu's or Kapur's methods [1]. Some new metaheuristics algorithms, such as cuckoo search [30], galaxy-based search algorithm [3], bat algorithm [23], electro-magnetism optimization [26], firefly algorithm [8], etc. are also used widely in image segmentation. Diego Oliva et al. proposed a method that combines the good search capabilities of the EMO algorithm with objective functions proposed by the popular MT methods of Otsu and Kapur.

In this chapter, a multi-level thresholding image segmentation method is proposed, based on the cricket chirping algorithm combined with Kapur's entropy criterion method and Otsu's between-class variance method. The proposed method takes the random solution from a feasible search space inside the image histogram. Such samples build each cricket (candidate solution) in the CCA context and its quality is evaluated considering the objective function that is employed by the Kapur's and Otsu's methods. Directed by this objective value, the set of candidate solutions are adapted using the CCA operators, and a search for the optimal solution in the search space proceeds. The rest of this chapter is organized as follows. In Section 2.2, the CCA is introduced. Section 2.3 gives a brief description of the Kapur's and Otsu's methods. Section 2.4 explains the implementation of the proposed algorithm. Section 2.5 discusses experimental results after testing the proposed method over a set of benchmark images and performing the comparison with its counterpart in Section 2.6. Finally, Section 2.7 concludes the chapter.

2.2 Cricket Chirping Algorithm

A cricket is an insect similar to a grasshopper, with a flattened body that makes a sound that is known as chirping. Many animals use acoustic signals for intraspecific (within species) communication. For example, birds sing, frogs croak, and crickets chirp, etc. The cricket uses this chirping song mainly for mating and aggression. Based on this chirping behavior of the cricket, a new cricket chirping algorithm (CCA) has been proposed by J. Deuri and S. S. Sathya [8, 9]. The chirping of a cricket is categorized as a calling chirp when the cricket chirps for mating, a courtship chirp when the male cricket attempts to mate with the female, a copulatory chirp after successful mating, and an aggressive chirp when the cricket combats with other crickets.

The chirping rate of a cricket depends on the outside temperature. Each cricket is assumed to be a solution in the search space and is characterized by its position. Out of the total cricket population, a few of them as determined by the user are randomly designated as the female population. For simplicity, the crickets are assumed to be in two phases: mating phase and aggression phase.

- Mating Phase: In this phase, the male cricket produces a calling chirp for mating. This calling chirp attracts the female cricket, and the other male crickets will move away. After mating, the male and female produce offspring, which means they are taken to new positions in the search space.

- Aggression Phase: In this phase, the cricket chirps for aggression by emitting an aggressive chirp. The other male crickets are warned and get attacked, and the female crickets will move away. But all crickets may not chirp for aggression. The probability of chirping for aggression is assumed to be Pagg, which is in between [0, 1]. The male crickets fight with other male crickets, and the winning cricket takes the place of the solution and removes the loser cricket.

Based on his chirping rate in a certain temperature, the cricket moves to a new position by emitting a mating song, and mates with females to produce offspring. The offspring represent a new position of the cricket. The relationship between air temperature and the chirping rate is calculated by Dolbear's law, developed by an American physicist and naturalist, Amos Emerson Dolbear [10]. The chirping of crickets is related to temperature as well as age and mating success. Dolbear expressed the relationship using Equation 14.1, which provides a way to estimate the temperature T_c in degrees Celsius from the number of chirps per minute C_n:

$$T_c = 10 + \frac{C_n - 40}{7} \qquad (2.1)$$

The chirping rate is derived by using Dolbear's law in a certain temperature T_c,

$$C_n = (T_c - 10) * 7 + 40 \qquad (2.2)$$

We assume that the chirping rate is the frequency of the cricket's chirp. From the frequency we calculate the velocity of each cricket as follows:

$$vel = C_n * \lambda \qquad (2.3)$$

Here, λ is the wavelength, the gap between one chirp and another chirp, which is uniformly drawn. From the velocity, we calculate the step size by using Equation 14.4

$$Step = Vel * (P - P_{best}) * \alpha \qquad (2.4)$$

where $\alpha = 0.01$ is a constant value that is used to control the movements of the cricket within a bounded space, P is the current position, and P_{best} is the best position ever encountered by the cricket. Then the cricket will move to the new position by using the following formula:

$$New_{pos} = pos + Step \qquad (2.5)$$

Equations 14.1 to 14.5 are used when the cricket chirps for mating, and they change their step size according to the chirping rate at a certain temperature. The mating process is done by adopting the crossover operator of the genetic algorithm.

When the cricket chirps for aggression, he moves to a new position using the random walk. By emitting an aggressive song, he fights with other male crickets and the winning crickets reach new positions in the search space. The cricket with the highest fitness will be selected as the winner cricket. The fitness of the male cricket is calculated based on his attractiveness, and the position of low fit cricket is replaced with a high fit cricket. The aim is to use the new and potentially better solutions (crickets) to replace a not-so-good solution. The attraction is based on the loudness of the chirping sound. Based on the chirping rates the cricket moves to the new position. The pseudo code of the cricket chirping algorithm appears below.

Algorithm [Pseudocode for Cricket Chirping Algorithm]
Begin
Input: $f_c(x)$: Objective function; n: number of crickets; T: Temperature; Pagg: Aggression rate; k: No. of female crickets, $1 < k < n/2$

1. Initialize the cricket population.

2. Randomly choose k crickets as female crickets.

3. Calculate the fitness of each cricket.

4. Assign best cricket f_{best} ← value of the best fit cricket, P_{best} ← position of the best cricket.

5. Set f_{best} as the current g_{best}. In initial generation, $g_{best} = f_{best}$.

6. Repeat till the stopping criteria not met.

    (a) Mating Phase: Chirp for mating (Call procedure calling-chirp()). Mate with female crickets (Call procedure mating()).

    (b) Agrression Phase: With probability $Pagg$, allow the male crickets to chirp for aggression (Call procedure aggr-chirp()).

    (c) Compute the fitness of the crickets produced by mating and aggression in the new positions.

    (d) Select f_{best}, from the new positions of the cricket.

    (e) If $f_{best} > g_{best}$, then update g_{best} with the current f_{best}.

7. End while

8. Return the global best cricket at termination.

End

Procedure calling-chirp()
Begin
For every male cricket
1. Calculate the frequency of Chirping and velocity using Equations 14.2 and 14.3.
2. Calculate the step size using Equation 14.4.
3. Move each cricket to the new position using Equation 14.5
4. Return crickets in new position.
End

Procedure mating()
Begin
1. For every male crickets Mi in their new position, randomly choose a female cricket Fi.
2. Randomly choose a cut point in both Mi and Fi.
3. Exchange the genetic materials of both Mi and Fi with reference to their cut to produce two new offspring.
4. Return the two offspring and the parents as the new cricket positions.
End

Procedure aggr-chirp()
Begin
1. If $rand \geq Pagg$
2.   Randomly walk to the new position.
3.   Fight with other male crickets.
4.   Return the winner cricket (position).
End

2.3 Multilevel Thresholding (MT)

In the thresholding process, the pixels of a gray scale image are divided into sets or groups depending on their intensity level l. For this classification, it is necessary to select a threshold value (θ) by following some simple rules, Where p is one of the $n \times m$ pixels of the gray scale image I_g that can be represented in l gray scale levels, $l = \{1, 2, \cdots, (l-1)\}$. G_0, G_1 are the groups in which the pixel p can be located.

$$\begin{aligned} G_0 \leftarrow p, & \quad if \ 0 \leq p \leq \theta \\ G_1 \leftarrow p, & \quad if \ \theta \leq p \leq l-1 \end{aligned} \qquad (2.6)$$

This can be extended for multilevel thresholding as follows:

$$
\begin{array}{ll}
G_0 \leftarrow p, & if\ 0 \leq p \leq \theta_1 \\
G_1 \leftarrow p, & if\ \theta_1 \leq p \leq \theta_2 \\
\ldots & \ldots \\
G_i \leftarrow p, & if\ \theta_i \leq p \leq \theta_{i+1} \\
\ldots & \ldots \\
G_n \leftarrow p, & if\ \theta_n \leq p \leq (l-1)
\end{array}
\tag{2.7}
$$

where, $\theta_1, \theta_2, \ldots \theta_{i+1}, \ldots \theta_n$, are different thresholds. The objective for both bi-level and multi-level threshold is to select the θ values that correctly identify the classes. Two most frequently used methods for determining such values are Otsu's and Kapur's methods, which propose a different objective function that must be maximized in order to find the optimal threshold values.

2.3.1 Kapur's Method (Entropy Criterion Method)

Kapur's method is based on the entropy and the probability distribution of the image histogram; it is also known as the entropy criterion method [16]. This method intends to obtain the optimal threshold value that maximizes the overall entropy of an image that measures the compactness and separability among classes or groups. In this regard, the entropy has the maximum value when the optimal value appropriately separates the classes. For the bi-level example, the objective function of the Kapur's problem can be defined as:

$$
I(\theta) = H_1^c + H_2^c
$$
$$
c = \begin{cases} 1,2,3 & if\ RGB\ image \\ 1 & if\ gray\ scale\ image \end{cases}
\tag{2.8}
$$

where H_1^c and H_2^c are entropies and computed by the following model:

$$
H_1^c = \sum_{i=1}^{thr} \frac{p_i^c}{w_0^c} \ln\left(\frac{p_i^c}{w_0^c}\right)
$$
$$
H_2^c = \sum_{thr=1}^{l} \frac{p_i^c}{w_1^c} \ln\left(\frac{p_i^c}{w_1^c}\right)
\tag{2.9}
$$

where p_i^c is the probability distribution of the intensity levels, which is obtained using [18]; w_0^c and w_1^c are probability distributions for c_1 and c_2. For multiple threshold values, it is necessary to divide the image into k classes using a similar number of thresholds. Under such conditions, the new objective function is defined as follows:

$$
F_{kapur}(TH) = max \sum_{i}^{k} H_i^c
$$
$$
c = \begin{cases} 1,2,3 & if\ RGB\ image \\ 1, & if\ gray\ scale\ image \end{cases}
\tag{2.10}
$$

where $TH = [\theta_1, \theta_2, \cdots, \theta_{k-1}]$ is a vector that contains the multiple thresholds. Each entropy is computed separately with its respective θ value. It is expanded for k entropies as follows:

$$H_1^c = \sum_{i=1}^{\theta_1} \frac{p_i^c}{w_0^c} ln\left(\frac{p_i^c}{w_0^c}\right)$$

$$H_2^c = \sum_{i=\theta_1+1}^{\theta_2} \frac{p_i^c}{w_1^c} ln\left(\frac{p_i^c}{w_1^c}\right) \quad (2.11)$$

$$\cdots$$

$$H_k^c = \sum_{i=\theta_{k+1}+1}^{\theta_2} \frac{p_i^c}{w_{k-1}^c} ln\left(\frac{p_i^c}{w_{k-1}^c}\right)$$

where $w_0(\theta) = \sum_{i=1}^{\theta_1} p_i$; $w_1(\theta) = \sum_{i=\theta_1+1}^{\theta_2} p_i$; $\cdots$ $w_{k-1}(\theta) = \sum_{i=\theta_{k+1}}^{l} p_i$.

2.3.2 Otsu's Method (Between-Class Variance)

The between-class variance is a non-parametric and unsupervised technique for thresholding proposed by Otsu [25], one that employs the maximum variance value of the different classes as a criterion to segment the image. Taking the intensity levels from a gray scale image or from each component of an RGB (red, green, and blue) image, the probability distribution of the intensity values is computed as follows:

$$p_i^c = \frac{h_i^c}{NP}; \sum_{i=1}^{NP} p_i^c = 1$$

$$c = \begin{cases} 1,2,3 & if \ RGB \ image \\ 1, & if \ Gray \ scale \ image \end{cases} \quad (2.12)$$

where i is a specific intensity level ($0 \leq i \leq (l-1)$), c is the component of the image that depends on the image type (gray scale or RGB), NP is the total number of pixels in the image, H_i^c (Histogram) is the number of pixels that corresponds to the i intensity level in c, and the histogram is normalized within a probability distribution p_i^c. For a bi-level segment, which is the simplest segmentation, two classes are defined as:

$$G_1 = \frac{p_1^c}{w_0^c}, \cdots, \frac{p_\theta^c}{w_0^c}$$

$$G_2 = \frac{p_{\theta+1}^c}{w_1^c}, \cdots, \frac{p_l^c}{w_1^c} \quad (2.13)$$

where w_0^c and w_1^c are probabilities distributions for G_1 and G_2, as is shown by Equation 13.14

$$w_0^c = \sum_{i=1}^{\theta} p_i^c; \quad w_1^c = \sum_{i=\theta+1}^{l} p_i^c \qquad (2.14)$$

The mean level is calculated as follows:

$$\mu_0 = \sum_{i=1}^{\theta} \frac{ip_i^c}{w_0^c} \mu_1 = \sum_{i=1}^{\theta} \frac{ip_i^c}{w_1^c} \qquad (2.15)$$

Then Otsu's variance between classes is calculated as follows:

$$\sigma^2 = \sigma_1 + \sigma_2 \qquad (2.16)$$

where σ_1 and σ_2 are variances of G_1 and G_2 that are calculated as follows:

$$\sigma_1 = w_0^c \left(\mu_0 + \mu_T\right)^2 \sigma_2 = w_1^c \left(\mu_1 + \mu_T\right)^2 \qquad (2.17)$$

where, $\mu_T = w_0^c \mu_0 + w_1^c \mu_1$ and $w_0^c + w_1^c = 1$. Based on the values, the objective function is defined as:

$$F_{Otsu}(\theta) = max\left(\sigma^2(\theta)\right)$$
$$0 \leq \theta \leq (l-1) \qquad (2.18)$$

where σ^2 is the Otsu's variance for a given θ value. The optimization problem is reduced to find the intensity levels (θ) that maximize Equation 13.18. This equation can be rewritten for multiple threshold values as follows:

$$F_{Otsu}(X) = max\left(\sigma^2(X)\right)$$
$$0 \leq \theta_i \leq (l-1), \quad i = 1, 2, \cdots, k \qquad (2.19)$$

where $X = [\theta_0, \theta_1, \cdots, \theta_{k-1}]$ is a vector containing thresholds and the variances are computed as:

$$\sigma^2 = \sum_{i=1}^{k} \sigma_i = \sum_{i}^{k} w_i \left(\mu_i + \mu_T\right)^2 \qquad (2.20)$$

Here i represent a specific class. w_i and μ_i are the probability of occurrence and the mean of a class, respectively. For MT, such values are obtained as follows:

$$w_0(\theta) = \sum_{i=1}^{\theta_1} p_i$$
$$w_1(\theta) = \sum_{i=\theta_1+1}^{\theta_2} p_i \qquad (2.21)$$
$$\cdots$$
$$w_{k-1}(\theta) = \sum_{i=\theta_{k+1}+1}^{l} p_i$$

and for mean values,

$$\mu_0 = \sum_{i=1}^{\theta_1} \frac{ip_i}{w_0(\theta_1)}$$

$$\mu_1 = \sum_{i=\theta_2+1}^{\theta_2} \frac{ip_i}{w_1(\theta_2)} \quad (2.22)$$

$$\cdots$$

$$\mu_{k-1} = \sum_{i=\theta_k+1}^{l} \frac{ip_i}{w_i(\theta_1)}$$

2.4 Multi-Level thresholding Using Cricket Chirping Algorithm

In this section, the cricket chirping algorithm is utilized to find the optimal threshold values for multilevel thresholding in image segmentation. The problem is viewed as an optimization problem and the methodology to apply CCA for optimizing threshold values is presented. Here the algorithm is implemented considering the two objective functions, namely Otsu's between class variance and Kapur's entropy criteria. The segmentation problem is represented as an optimization problem as follows:

$$Maximize, \; F(x)$$

$$F(x) = \left[\int_{kapur}(TH) \, or \int_{otsu}(TH) \right] \quad (2.23)$$

$$TH = [\theta_1, \theta_2, \cdots, \theta_k]$$

$$Subject \; to, TH \in X$$

where $X = \{TH \in \theta_i | \; 0 \leq \theta_i \leq 255, i = 1, 2, \cdots, k\}$ is the bounded feasible region constrained by the interval [0-255].

In the cricket population, each cricket uses k different decision variables within the optimization algorithm, each of which represents different threshold points that are used for segmentation. The complete cricket population is represented as: $S = [TH_1, TH_2, \cdots, TH_N], TH_i = [\theta_1, \theta_2, \cdots, \theta_k]^T$ with the search space boundary having lower bound $lb = 0$, upper bound $ub = 255$, and N being the total cricket population. The implementation of the proposed method is summarized below.

Pseudo code for multilevel image segmentation using CCA

Begin

1. Read the image Ig or Irgb.
2. Obtain histograms, for RGB images hr, hg, hb, and for gray scale images hgr.
3. Calculate the probability distribution using Equation 14.12 and obtain the histograms.
4. Initialize the CCA parameters: $T, k, Cr, Pagg$.
5. Initialize the crickets.
6. Evaluate the fitness using Kapur's or Otsu's methods.
7. While ($i \leq max - iter$) or stopping criteria not met:
    (a) Allow the crickets to chirp for mating.
        i. Compute the $Cn, Vel, Step$ using Equations 14.2, 14.3, and 14.4.
        ii. Let them mate with female crickets.
    (b) Allow the crickets to chirp for aggression.
        i. Let them fight with other male crickets
    (c) Calculate the fitness using Kapur's or Otsu's methods.
    (d) Select the cricket that has the best fit objective function value.
8. Apply the threshold values to the image Ig or Irgb after computing the best fit objective function value.

End

The performance of CCA depends on the parameter settings. In this implementation, we set the aggression rate as 0.45, mating rate 0.80, and population size 50, as shown in Table 2.1. The temperature of the outside environment is chosen randomly from [10 to 100] degrees Celsius. The female population k is chosen randomly such that $k \leq n/2$, and one-point crossover is used for mating.

TABLE 2.1: Control parameters of CCA.

Population Size(n)	Female pop. Size(k)	Aggression rate	Mating rate
50	25	0.45	0.80

2.5 Experiment Results

The proposed CCA algorithm for multi-level thresholding has been tested for 8 benchmark test images, which are in JPEG format and provided by the Berkeley segmentation data set [11, 4]. The experiments are carried out on an Intel Core^{TM} i5 CPU and 4 GB RAM, running under the Windows 7.0 operating system. The algorithms were implemented using MATLAB. The required parameters are set as in Table 2.1 for all the test images. The program was run 50 times for each image separately, and the performance of the proposed method was evaluated using well-known parameters such as peak signal-to-noise ratio (PSNR) and structural similarity indices (SSIM) [2, 32]. PSNR is used to assess the similarity of the segmented image against a reference image (original image) based on θ. Both PSNR and MSE are defined as:

$$SNR = 20 \log_{10}\left(\frac{255}{\sqrt{MSE}}\right)$$
$$MSE = \frac{\sum_{i=1}^{r}\sum_{j=1}^{c}\left[I_0(i,j) - I_s(i,j)\right]^2}{r \times c} \quad (2.24)$$

where, I_0 is the original image and I_s is the segmented image, r and c are the total number of rows and columns of the image, respectively. The structural similarity (SSIM) index is a full reference metric used for measuring the similarity between two images. It is the measurement or prediction of image quality based on an initial uncompressed or distortion-free image as a reference [2]. SSIM evaluates the visual similarity between the original image x and the segmented image y, both of size $N \times N$, as:

$$SSIM(x,y) = \frac{(2\mu_x\mu_y + c_1)(2\sigma_{xy} + c_2)}{(\mu_x^2 + \mu_y^2 + c_1)(\sigma_x^2 + \sigma_y^2 + c_2)} \quad (2.25)$$

where μ_x is the average of x, μ_y is the average of y, σ_x^2 the variance of x, σ_y^2 the variance of y, σ_{xy} the covariance of x and y. $C_1 = (k_1 l)^2, C_2 = (k_2 l)^2$ are two variables to stabilize the division with the weak denominator l the dynamic range of the pixel values (typically this is $2^{bitperpixel} - 1$), $k_1 = 0.01$ and $k_2 = 0.03$ by default.

2.5.1 Image Segmentation Using CCA and Kapur's Method

In this section, the CCA is run considering Kapur's entropy method as the objective function that is given in Equation 14.10. The approach is applied to the complete set of benchmark images considering four different threshold points, 2, 3, 4, 5, and the results such as PSNR, SSIM, execution time, etc., are calculated. The original image and segmented image with a different threshold value and their histogram and fitness graph are shown in Figures 2.1 to 2.4.

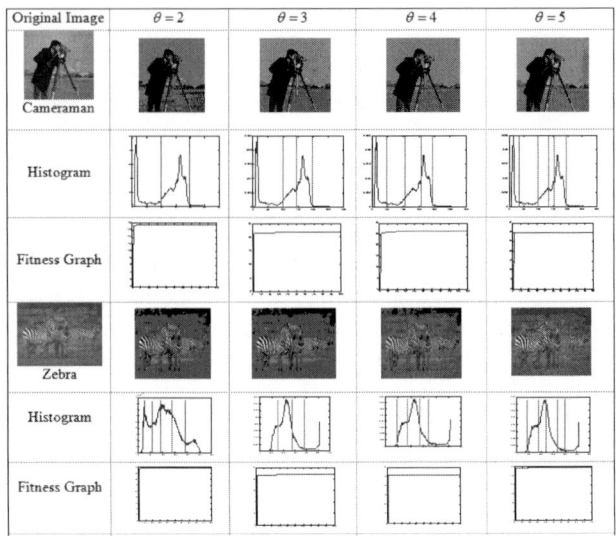

FIGURE 2.1: Resultant images after applying the CCA to the set of benchmark images (Cameraman and Zebra) using Kapur's function

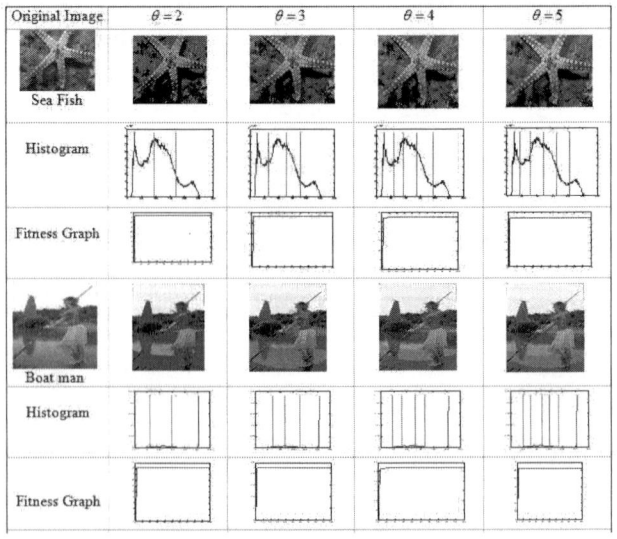

FIGURE 2.2: Resultant images after applying the CCA to the set of benchmark images (Sea Fish and Boat Man) using Kapur's function

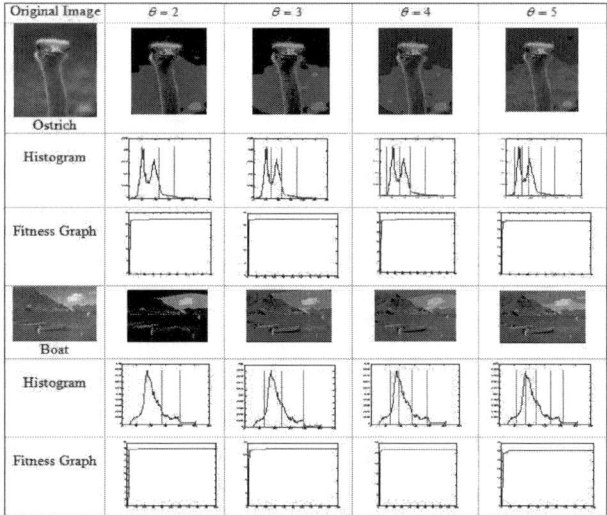

FIGURE 2.3: Resultant images after applying the CCA to the set of benchmark images (Ostrich and Boat) using Kapur's function

From the results, it is evident that the PSNR and SSIM values increment their magnitude as the number of threshold points increases. The best result found using CCA with Kapur's function in the 50 runs for each image with the different threshold value is shown in Table 2.2.

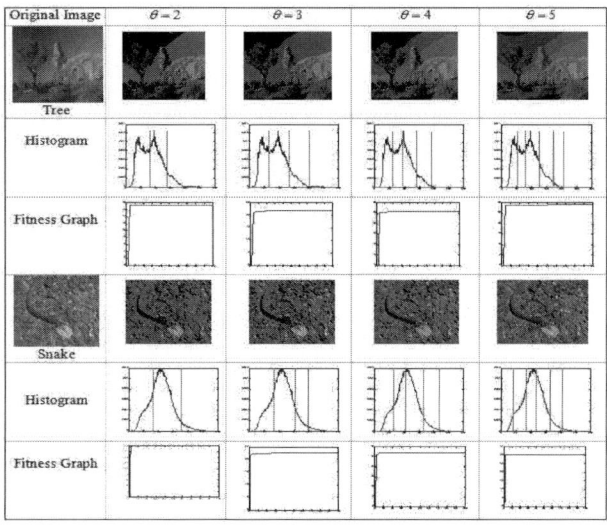

FIGURE 2.4: Resultant images after applying the CCA to the set of benchmark images (Tree and Snake) using Kapur's function

TABLE 2.2: Best results after applying the CCA using Kapur's function to the set of benchmark images.

Image	k	Best fitness	PSNR	SSIM	Time(sec)	Threshold value
Cameramen	2	17.5526	14.1796	0.5394	0.4604	120 193
	3	20.8251	17.9649	0.6351	0.5469	100 137 216
	4	25.8039	18.0425	0.6310	0.6300	36 82 155 199
	5	30.2333	21.8695	0.7037	0.7115	31 91 126 149 194
Starfish	2	18.7536	14.4390	0.4081	0.4665	90 170
	3	23.2861	17.0826	0.5248	0.5484	65 127 179
	4	27.4114	19.2181	0.6367	0.6410	58 93 138 193
	5	31.4256	20.4641	0.7104	0.7221	43 76 121 165 207
Zebra	2	17.8065	14.4871	0.4552	0.4592	92 160
	3	22.1330	16.0182	0.5352	0.5602	78 138 189
	4	25.8618	19.1451	0.5639	0.6451	44 87 136 168
	5	29.8213	21.2276	0.6913	0.7214	44 88 122 165 197
Boat Man	2	18.0242	16.0707	0.6066	0.4620	61 147
	3	22.6285	19.2583	0.7321	0.5506	66 115 177
	4	26.9793	21.1183	0.7657	0.6339	45 82 131 166
	5	30.9901	22.6232	0.7820	0.7164	48 78 119 148 184
Ostrich	2	22.5240	16.2691	0.0771	0.4633	64 125 186
	3	22.5229	16.1868	0.4785	0.5496	73 116 176
	4	26.7208	19.9436	0.7018	0.6325	27 79 121 180
	5	30.4907	22.7053	0.7524	0.7199	32 64 92 141 192
Boat	2	18.0903	9.2283	0.1788	0.4734	138 202
	3	22.7516	17.5315	0.5702	0.5494	66 121 196
	4	26.7387	19.8912	0.6812	0.6466	65 94 139 192
	5	30.8757	21.2346	0.7143	0.7208	57 89 121 168 209
Tree	2	17.2734	15.5238	17.2734	0.2463	81 141
	3	21.7707	16.5785	21.7707	0.2905	68 127 196
	4	25.6561	18.6202	25.6561	0.3299	57 102 141 187
	5	29.2947	21.4849	29.2947	0.3922	40 79 104 149 201
Snake	2	17.9342	14.6194	0.5222	0.4603	84 176
	3	22.5789	15.4028	0.5948	0.5548	84 154 199
	4	26.7863	18.4103	0.7405	0.6347	66 104 165 220
	5	30.5564	21.0050	0.8358	0.7187	42 86 119 164 207

TABLE 2.3: Best results of applying the CCA using Otsu's function to the set of benchmark images.

Image	k	Best fitness	PSNR	SSIM	Time (sec)	Threshold value
Cameramen	2	3.6517e+003	18.5827	0.5738	0.3777	70 145
	3	3.7269e+003	21.1954	0.6444	0.4221	58 127 157
	4	3.7733e+003	22.6114	0.686	0.4691	43 98 135 159
	5	3.8033e+003	23.6897	0.6981	0.5438	52 103 130 151 175
Sea Fish	2	2.5466e+003	15.5738	0.4545	0.3466	84 155
	3	2.7757e+003	17.8468	0.5645	0.3857	66 119 171
	4	2.8594e+003	19.2581	0.6483	0.4147	52 98 132 183
	5	2.8942e+003	21.2615	0.7152	0.4520	39 77 119 141 188
Zebra	2	1.3940e+003	13.9618	0.4453	0.3568	98 171
	3	1.5184e+003	16.3846	0.5785	0.5267	81 121 205
	4	1.5648e+003	18.6222	0.6714	0.6239	72 98 137 208
	5	1.5780e+003	20.1544	0.7101	0.6898	64 81 104 154 203
Boat Man	2	5.0744e+003	14.2071	0.5385	0.3552	107 193
	3	5.2342e+003	17.5607	0.6743	0.3840	83 134 193
	4	3080e+003	20.2656	0.7661	0.4268	59 107 153 234
	5	5.3443e+003	21.3697	0.7818	0.4473	68 108 135 159 205
Ostrich	2	1.0729e+003	16.2329	0.4439	0.3614	73 133
	3	1.1352e+003	17.3580	0.4815	0.3907	63 96 136
	4	1.1716e+003	18.3364	0.5447	0.4205	57 84 133 180
	5	1.1958e+003	22.7184	0.7019	0.4480	47 68 90 123 163
Boat	2	1.2643e+003	12.3241	0.3490	0.3654	100 153
	3	1.3699e+003	18.6936	0.6477	0.3915	64 107 162
	4	1.4299e+003	18.3164	0.6735	0.4215	62 99 138 190
	5	1.4689e+003	21.0115	0.7260	0.2464	69 91 113 139 182
Tree	2	1.1887e+003	16.7795	0.5392	0.3101	72 119
	3	1.2672e+003	19.1352	0.6553	0.3674	56 92 129
	4	1.3031e+003	20.9931	0.7367	0.4197	47 80 111 153
	5	1.3220e+003	24.4303	0.8341	0.4828	36 56 85 112 136
Snake	2	1.1178e+003	15.7850	0.6250	0.3233	87 132
	3	1.2271e+003	18.7065	0.7594	0.3748	74 110 145
	4	1.2727e+003	20.5848	0.8205	0.46164	63 96 132 156
	5	1.3070e+003	22.0542	0.8619	0.5188	54 84 103 136 168

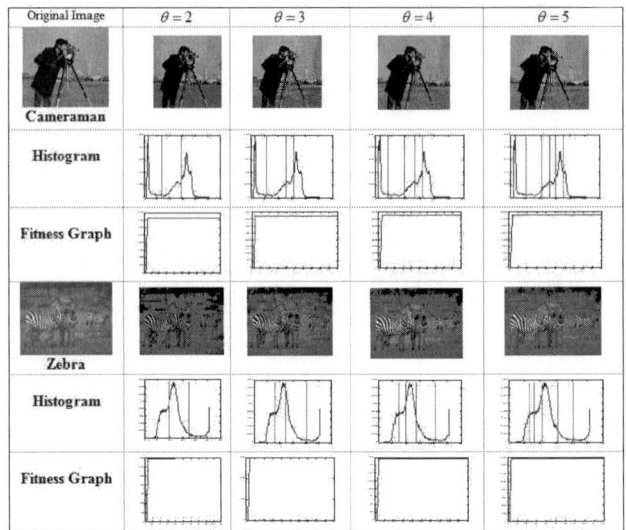

FIGURE 2.5: Resultant images after applying the CCA to the set of benchmark images (Cameraman and Zebra) using Otsu's function

2.5.2 Image Segmentation Using CCA and Otsu's Method

In this section, the CCA is run considering Otsu's between-class variances method as the objective function that is given in Equation 13.19. The approach is applied to the complete set of benchmark images considering four different threshold points, 2, 3, 4, 5, and the best results such as PSNR, SSIM, time, etc., are stated in Table 2.3. The segmented image, histogram, and fitness graph are shown in Figures 2.5 to 2.8. From the results, it is shown that the PSNR and SSIM values increment their magnitude as the number of threshold points increases.

2.6 Comparisons and Statistical Analysis

In this section, three different comparisons are done to analyze the results of the proposed method. The first comparison is executed between the two versions of CCA, via Kapur's function and Otsu's criterion. The second one analyses the comparison of CCA with the other state-of-the-art approaches. The third one is the statistical analysis of the obtained results of CCA and MTEMO (multilevel thresholding using electro-magnetism optimization) in order to verify its performance and computational effort.

Multilevel Thresholding for Image Segmentation Using CCA 49

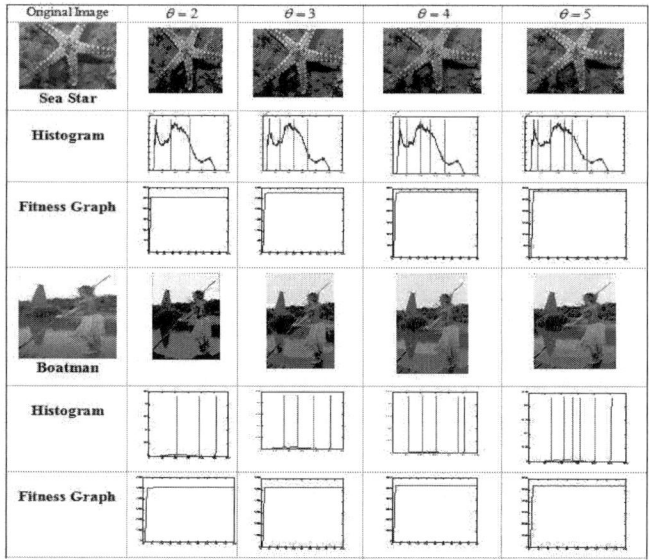

FIGURE 2.6: Resultant images after applying the CCA to the set of benchmark images (Sea Star and Boat Man) using Otsu's function

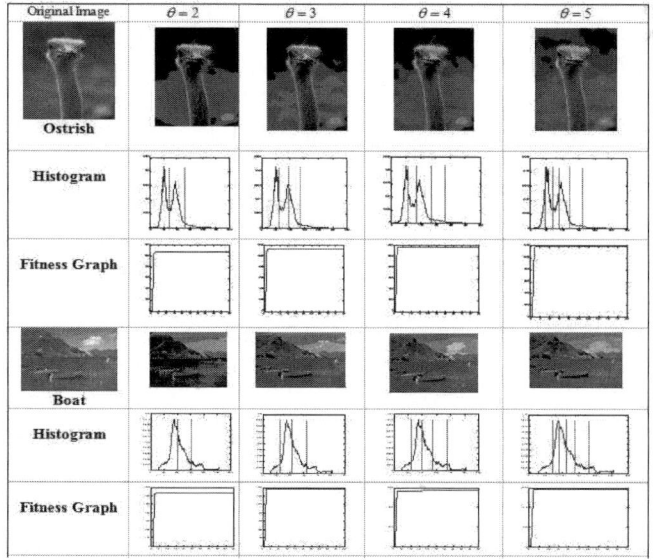

FIGURE 2.7: Resultant images after applying the CCA to the set of benchmark images (Ostrich and Boat) using Otsu's function

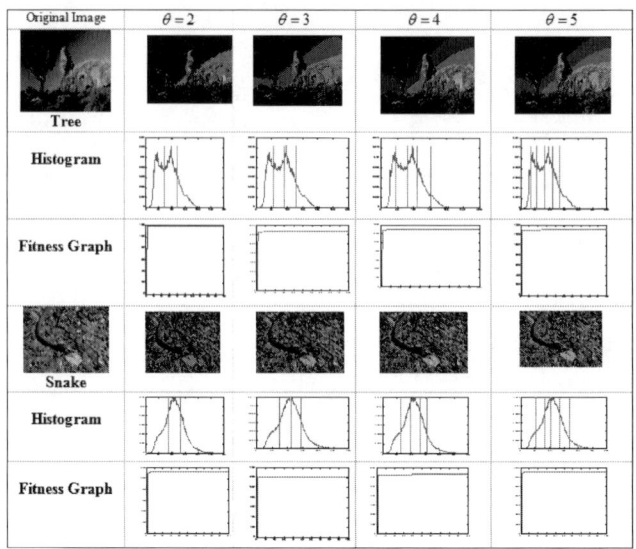

FIGURE 2.8: Resultant images after applying the CCA to the set of benchmark images (Tree and Snake) using Otsu's function

All the algorithms are run 50 times over each selected image. The images used for this test are the same as those selected in Section 2.5. For each image, the PSNR, SSIM, time, and the mean of the objective function values are calculated, wherein the entire test is performed using both Otsu's and Kapur's objective functions. We consider the MTEMO algorithm for comparison as presented by Oliva et al. [23]. The other methods, like genetic algorithms (GA), particle swarm optimization (PSO), and bacterial foraging (BF) are not considered for comparison as they were already compared with MTEMO [23].

2.6.1 Comparison between Otsu and Kapur in CCA

In order to statistically compare the results of Otsu and Kapur, a nonparametric significance test known as the Wilcoxon's rank test [21] for 50 independent samples has been conducted. Such proof allows assessing result differences among two related methods. The analysis is performed considering a 5% significance level over the peak-to-signal ratio (PSNR) data corresponding to the test images with two to five threshold points. Table 2.4 reports the p values produced by Wilcoxon's test for a pairwise comparison of the PSNR values between the Otsu and Kapur objective functions. The h represents the hypothesis. In this analysis, the null hypothesis is assumed that there is no difference between the values of the two objective functions. The alternative hypothesis considers a significant difference between the values of both approaches. All p values reported in Table 2.4 are less than 0.05 (5% significance level), which is strong evidence against the null hypothesis. So, it

TABLE 2.4: p-values produced by Wilcoxon's test comparing Otsu vs. Kapur over PSNR.

Image	θ	P-value: Otsu vs. Kapur	H
Cameramen	2	0.00	1
	3	0.00	1
	4	0.00	1
	5	0.00	1
Sea Star	2	0.00	1
	3	0.00	1
	4	0.00	1
	5	0.00	1
Zebra	2	0.00	1
	3	0.00	1
	4	0.00	1
	5	0.00	1
Boat Man	2	0.00	1
	3	0.00	1
	4	0.00	1
	5	0.00	1
Ostrich	2	0.00	1
	3	0.00	1
	4	0.00	1
	5	0.00	1
Boat	2	0.00	1
	3	0.00	1
	4	0.00	1
	5	0.00	1
Tree	2	0.00	1
	3	0.00	1
	4	0.00	1
	5	0.00	1
Snake	2	0.00	1
	3	0.00	1
	4	0.00	1
	5	0.00	1

is concluded that there is a significant difference between the values of the two objective functions. It means Otsu's PSNR mean values for the performance are statistically better than Kapur's PSNR values.

2.6.2 Comparison of CCA with MTEMO

The proposed method is compared with another metaheuristic method, MTEMO (multilevel thresholding using electro-magnetism optimization). Both the algorithms are run 50 times over each selected image. For each image, the PSNR, SSIM, time and the mean of the objective function values are

TABLE 2.5: Comparison of CCA and MTEMO using Kapur's method.

Image	θ	MTEMO			CCA			
		PSNR	Mean	SSIM	PSNR	Mean	SSIM	Time
Cameramen	2	13.626	17.5842	0.5202	14.1255	17.5663	0.5394	0.4604
	3	14.4602	21.976	0.5966	17.4939	21.9109	0.6351	0.5469
	4	20.1531	26.586	0.6672	20.1656	26.2674	0.6310	0.6300
	5	20.661	30.506	0.6850	21.8695	30.2333	0.7037	0.7115
Sea Star	2	14.3982	18.7542	0.4008	14.5611	18.7503	0.4081	0.4665
	3	16.987	23.3233	0.5257	17.0149	23.2790	0.5248	0.5484
	4	18.304	27.5817	0.5901	18.9300	27.4544	0.6367	0.6410
	5	20.165	31.5626	0.6697	20.7114	31.1217	0.7104	0.7221
Zebra	2	13.8051	17.8802	0.4455	14.2625	17.7929	0.4552	0.4592
	3	15.0013	22.3129	0.5162	15.8788	22.1162	0.5352	0.5602
	4	15.5085	26.5212	0.5440	16.2427	26.0255	0.5639	0.6451
	5	20.2723	30.2664	0.7052	20.5283	29.5575	0.6913	0.7214
Boat Man	2	15.8962	18.0625	0.6238	16.0127	18.0066	0.6258	0.4620
	3	18.1776	22.8079	0.7079	19.1801	22.6660	0.7498	0.5708
	4	20.1480	27.2968	0.7738	21.0354	26.8740	0.7852	0.6469
	5	22.2361	31.3337	0.8137	22.6732	30.9901	0.8355	0.7264
Ostrich	2	10.8341	18.1959	0.0804	10.9067	18.1839	0.0840	0.4833
	3	16.0263	22.5932	0.4415	16.3296	22.5554	0.4785	0.5496
	4	16.2206	26.8341	0.4528	19.1780	26.6143	0.7018	0.6325
	5	20.7745	31.0419	0.7763	21.5161	30.4390	0.7825	0.7220
Boat	2	9.3556	18.0709	0.1629	9.7614	18.0781	0.1963	0.4705
	3	17.1017	22.9471	0.5558	17.4877	22.8050	0.5702	0.5494
	4	18.9648	27.1648	0.6253	19.7622	26.8680	0.6812	0.6466
	5	19.8263	31.0884	0.6556	20.5056	30.5153	0.7143	0.7208
Tree	2	15.6147	17.2739	0.4821	15.8236	17.2706	0.4962	0.4776
	3	15.7403	21.8542	0.4905	16.3621	21.7709	0.5507	0.5578
	4	16.9652	26.0243	0.5553	18.4098	25.8224	0.6246	0.6436
	5	18.5409	29.9408	0.6343	21.4849	29.4536	0.7698	0.7108
Snake	2	14.7184	17.9398	0.5357	14.8246	17.9321	0.5431	0.4748
	3	15.4042	22.6388	0.5879	16.2885	22.5853	0.5604	0.6482
	4	17.5795	26.9963	0.7205	18.4103	26.7863	0.7405	0.6347
	5	19.7672	31.0221	0.8011	21.0050	30.5564	0.8358	0.7187

reported. However, the entire test is performed using both Otsu's and Kapur's objective functions. Table 2.5, Table 2.6, and Table 2.7 show the comparison of CCA and MTEMO for Kapur's and Otsu's methods, respectively.

2.6.3 Statistical Analysis

For statistical analysis of the results shown in Tables 2.5 and 2.6, a one-way ANOVA test has been done. The analysis is performed considering a 5% significance level over the execution time data corresponding to the test images

TABLE 2.6: Comparison of CCA with MTEMO using Otsu's method.

Image	θ	MTEMO			
		PSNR	Mean	SSIM	Time
Cameramen	2	17.2474	3.6519e+03	0.5964	2.9145
	3	20.2114	3.7270e+03	0.6415	4.5650
	4	21.5328	3.7824e+03	0.6648	6.5646
	5	23.2235	3.8119e+03	0.6951	8.6998
Starfish	2	14.8158	2.5469e+03	0.4230	3.6676
	3	17.3301	2.7799e+03	0.5485	5.8401
	4	19.1259	2.8657e+03	0.6386	8.1728
	5	20.7674	2.9128e+03	0.7092	10.7722
Zebra	2	13.4728	1.3947e+03	0.4310	3.8425
	3	15.2286	1.5263e+03	0.5444	5.9443
	4	16.8718	1.5825e+03	0.6509	8.3692
	5	18.2373	1.6105e+03	0.7101	10.8720
Boat Man	2	12.6309	5.0750e+03	0.5422	3.5575
	3	15.0155	5.2399e+03	0.6383	5.4001
	4	17.6208	5.3169e+03	0.7547	7.4238
	5	18.8359	5.3559e+03	0.7973	9.5893
Ostrich	2	15.6925	1.0735e+03	0.4219	3.7034
	3	16.8151	1.1392e+03	0.4534	5.7863
	4	17.4478	1.1786e+03	0.4870	8.2029
	5	18.7970	1.2037e+03	0.5584	10.2127
Boat	2	12.3263	1.2645e+03	0.3492	3.8212
	3	17.8963	1.3749e+03	0.5972	6.7030
	4	19.1961	1.4373e+03	0.6504	9.0585
	5	20.9881	1.4794e+03	0.7201	11.8090
Tree	2	16.7057	1.1890e+03	0.5241	4.1409
	3	18.6252	1.2693e+03	0.6271	6.4711
	4	20.2430	1.3135e+03	0.7005	8.5807
	5	21.7758	1.3344e+03	0.7497	11.1699
Snake	2	15.6662	1.1186e+03	0.6226	3.8269
	3	17.9818	1.2313e+03	0.7354	5.9151
	4	19.6907	1.2865e+03	0.8005	8.0355
	5	20.8990	1.3170e+03	0.8395	10.8160
		CCA			
		PSNR	Mean	SSIM	Time
Cameramen	2	18.1827	3.6519e+03	0.6022	**0.3777**
	3	21.0693	3.7227e+03	0.6512	**0.4221**
	4	22.6114	3.7733e+03	0.6599	**0.4691**
	5	23.5897	3.8033e+03	0.7058	**0.5438**
Starfish	2	15.5738	2.5466e+03	0.4545	**0.3466**
	3	17.8468	2.7757e+03	0.5645	**0.3857**
	4	19.2581	2.8594e+03	0.6483	**0.4147**
	5	21.2615	2.8942e+03	0.7152	**0.4520**

(continued)

TABLE 2.6: (continued) Comparison of CCA with MTEMO using Otsu's method.

Image	θ	CCA			
		PSNR	Mean	SSIM	Time
Zebra	2	13.9618	1.3940e+03	0.4453	**0.3568**
	3	16.3846	1.5184e+03	0.5785	**0.5267**
	4	18.6222	1.5648e+03	0.6714	**0.6239**
	5	20.1544	1.5780e+03	0.7101	**0.6898**
Boat Man	2	14.0071	5.0744e+03	0.5385	**0.3552**
	3	17.3007	5.2342e+03	0.6743	**0.3840**
	4	20.0656	5.3080e+03	0.7661	**0.4268**
	5	21.0697	5.3443e+03	0.7818	**0.4473**
Ostrich	2	16.0329	1.0729e+03	0.4439	**0.3614**
	3	17.3580	1.1352e+03	0.4815	**0.3907**
	4	18.3164	1.1716e+03	0.5447	**0.4205**
	5	22.1184	1.1958e+03	0.7019	**0.4480**
Boat	2	12.3241	1.2643e+03	0.3490	**0.3654**
	3	18.6936	1.3699e+03	0.6477	**0.3915**
	4	18.3164	1.4299e+03	0.6735	**0.4215**
	5	20.9115	1.4689e+03	0.7260	**0.2464**
Tree	2	16.7595	1.1886e+03	0.5392	**0.3101**
	3	19.1352	1.2663e+03	0.6553	**0.3674**
	4	20.9931	1.3038e+03	0.7367	**0.4197**
	5	24.0303	1.3225e+03	0.8341	**0.4828**
Snake	2	15.7050	1.1182e+03	0.6250	**0.3233**
	3	18.4765	1.2275e+03	0.7594	**0.3748**
	4	20.3848	1.2770e+03	0.8205	**0.46164**
	5	22.0442	1.3032e+03	0.8619	**0.5188**

TABLE 2.7: ANOVA test over the CCA and MTEMO based on Kapur's method.

	ANOVA				
	Sum of Squares	Df	Mean Square	F	Sig.
Between Groups	422.678	1	422.678	220.149	.000
Within Groups	119.037	62	1.920		
Total	541.715	63			

with two to five threshold points. The hypothesis made for the analysis is as follows.

Null hypothesis H0: There is no significant difference in the execution time between the two methods MTEMO and CCA.

Alternative hypothesis H1: There is a significant difference in the execution time between the two approaches MTEMO and CCA.

The ANOVA test is conducted using the SPSS tool and the results found in the experiment are shown in Table 2.7 and Table 2.8. Table 2.7 presents

TABLE 2.8: ANOVA test over the CCA and MTEMO based on Otsu's method.

	ANOVA				
	Sum of Squares	Df	Mean Square	F	Sig.
Between Groups	695.128	1	695.128	195.789	.000
Within Groups	220.124	62	3.550		
Total	915.253	63			

the results of the ANOVA test regarding execution time that is obtained from Table 2.5, and Table 2.8 presents the results of the ANOVA test regarding execution time that is obtained from Table 2.6. The p values are less than 0.05 (5% significance level), which is strong evidence against the null hypothesis. Hence, it is concluded that there is a significant difference in the execution time between MTEMO and CCA.

2.7 Conclusion

In this chapter, an efficient optimization algorithm, namely, the cricket chirping algorithm (CCA), has been used for solving the multilevel thresholding (MT) problem in image segmentation. The approach combines the good searching capabilities of the CCA algorithm with the two objective functions proposed by the popular MT methods of Otsu class variance and Kapur's entropy method. In order to measure the performance of the proposed approach, the peak signal-to-noise ratio (PSNR), which assesses the segmentation quality, considering the coincidences between the segmented and the original images, SSIM (Structural Similarity), and computational time are used. To see the effect of Otsu's and Kapur's methods with CCA for multilevel thresholding of image segmentation, the Wilcoxon rank test is used. The proposed algorithm is then compared with MTEMO, another popular algorithm, and statistically analyzed using a one-way ANOVA test. Finally, it is concluded that CCA is better than its counterparts in terms of fitness value, PSNR, SSIM, and computational times. In the future, CCA could be extended for other multi-objective optimization problems in image processing.

Bibliography

[1] Sanjay Agrawal, Rutuparna Panda, Sudipta Bhuyan, and Bijaya K. Panigrahi. Tsallis entropy based optimal multilevel thresholding using

cuckoo search algorithm. *Swarm and Evolutionary Computation*, 11:16–30, 2013.

[2] Bahriye Akay. A study on particle swarm optimization and artificial bee colony algorithms for multilevel thresholding. *Applied Soft Computing*, 13(6):3066–3091, 2013.

[3] Adis Alihodzic and Milan Tuba. Improved bat algorithm applied to multilevel image thresholding. *The Scientific World Journal*, 2014.

[4] George E.P. Box, William Gordon Hunter, and J. Stuart Hunter. *Statistics for experimenters: an introduction to design, data analysis, and model building*, volume 1. JSTOR, 1978.

[5] Sebastien Chabrier, Christophe Rosenberger, Bruno Emile, and Hélene Laurent. Optimization-based image segmentation by genetic algorithms. *Journal on Image and Video Processing*, 2008:10, 2008.

[6] Akhilesh Chander, Amitava Chatterjee, and Patrick Siarry. A new social and momentum component adaptive pso algorithm for image segmentation. *Expert Systems with Applications*, 38(5):4998–5004, 2011.

[7] M. Portes De Albuquerque, Israel A. Esquef, and A.R. Gesualdi Mello. Image thresholding using tsallis entropy. *Pattern Recognition Letters*, 25(9):1059–1065, 2004.

[8] Jonti Deuri and S. Siva Sathya. A novel cricket chirping algorithm for engineering optimization problem. *Advances in Natural and Applied Sciences*, 9(6 SE):397–403, 2015.

[9] Jonti Deuri and S. Siva Sathya. Cricket chirping algorithm: an efficient metaheuristic for numerical function optimisation. *International Journal of Computational Science and Engineering*, in press.

[10] A.E. Dolbear. The cricket as a thermometer. *The American Naturalist*, 31(371):970–971, 1897.

[11] Salvador García, Daniel Molina, Manuel Lozano, and Francisco Herrera. A study on the use of non-parametric tests for analyzing the evolutionary algorithms behaviour: a case study on the cec2005 special session on real parameter optimization. *Journal of Heuristics*, 15(6):617, 2009.

[12] Kamal Hammouche, Moussa Diaf, and Patrick Siarry. A multilevel automatic thresholding method based on a genetic algorithm for a fast image segmentation. *Computer Vision and Image Understanding*, 109(2):163–175, 2008.

[13] Kamal Hammouche, Moussa Diaf, and Patrick Siarry. A comparative study of various metaheuristic techniques applied to the multilevel thresholding problem. *Engineering Applications of Artificial Intelligence*, 23(5):676–688, 2010.

[14] Kazim Hanbay and M. Fatih Talu. Segmentation of sar images using improved artificial bee colony algorithm and neutrosophic set. *Applied Soft Computing*, 21:433–443, 2014.

[15] Ming-Huwi Horng. Multilevel thresholding selection based on the artificial bee colony algorithm for image segmentation. *Expert Systems with Applications*, 38(11):13785–13791, 2011.

[16] Jagat Narain Kapur, Prasanna K. Sahoo, and Andrew K.C. Wong. A new method for gray-level picture thresholding using the entropy of the histogram. *Computer vision, graphics, and image processing*, 29(3):273–285, 1985.

[17] Muhammad Waseem Khan. A survey: Image segmentation techniques. *International Journal of Future Computer and Communication*, 3(2):89, 2014.

[18] Yi Liu, Caihong Mu, Weidong Kou, and Jing Liu. Modified particle swarm optimization-based multilevel thresholding for image segmentation. *Soft Computing*, 19(5):1311–1327, 2015.

[19] Miao Ma, Jianhui Liang, Min Guo, Yi Fan, and Yilong Yin. Sar image segmentation based on artificial bee colony algorithm. *Applied Soft Computing*, 11(8):5205–5214, 2011.

[20] Madhubanti Maitra and Amitava Chatterjee. A hybrid cooperative–comprehensive learning based pso algorithm for image segmentation using multilevel thresholding. *Expert Systems with Applications*, 34(2):1341–1350, 2008.

[21] David Martin, Charless Fowlkes, Doron Tal, and Jitendra Malik. A database of human segmented natural images and its application to evaluating segmentation algorithms and measuring ecological statistics. In *Computer Vision, 2001. ICCV 2001. Proceedings. Eighth IEEE International Conference on*, volume 2, pages 416–423. IEEE, 2001.

[22] Kamal E. Melkemi, Mohamed Batouche, and Sebti Foufou. A multiagent system approach for image segmentation using genetic algorithms and extremal optimization heuristics. *Pattern Recognition Letters*, 27(11):1230–1238, 2006.

[23] Diego Oliva, Erik Cuevas, Gonzalo Pajares, Daniel Zaldivar, and Valentín Osuna. A multilevel thresholding algorithm using electromagnetism optimization. *Neurocomputing*, 139:357–381, 2014.

[24] Diego Oliva, Erik Cuevas, Gonzalo Pajares, Daniel Zaldivar, and Marco Perez-Cisneros. Multilevel thresholding segmentation based on harmony search optimization. *Journal of Applied Mathematics*, 2013.

[25] N. Level Otsu. A threshold selection method from gray-level histogram. *IEEE Transactions on Systems, Man and Cybernetics*, 9(1):62–66, 1979.

[26] N. Raja, V. Rajinikanth, and K. Latha. Otsu based optimal multilevel image thresholding using firefly algorithm. *Modelling and Simulation in Engineering*, 2014:37, 2014.

[27] P.D. Sathya and R. Kayalvizhi. Optimum multilevel image thresholding based on tsallis entropy method with bacterial foraging algorithm. *International Journal of Computer Science*, 7(5):336–343, 2010.

[28] P.D. Sathya and R. Kayalvizhi. Modified bacterial foraging algorithm based multilevel thresholding for image segmentation. *Engineering Applications of Artificial Intelligence*, 24(4):595–615, 2011.

[29] P.D. Sathya and R. Kayalvizhi. Optimal multilevel thresholding using bacterial foraging algorithm. *Expert Systems with Applications*, 38(12):15549–15564, 2011.

[30] Hamed Shah-Hosseini. Multilevel thresholding for image segmentation using the galaxy-based search algorithm. *International Journal of Intelligent Systems and Applications*, 5(11):19, 2013.

[31] Constantino Tsallis. Entropic nonextensivity: a possible measure of complexity. *Chaos, Solitons & Fractals*, 13(3):371–391, 2002.

[32] Zhou Wang, Alan C. Bovik, Hamid R. Sheikh, and Eero P. Simoncelli. Image quality assessment: from error visibility to structural similarity. *IEEE Transactions on Image Processing*, 13(4):600–612, 2004.

[33] Yong Zhang, Dan Huang, Min Ji, and Fuding Xie. Image segmentation using pso and pcm with mahalanobis distance. *Expert Systems with Applications*, 38(7):9036–9040, 2011.

Chapter 3

Algorithms for Drawing Graphics Primitives on a Honeycomb Model-Inspired Grid

M. Prabukumar
VIT University, Vellore, India

3.1	Introduction	60
3.2	Literature Survey	61
3.3	Hexagonal Grid Simulator	67
	3.3.1 Hexagon Drawing	68
	3.3.2 Mapping Techniques	70
	3.3.3 Deviation Calculation	72
3.4	Rasterization Algorithms on a Hexagonal Display	73
3.5	Results and Discussion	76
	3.5.1 Analysis of Deviations in Sampling Lines in Both Square Grids and Hexagonal Grids	77
	3.5.2 Analysis of Deviations in Sampling Circles in Both Square Grids and Hexagonal Grids	79
	3.5.3 Analysis of Deviations in Sampling Ellipses in both Square and Hexagonal Grids	81
	3.5.4 Character Generation	83
3.6	Conclusion	87
	Bibliography	87

In this chapter, the relative benefits that the lattices provide for representing lines and curves is investigated. The hexagonal simulator to rasterize the basic output primitives such as line, circle, and ellipse on square and hexagonal grids is introduced. The pros and cons of the hexagonal grid-based rendering algorithms over conventional square grid algorithms are demonstrated. The performance of the hexagonal output primitive drawing algorithm is compared quantitatively by computing average error (AE), mean squares error (MSE), and maximum errors, and compared in the same way with square output primitive drawing algorithms. Experimental results show that the hexagonal grid produces less jaggies on regions of higher curvature than the square grid, and the regions with low curvature are found to have more jaggies on hexagonal grids than on the square grid. This feature of hexagonal grids is reflected in character drawing, too.

3.1 Introduction

As mentioned in many scientific papers and research projects the hexagonal grid is an alternative image sampling technique, which makes use of the structural attributes of the hexagon pixel as an advantage to present an image [32]. The hexagonal sampling is known for better image quality and different implementation approaches, and the latter one made it very difficult to find any working hardware that applies the hexagonal sampling concept [9]. The primary motivation behind the use of a hexagonal pixel display system over a conventional square pixel display system is that it is highly advantageous in many ways. Some of these include sampling efficiency ways as it reduces the number of pixels being used, thus rendering the same resolution at a much better economic level; sixfold symmetry offers superior symmetry, in due course reducing the computational time for processing. The well-behaved connectivity provides less ambiguous identity of boundaries and regions corresponding to a particular pixel and thereby defines a definite neighborhood. Another interesting advantage is that the retina of the human eye closely resembles a hexagonal display system, thus a better view of the object is achieved with the hexagonal display system [3].

In this chapter a survey of output primitives generation on hexagonal grids is presented. The hexagonal simulator for a rasterizing the basic output primitives such as line, circle, and ellipse on hexagonal grid is introduced. Experimental results prove that the hexagonal grid produces less jaggies on regions of higher curvature than the square grid, and the regions with low curvature are found to have more jaggies on the hexagonal grid than on the square grid. This feature of hexagonal grids is reflected in character drawing, too. This optimality arises because of the geometric properties of the grid. Owing to this approach, the objects in the CAD tool can be modeled efficiently. The chapter is organized as follows: In Section 2 the literature survey of hexagonal sampling is discussed. In Section 3, details of the hexagonal simulator are given. Rasterization algorithms on a hexagonal grid are presented in Section 4. The performance analysis of the rasterization algorithm on both square and hexagonal grids is made in Section 5, and finally, a conclusion is drawn in Section 7.

3.2 Literature Survey

A large number of researchers have studied and scrutinized the problem of rasterization of curves in general, and lines and conic sections in particular. Algorithms already exist for drawing arbitrary curves and for plotting conic sections or equations of the second order. The general equation of a line is simpler than that of a curve. This explains why there is a large body of literature that examines the problem of efficient line drawing on raster displays. The most fundamental of these is the Bresenham algorithm [3], which uses an incremental technique, similar to solving differential equations. Bresenham also designed an algorithm for drawing circular arcs [3] and showed that his lines and circles are the best-fit curves. The error is that it overlooked the fact that the vertical distance between the selected pixel and the analog line is always less than or at the most equal to 0.5. Bresenham's technique does not generalize easily to arbitrarily selected curves.

Wu and Rokne [47] suggested a double-step algorithm for incremental generation of lines as well as circles, selecting two pixels per iteration. While this technique leads to algorithms that require the same amount of integer arithmetic per iteration as the single-step algorithms, this significantly reduced to half the number of iterations that are required. Wu and Rokne [48] also showed the use of a double-step technique for pixel selection for ellipses. Bresenham [3] suggested a similar but more complex algorithm using quadruple step patterns. Rokne et. al [32] developed a fast scan conversion algorithm for line drawing based on Bresenham's algorithm combined with double-step and symmetry [32]. Bresenham also probably observed symmetry on the digital line. Danielsson [6] presented an incremental method for curve generation using parametric equations. It not only deals with lines but also with other curves. In the same paper, he also discussed curve generation using an equation without parameters.

Pitteway and Watkinson [23] addressed the problem of line drawing with gray scale. This results in pleasing visual effects with display devices that permit multiple levels of intensities. Dan Field [7] presented two incremental linear interpolation algorithms and analyzed their speed and accuracy. The first of these algorithms is a simple digital differential algorithm relying on fixed-point arithmetic, and the second makes use of integer arithmetic and is a generalization of Bresenham's line drawing algorithm. The algorithm is accurate and faster than the fixed-point algorithm depending upon the underlying processor capability. Sproull [37] gave an elegant derivation of a line drawing algorithm as a series of program transformations from the original brute-force algorithm. Tran-Thong [40] proposed a symmetric algorithm for line drawing that is independent of the order of the end points. Mcilroy [54] presented an algorithm for generation of circles. Van Aken [43] presented an algorithm for drawing an ellipse. Kappel [55], reviewed a number of the then-existing algorithms and the different methods used in their creation, and then presented an ellipse-plotting algorithm that he claimed embodies the best features from each of the algorithms that he reviewed.

Algorithms designed specifically for the generation of hyperbolas and parabolas are rare and harder to find. Surany [56] addressed the generation of the hyperbola rather briefly. He presented an algorithm for the generation of parabolas, but Jordan [57] detected flaws and called it inefficient. The method of deviation was introduced in [37] for drawing an implicit curve. The basic idea involved in the method of deviation is finding a pixel closest to an analog curve. The method succeeds to draw a curve despite self intersection. The method was subsequently used in [31] to draw lines and conic sections.

The application of hexagonal sampling is not a new idea. Over the past forty years many researchers have examined various aspects of this field. Theoretical studies of a possible alternative sampling system for a 2-D Euclidean space starts with [22]. He concluded that the most efficient sampling can be obtained in hexagonal grids. The study of distance functions on digital pictures using a variety of coordinate systems was done by [33]. He evaluated morphological operators in these coordinate systems via simple applications. Finally, he came up with the suggestion of displaying the hexagonal images by arranging the individual square pixels like a brick wall. The hexagonal structures-based parallel computing systems were developed by [58]. Later the computational savings offered by hexagonal sampling consolidated the earlier work by formulating the hexagonal sampling theorem motivated by [59]. This was used to derive the discrete hexagonal Fourier transform. He also proved that the sampling on a hexagonal grid achieves greater efficiency than on a square lattice due to less aliasing.

The motivation to find efficient storage schemes for an image using tree and pyramid data structures was performed by [4]. The sept-tree, which was composed of a central hexagonal cell surrounded by six others, was advocated as the most useful structure, as it was roughly hexagonal in shape. Around the same epoch of time, the problem of conversion of map data to other forms for graphical information systems was studied by [60]. They devised a data structure known as the generalized balanced ternary (GBT). The GBT consists of a hierarchy of cells, with the cells at each level constructed from those at a previous level using an aggregation rule. As each level consists of clusters of seven cells, any GBT-based structure can be represented by unique, base-seven indices.

Whitehouse and Phillips established the superiority of a hexagonal sampling lattice over the square lattice for cases of high data density at the theoretical level. This was done on the basis of a quantitative comparison of the three regular tilings (with triangles, squares, and hexagons) of the plane and their associated surfaces. Later consolidated and extended versions of the earlier work on mathematical morphology were developed by [35]. Serra evaluated a variety of problems, and in each one of them, chose a suitable coordinate system. Generally, a preference for hexagonal lattices has been shown in the literature due to their consistent connectivity. Watson and Ahumada [53] proposed a hexagonal orthogonal-oriented pyramid structure for image coding. Drawing from biological precepts, they generated a structure that was similar to the pyramid of Burt and the GBT. Groups of 7 pixels were used as kernels to represent the image. The image coding regime derived from this showed a high degree of efficiency. Again, the coordinate system used was based upon a skewed axis, in this case the y-axis being at 60°.

Staunton [38] and Staunton and Storey [39] performed some work on a practical, hardware-based system for hexagonal-image processing. First, a conventional frame grabber was modified to perform pseudo-hexagonal sampling by offsetting alternate rows by half an element. The resulting data was processed using a pipeline computer system. Simple algorithms such as thresholding were examined, and local operators and the processed images were displayed on a computer screen using a brick wall approach. An algorithmic comparison of square and hexagonal lattices was performed by [49]. Again, hexagonal coordinates were represented on a skewed coordinate system; this time the x-axis was rotated by 60° in the anti-clockwise direction. The line and circle generation algorithms were implemented and compared using both square and hexagonal lattices. An algorithmic comparison was done using a simulator, which represented a hexagonal pixel as an accumulation of square pixels. The results confirmed the strong similarity between the hexagonal and square-based systems. Based on many years of experience in modeling the limits of human visual perception, Overington [21] proposed a matrix model for early human vision. The sampling step involved a simple reordering of the input image data, reducing every eight rows to seven to an approximate ideal hexagonal sampling. The data was displayed on an offset brick wall. Low-level image operations were tested within this framework and the model appeared consistent with many aspects of the early human visual system.

Her [9] and Her and Yuan [10] were interested in representing hexagonal data, and focused on two different aspects. The first was re-sampling a square-image to a hexagonal image. The effectiveness of different sampling kernels were examined in the re-sampling process. Display of the hexagonal image again used the brick wall approach. He also proposed a new three-coordinate system to represent hexagonal data and used a three-tuple approach to represent the displacement from each of the major axes of the hexagon. A variety of operators and translations were defined on this framework. Lester and Sandor [15] discussed the processing of rectangular and hexagonal images. It was shown that despite the differences in their mathematical foundation, their algorithmic implementations are similar. A series of test patterns are presented and analyzed. They concluded that the hexagonal processing results in a greater reduction in aliasing in regions of vertical/near vertical features of images than is achieved by rectangular processing, without any degradation in other regions, and with negligible additional computational effort.

Yong-Kui [51] used only integer arithmetic in order to develop an algorithm for generation of straight lines on a hexagonal grid. He categorized line

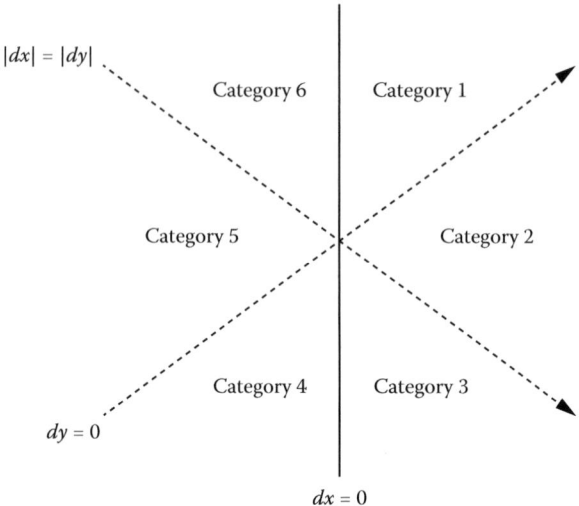

FIGURE 3.1: Line on hexagonal grid

segments on a hexagonal grid into six categories. He considered two endpoints of the straight lines $P(x_1, y_1)$ and $Q(x_2, y_2)$ in hexagonal coordinates, and let $dx = x_2 - x_1$ and $dy = y_2 - y_1$. Then he transformed two endpoints of the line into cartesian coordinates using the transformation function $X = (\sqrt{3}/2)x$ and $Y = y + (1/2)x$. The algorithm was further processed in terms of the cartesian coordinates. The equation of the straight line in terms of the cartesian coordinates is of the form $Y = AX + B$, where $A = (2dy + dx)/(\sqrt{3}dx)$ and $B = (x_2 y_1 - x_1 y_2)/dx$. The straight lines on the hexagonal grid are categorized into six categories using three lines, such as $dx = 0$, $dy = 0$, and $|dx| = |dy|$, as shown in Figure 3.1.

Two candidate pixels are chosen based on the type of line to be drawn. Among the two candidate pixels, the one that is closer to the true line is determined by using the point-to-line distance approach. Then the coordinate of the chosen pixel is converted back to a hexagonal coordinate for plotting. Yong-Kui [51] found the closest integer coordinates to the actual circular arc by using only integer arithmetic, and designed two algorithms that can generate circular arcs on hexagonal grids. In his approach the circle is divided into six equal parts using three lines, $x - y = 0$, $x + 2y = 0$, and $2x + y = 0$, as shown in Figure 3.2. The algorithm finds the pixels belonging to the first half of the circle, and using symmetry, the remaining pixels may be generated.

Tytkowski [42] discusses some advantages of a hexagonal lattice over conventional propositions, and the construction of hexagonal mesh hardware for a graphics display unit, and also presents a new approach for the representation of pixels in raster graphics. The properties and advantages of grids based on rhombic truncated octahedral tilings are given in a study by Miller [17]. Better perimeter estimates using local counting algorithms can be obtained using

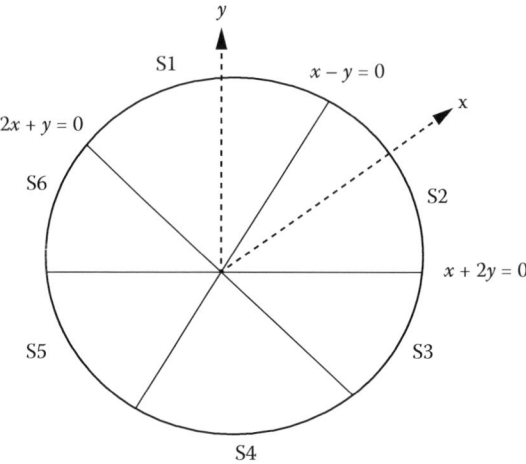

FIGURE 3.2: Circle on hexagonal grid

hexagonal or triangular grids. A method to trace lines in non-orthogonal grids in any dimension using only additions during the line tracing process was proposed by Luis et. al [11] and was found to achieve good performance. Pitteway [23, 24, 25] presented an algorithm that could outline an ellipse, a circle, or any of the other conic sections on a hexagonal lattice. The basic algorithm requires just one test and three addition operations in the inner loop, and an additional test to detect a change in the overall direction between the two adjacent secants.

Condat and Ville [5] introduced a spline model for different orders, both for orthogonal and hexagonal lattices. They derived an expression for a least-squares approximation that can be applied to convolution-based resampling. They also demonstrated the feasibility of the proposed method and compared them against the standard approach for a printing application. Ville et al. [44] constructed a new family of hex-splines that are specifically designed for hexagonal lattices, and used these splines to derive the least square reconstruction function. Hex-splines are a new type of bivariate splines that are designed especially for hexagonal lattices. Inspired by the indicator function of the Voronoi cell, they are able to preserve the isotropy of the hexagonal lattice (as opposed to their B-spline counterparts). They are piecewise-polynomial (on a triangular mesh) and can be constructed for any order. Analytical formulas have been worked out in both spatial and Fourier domains. For orthogonal lattices, the hexsplines revert to the classical tensor-product B-splines. The standard approach to represent two-dimensional data uses orthogonal lattices even though hexagonal lattices provide several advantages, including a higher degree of symmetry and a better packing density.

Sproull [14] investigated the performance of image reconstruction on the hexagonal grid. Conventional image reconstruction methods are implemented on a square structure. However, on a square grid, the distance between adjacent pixels is different in the horizontal (or vertical) direction from that in the diagonal direction. This difference introduces inconsistency when neighboring pixel values are interpolated with a spherically symmetric weighting function, of which weight depends on the distance between a given position and the central pixel. He compared the accuracy of the reconstructed images and compared the results with those obtained on a square grid. His experimental results on disc-shaped images showed a better reconstruction quality on hexagonal grids than on rectangular grids. Condat et al. [5] proposed a new family of bivariate, non-separable splines, called hex-splines, specially designed for hexagonal lattices. The starting point of the construction is the indicator function of the Voronoi cell, which is used to define, in a natural way, the first-order hex-spline. Higher-order hex-splines are obtained by successive convolutions. A mathematical analysis of this new bivariate spline family is presented. In particular, the authors derived a closed form for a hex-spline of arbitrary order. They also discussed important properties, such as Fourier transform and formulated Riesz basis. They also highlight the order of approximation. For conventional rectangular lattices, hex-splines revert to classical separable tensor-product B-splines. Finally, some prototypical applications and experimental results demonstrate the usefulness of hex-splines for handling hexagonally sampled data.

Sheridan et al. [36] proposed a new grid conversion method to resample between two 2-D periodic lattices with the same sampling density. They also showed that the method is fast and provides high-quality resampled images. Kimuro and Nagata [14] used a spherical hexagonal pyramid to analyze the output from a multi-focus camera. The results were promising and were better than such systems designed using square images. Staunton and Storey [39] revisited the work of Gibson and Lucas and labeled his version of the generalized balanced ternary as the spiral honeycomb mosaic, and studied modulo arithmetic upon this structure. Texture characterization using co-occurrence matrices has also been studied on the hexagonal lattice. It was found that there were benefits for textures that have information at multiples of 60°. Additionally, texture synthesis using Markov random fields has been proposed which permit classes of texture that are not simple to generate using square images. Quantization error measures and profiles have been developed and used to compare the square and hexagonal sampling in [12] and [13]. The results obtained for line representation show the hexagonal grid somewhat favorably. However, it is cautioned that the specific algorithm to be used should be taken into consideration for an accurate evaluation.

Almansa [1] presents a quantitative means to measure the elective resolution of image acquisition systems, one that can be used as a basis of comparison of square and hexagonal sampling as well as for improving image resolution. Wu [46] presented a hexagonal discrete cosine transform (HDCT) for encoding the hexagonally sampled signals. He concluded that hexagonal sampling is the optimal sampling strategy for two-dimensional signals in the sense that exact reconstruction of the waveform requires a lower sampling density than those with alternative schemes. An efficient 3D line generation algorithm on a hexagonal prism grid is proposed by [8]. The algorithm is based on the adjunct parallelopiped grid and the 3D cubic Bresenham's line drawing algorithm. The algorithm can be implemented using only integer arithmetic operations. So, it is faster, and the accumulation of rounding errors is eliminated completely.

A display panel having a novel geometrical configuration was patented by [41]. The proposed geometrical configuration can be viewed as a hexagon. He claims that the hexagonal display improves the legibility of characters as compared to the conventional square display. Moreover it greatly improves the average light intensity of diagonal and curvilinear lines. So, a hexagonal display offers substantial advantages for displaying oriental languages. Use of hexagonal grids in fiber optic displays [41, 34] would also prove to be much more efficient and profitable than using square grids. This is due to the reason that the shape of the fiber section is close to a circle and thus further use of square grids can cause serious technological problems. Wu and Rokne [47] used neighborhood sequences in triangular grids. He presented an algorithm that provides the shortest path between two arbitrary triangles by using the given neighborhood sequences. He also discussed properties of triangular grids. He showed that the distance based on neighbourhood sequences in triangular grids may not be symmetric. Wu and Rokne [47] and Wuthrich and Stucki [49] proposed a method for drawing digital circles on a triangular grid, and also proved that circles on a triangular grid are better approximations to the Euclidean circles than the ones in a square grid. Prabukumar et al. [16, 30, 27, 26, 29, 28] proposed drawing graphics primitives such as lines, circles, ellipses, parabolas, and hyperbolas on hexagonal grids using two different approaches, namely mid-point and method of deviation. They compared the same with a square grid-based algorithm and found the pros and cons of the hexagonal grid over conventional square grid algorithms.

Two points can be concluded from the above review. First, the merits of hexagonal sampling in general and for computer graphics in particular have been widely acknowledged. However, a solution for a complete practical system for representing hexagonal objects is lacking. Second, graphics primitives other than lines and circles on a hexagonal grid have not been studied by and large in the context of hexagonal sampling. We address these points in the forthcoming discussions.

3.3 Hexagonal Grid Simulator

In this section, theoretical study of drawing the hexagonal pixel, the routine for mapping Cartesian coordinates to hexagonal coordinates, the routine for mapping hexagonal coordinates to Cartesian coordinates, pixel drawing and storing, deviation calculation, and GUI design are discussed in detail.

3.3.1 Hexagon Drawing

A hexagonal grid can be derived from a square grid. Since designing a coordinate system for a square grid is straightforward, the derivation will guide us in designing coordinate systems for hexagonal grids. A regular hexagon has 6 edges of equal length. As shown in Figure 3.3, if the upper edge is d, then the middle edge is $2d$. So the width and height of a hexagon is $2d$, $\sqrt{3}d$, respectively. The ratio of height to width is approximately equal to $0.8660 : 1$. Thus, we should construct the hexagon in a such way that the ratio is close to this value. Since the hardware for a hexagonal grid is not available at present, we can use regular square pixel to construct a hexagonal pixel (Figure 3.4).

To create a single hexagonal pixel, we need to use multiple square pixels. At the same time, the total number of square pixels available is found to be less in a fixed-size frame because of hardware limitations. In order to compromise both the constraints, the ratio is increased to 13:15, which is 0.8667:1; it is relatively very close to the ratio (0.8660:1). Here the total number of pixels used is still acceptable. More importantly, if the ratio is odd:odd, it will be very easy to locate the center pixel, as shown in Figure 3.5.

With a collection of this kind of hexagonal pixels, we can make a hexagonal grid as shown in Figure 3.6. This grid has a gap between adjacent pixels that can be removed by slightly modifying the pixel structure, as shown in Figure 3.7. This will not change the height–width ratio of 13:15, and after this modification, all six neighbor pixels can be perfectly combined together as shown in Figure 3.8. Finally, the hexagonal grid is constructed as shown in Figure 3.9.

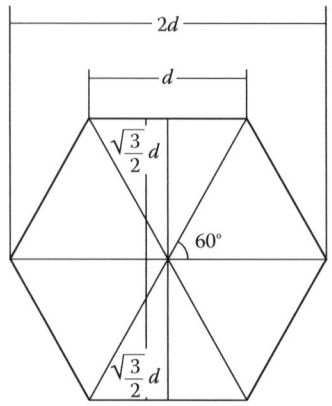

FIGURE 3.3: Structure of a real hexagonal pixel

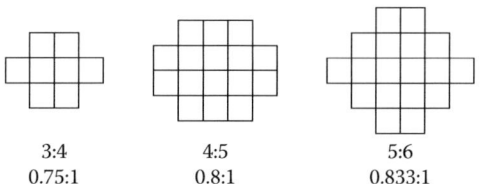

FIGURE 3.4: Different hexagonal structures presented using square grids

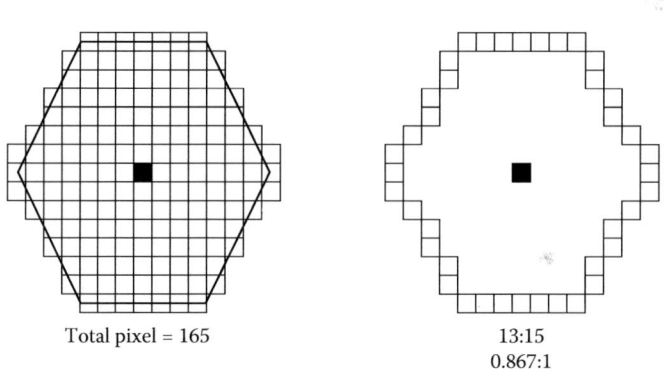

FIGURE 3.5: The selected hexagonal pixel structure

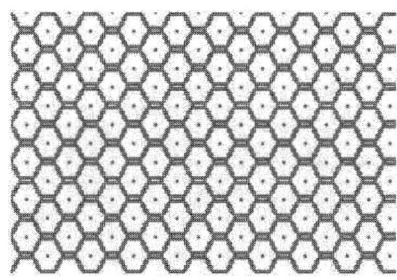

FIGURE 3.6: Hexagonal grids constructed by the selected hexagonal pixels

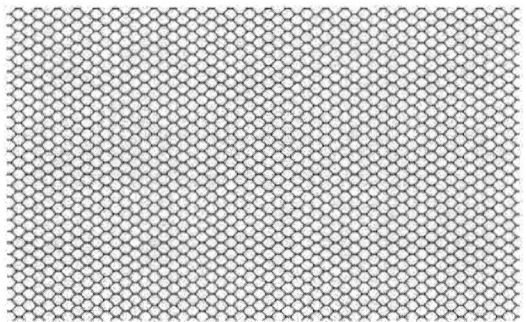

FIGURE 3.7: Modified hexagonal pixel structure

3.3.2 Mapping Techniques

In order to execute the existing hexagonal rendering algorithms, the coordination system is modified as shown in Figure 3.10. When the pixel is plotted, the position is located with the help of its actual position in the Cartesian coordination system in Figure 3.11.

It is obvious that as X increases, B and A both increase; while as y increases, A decreases and B increases. The angle between B and the y axis is $30°$ and the angle between A and y is $60°$. We can calculate the increment and decrement of A and B as x and y increase or (decrease). Increasing one unit vector on y reflects a unit increment on B and a $\sqrt{3}/2$ unit decrement on A.

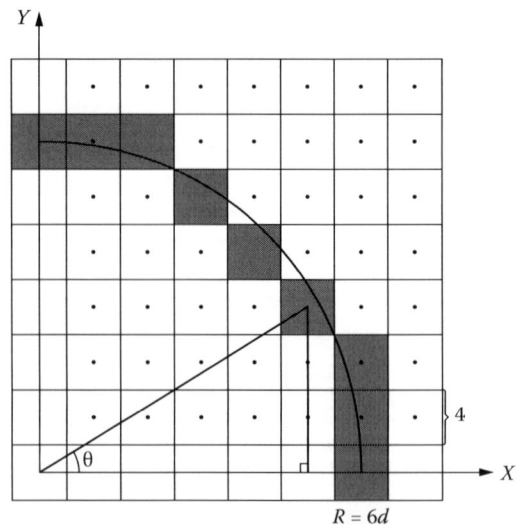

FIGURE 3.8: Six neighbor pixels using modified hexagonal pixel structure

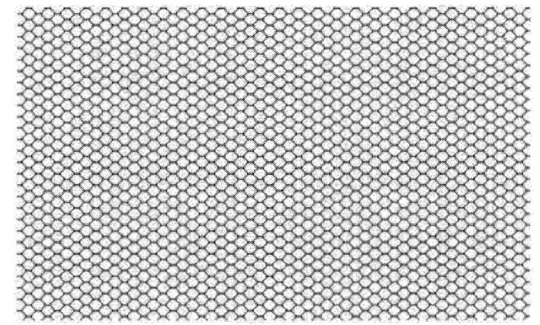

FIGURE 3.9: The hexgonal grids after modification

Similarly increasing one unit vector on x reflects a unit increment on A and a $\sqrt{3}/2$ unit increment on B. So, we can derive the relationship between the Cartesian coordinate system (A, B) and hexagonal the coordination system (x, y) using the followng transformation.

If we know the point (x, y) on the hexagonal coordinate system, the corresponding point on the Cartesian coordination system (A, B) is computed using the following transformation:

$$A = x/\sqrt{3} - y \quad (3.1)$$

$$B = x/\sqrt{3} + y \quad (3.2)$$

Similarly, if we know the point (A, B) on the Cartesian coordination system, the corresponding point (x, y) on the hexagonal coordination system is computed using the following transformation:

$$x = (\sqrt{3}A + \sqrt{3}B)/2 \quad (3.3)$$

$$y = (B - A)/2 \quad (3.4)$$

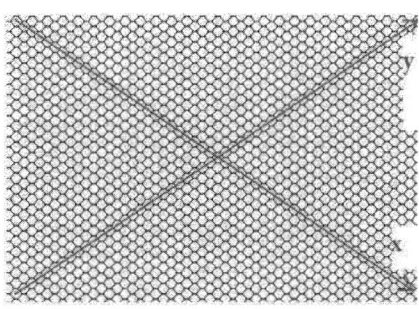

FIGURE 3.10: Coordinate system in hexagonal grids

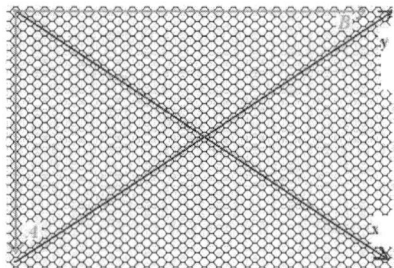

FIGURE 3.11: The actual coordinate system on the monitor (gray) and the applied coordinate system (dark gray)

But actually, the starting position of each column of pixels is different. Referring to Figure 3.12, it can be seen that pixels (B, A) with even B value (gray ones) have a starting position that is beyond that of the ones with odd B value (dark gray ones). To overcome this difference, pixels have to be handled in two categories according to their B value. Each hexagonal pixel size is 15×13, which will cover the hexagonal grids in order to create the visual effect as if the specific grid is plotted. So when a pixel is at the bottom of another, then the distance is 12 pixels in A direction (Figure 3.13). The pixel that is situated in the bottom-right is also 12 pixels (Figure 3.14).

So, every time the movement has a component along the B axis, the component movement should be carefully handled; if the movement is of odd unit vector, the pixel will move in both the A and the B directions, otherwise it will move in the B direction only. The above movement occurs along the x axis; the movement along the y axis can also be calculated in the same way.

3.3.3 Deviation Calculation

The deviation will be calculated as the distance between each selected pixel and the actual image. The deviation calculation of the coordinates in a hexagonal grid is different from that of square grid. In a square grid, with a given arc, the closest pixels are selected and plotted to present the arc in an analog manner. Here is an example that takes the width of a square pixel as d, the

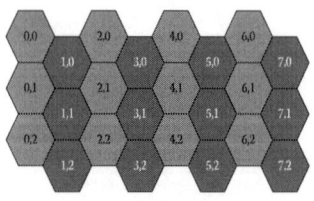

FIGURE 3.12: Different initial positions in various columns

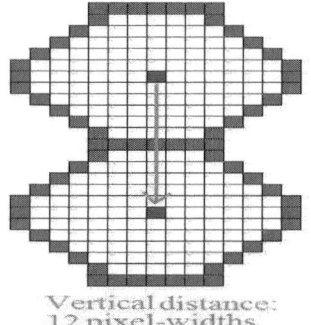

FIGURE 3.13: Distance between a pixel and its above pixel

radius of the arc is 6d. Then the actual distance between the selected pixel and the orientation is calculated as shown in Figure 3.15.

From Figure 3.16 it is clear that the analog radius is equal to 5.8309d, and the deviation of the arc is equal to 0.1691d. Since in the hexagonal grid the angle between two principle axes is not orthogonal, finding the deviation of the arc in the hexagonal grid in the same way is not possible. The distance between selected pixels and the orientation is calculated in the hexagonal grid as shown in Figure 3.17.

3.4 Rasterization Algorithms on a Hexagonal Display

In typical computer graphics, describing a scene in terms of the basic geometric structure is known as output primitive. The most commonly known output primitives in computer graphics are points, lines, circles, and ellipses.

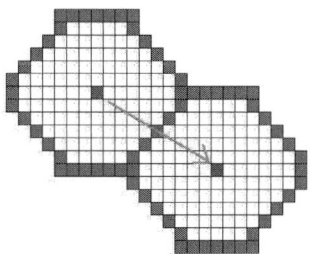

Distance of the center points of neighbor pixels:

$\sqrt{7^2 + 11^2}$ -1 = 13.03840 -1 = 12.03840

FIGURE 3.14: Distance between a pixel and its right bottom pixel

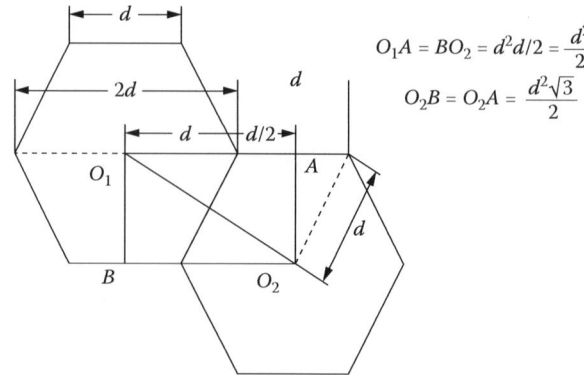

FIGURE 3.15: The sampling of a curve in square grids

To rasterize the output primitives on a square display, numerous algorithms have been proposed. Those algorithms fail to produce output primitives that resemble their analog counterparts. To minimize rasterization error, computer graphics algorithms are executed on a hexagonal display. The related work on hexagonal displays for computer graphics is summarized as follows:

Wuthrich and Stucki [49] provided a systematic proof for the similarity in the characteristics of digitized curves on square and hexagonal grids. Efficient hexagonal grid implementation of algorithms was pointed out to give high standards for the possibility of building graphics devices on hexagonal grids.

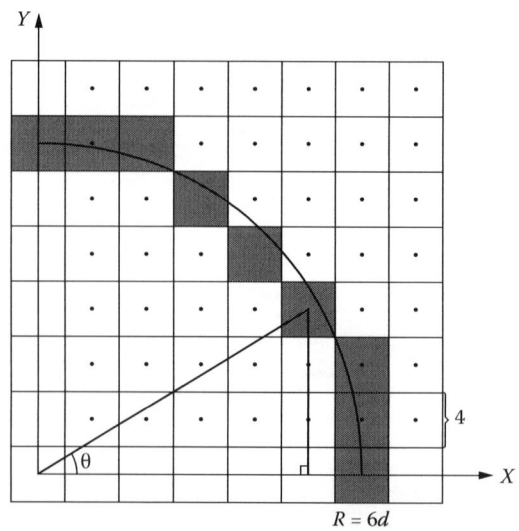

FIGURE 3.16: Approach to calculating the deviation in square sampling

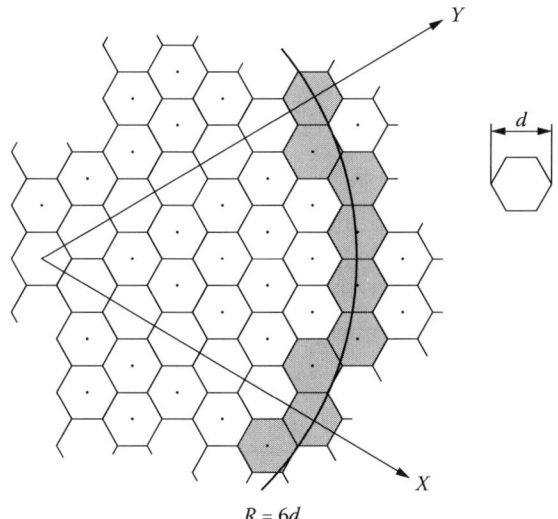

FIGURE 3.17: Approach to calculating the deviation in hexagonal grids

Liu Yong-Kui [51] used only integer arithmetic and developed an algorithm for the generation of straight line on a hexagonal grid. Liu Yong-Kui [50] found the closest integer coordinates to the actual circular arc by using only integer arithmetic and designed two algorithms that can generate circular arcs on hexagonal grids. Tytkowski [42], having discussed some advantages of hexagonal lattices over conventional ones, proposed the construction of hexagonal mesh hardware for a graphics display unit and also presented a new approach for the representation of a pixel in raster graphics. The properties and advantages of grids based on rhombic truncated octahedral tilings were given in a study by Miller [17]. Using local counting algorithms, better perimeter estimates can be obtained using hexagonal or triangular grids. A method for tracing lines in non-orthogonal grids in any dimension using only additions during the line tracing process was proposed by Luis Ibez et al. [11], and achieved good performance. L. V. Pittway [24] presented an algorithm that could outline ellipses, circles, or any of the other conic sections on a hexagonal lattice.

The basic algorithm requires just one test and three addition operations in the inner loop, and an additional test to detect a change in the overall direction between the two adjacent secants. It has geometric symmetries that lead to better topological properties. Prabukumar [16] proposed an algorithm that generates lines on a hexagonal grid. The qualitative and quantitative comparison of the algorithm with the conventional mid-point algorithm on a square grid was also presented. The idea of the improved mid-point circle drawing algorithm on a hexagonal grid was proposed by Prabukumar [26]. It made use of the advantages of hexagonal sampling that can scan-convert a circle on the raster display with significantly less scan-conversion error. Prabukumar and Ray [27] presented the idea of a mid-point ellipse drawing algorithm on a hexagonal grid. The qualitative analysis and execution time show that the proposed algorithm performs better than the conventional ellipse drawing algorithm on a square grid.

The transformation formula $(x, y) \rightarrow (u, v) = (x\sqrt{(3)}/2, y + x/2)$ must be used in order to virtually transform the architecture of the frame buffer to a hexagonal fashion such that it is on par with the hexagonal alignment of the screen. The spatial relationship between an arbitrary point (u, v) and a line of slope m is computed on a hexagonal grid using the equation

$$f(u, v) = (m - 1)u - v/2 + b \qquad (3.5)$$

The spatial relationship between an arbitrary point (u, v) and a circle of radius r centered at the origin is computed on a hexagonal grid using the equation

$$f(u, v) = u^({2}) + v^2 + uv - r^2 \qquad (3.6)$$

The spatial relationship between an arbitrary point (u, v) and an ellipse of major and minor axes, respectively a, b centered at the origin, is computed on a hexagonal grid using the equation

$$f(u, v) = 0.75u^2b^2 + a^2v^2 + (0.25a^2u)^2 + a^2uv - a^2b^2 \qquad (3.7)$$

3.5 Results and Discussion

In this section, analysis of deviation in sampling circles with different radii, lines in different orientations, and ellipses with different major and minor axis radii are discussed. The scenes in nature are continuous, while images displayed on the raster monitors are presented by discrete points, so that the sampling disciplines must be the kinds that will make the original image as detailed as possible, or reduce the gap between sampling points as much as possible. Surely this can be achieved by increasing the amount of sampling points. But the number of pixels (resolution) of each raster monitor is fixed, which is determined by the size of the raster points. The raster points cannot be minimized to infinity since this is limited by the manufacturing process, thus in the condition that the size of each raster point is fixed, only by applying a reasonable arrangement can the number of sampling points increase. In fact, raster points are the cycles generated by the impact of electron beams on the screen. At this point, hexagon a compared to square is closer to the actual shape of the raster point. Let the diameter of the cycle (or length of a pixel) be 1, take a screen of area mn; if square grid distribution is applied, mn

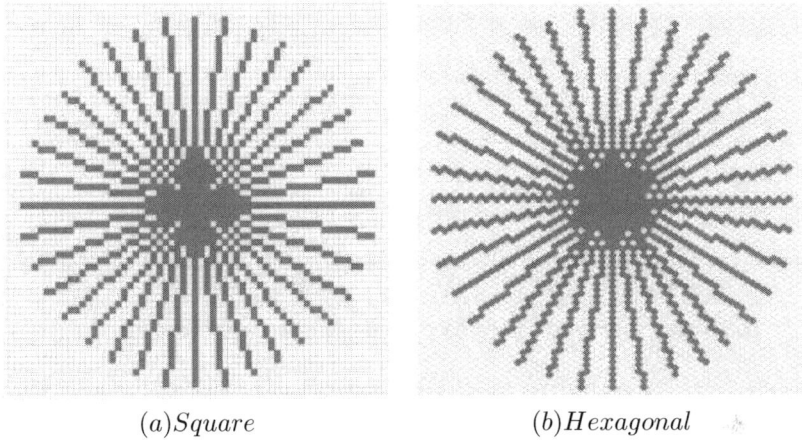

(a)Square (b)Hexagonal

FIGURE 3.18: Scan-converted lines of angle from 0 to 350 degrees

points are contained; if hexagonal grid distribution is applied, mn points will cover 86.6% of the area. In this case the distance between two points is not 1 but 0.866, thus the whole screen can contain $(1.155m * n)$ points, with the number of points increased by 15.5%. This indicates the intensity of points has increased, and the image can be presented in a more detailed manner.

3.5.1 Analysis of Deviations in Sampling Lines in Both Square Grids and Hexagonal Grids

The lines considered to perform the qualitative analysis are in the orientation from 0° to 350° with respect to the x-axis. The comparison is performed qualitatively by plotting lines on the screen as shown in Figure 3.18, and quantitatively by computing average error (AE), mean square error (MSE), and maximum error, as shown in Table 3.1.

The table values we infer are based on the deviation of a line on a hexagonal grid being slightly larger than that on a square grid, but still the increment is too small. In an attempt to compute a single statistical measure for quantitative comparison between quality of output on square and hexagonal grids, the overall average (mean) and standard deviation (SD) of the average error, maximum error, and mean square error for different orientations of lines (shown in Table 3.1) are computed and shown in the last two rows of the table. As seen from these results, the approximation error for lines on hexagonal grids is higher than that on the square grid, and the overall variation of approximation error (last row of the table) across orientation of the lines is also higher on the hexagonal grid than on the square grid.

But from the visual inspection, we observe some important characteristics of lines drawn on a hexagonal grid and on a square grid. The 0^0 line on a hexagonal grid appears to be less smooth than the one on a square grid due

TABLE 3.1: Deviation of lines in different angles in both square and hexagonal grids.

Orientation	Lines on square grid			Lines on hexagonal grid		
	Average error	MSE	Maximum error	Average error	MSE	Maximum error
0	0.00	0.00	0.00	0.49	0.49	1.00
10	0.25	0.08	0.48	0.37	0.19	0.73
20	0.25	0.08	0.50	0.29	0.12	0.56
30	0.24	0.08	0.40	0.00	0.00	0.00
40	0.25	0.08	0.50	0.29	0.12	0.56
50	0.25	0.08	0.50	0.37	0.19	0.73
60	0.24	0.08	0.40	0.00	0.00	0.00
70	0.25	0.08	0.50	0.37	0.19	0.73
80	0.25	0.08	0.48	0.29	0.12	0.56
90	0.00	0.00	0.00	0.00	0.00	0.00
100	0.25	0.08	0.48	0.29	0.12	0.56
110	0.25	0.08	0.50	0.37	0.19	0.73
120	0.25	0.08	0.48	0.49	0.49	1.00
130	0.25	0.08	0.50	0.37	0.19	0.73
140	0.25	0.08	0.50	0.29	0.12	0.56
150	0.24	0.08	0.40	0.00	0.00	0.00
160	0.25	0.08	0.50	0.29	0.12	0.56
170	0.25	0.08	0.48	0.37	0.19	0.73
180	0.00	0.00	0.00	0.49	0.49	1.00
190	0.25	0.08	0.48	0.37	0.19	0.73
200	0.25	0.08	0.50	0.29	0.12	0.56
210	0.25	0.08	0.48	0.00	0.00	0.00
220	0.25	0.08	0.50	0.29	0.12	0.56
230	0.25	0.08	0.50	0.37	0.19	0.73
240	0.24	0.08	0.40	0.49	0.49	1.00
250	0.25	0.08	0.50	0.37	0.19	0.73
260	0.25	0.08	0.48	0.29	0.12	0.56
270	0.00	0.00	0.00	0.00	0.00	0.00
280	0.25	0.08	0.48	0.29	0.11	0.56
290	0.25	0.08	0.50	0.37	0.19	0.73
300	0.25	0.08	0.48	0.49	0.49	1.00
310	0.25	0.08	0.50	0.37	0.19	0.73
320	0.25	0.08	0.50	0.29	0.12	0.56
330	0.24	0.08	0.40	0.00	0.00	0.00
340	0.25	0.08	0.50	0.29	0.12	0.56
350	0.25	0.08	0.48	0.37	0.19	0.73
Mean	0.22	0.07	0.42	0.30	0.19	0.59
SD	0.08	0.03	0.16	0.15	0.15	0.31

to the shape of the hexagonal pixel. In all other cases, lines on a hexagonal grid appear to display less jaggies than those on a square grid, except the line with an angle of 45^0. The overall visual perception of lines on a hexagonal grid is better than that on a square grid. Due to better connectivity in the case of a hexagonal grid, discretized lines are approximated by small polylines, whereas on a square grid, discretized lines are approximated by pixels. Single pixel perception may be disturbing the impression of continuity of the discretized line. This is due to the fact that in the square grid, not all neighbors of a pixel are placed at the same distance. Moreover, two diagonal neighbors in the square grid have only one point in common, which is two horizontal or vertical neighbors of the square grid. On the other hand, all the neighbors of a pixel in the hexagonal grid have one segment in common with their neighbor. This fact produces thickness variations in square digitization, leading to greater edge busyness and to a thinner average width in a line digitization. So, the quality of the lines produced by the hexagonal grid algorithm is better than that of the existing line drawing algorithms [31].

$$AE = \sum_x |y - y_a|/n \qquad (3.8)$$

$$MSE = \sum_x (y - y_a)^2/n \qquad (3.9)$$

$$MaxE = Max_x |y - y_a| \qquad (3.10)$$

Where y is the value of the y coordinate obtained from an analog line equation for a different x value, y_a is the value of the y coordinate obtained from the algorithm, and n is the total number of pixels.

3.5.2 Analysis of Deviations in Sampling Circles in Both Square Grids and Hexagonal Grids

In this section, the output of the conventional circle-drawing algorithm on a hexagonal grid is compared with the circle-drawing algorithm on a square grid. Circles of radii $1, 5, 10, 15, 20, 25$, and 30 are drawn using a hexagonal display simulator. These drawings are shown in Figure 3.19.

Since a hexagonal grid offers greater angular resolution, the nearest neighboring pixels of the same type are separated by $60°$ instead of $90°$. In a square grid, pixels often disturb the perception of continuity of a circle. But due to better pixel connectivity in a hexagonal grid, circles are approximated by tiny polylines. The brightness of the circle is also found to be constant along its circumference (the is not so in a square grid) because of equidistant pixel neighbors. These result in circles of better quality than that in a square grid. In particular, the vertical and horizontal regions of the circles drawn by a square grid algorithm appear as a thick straight line. The overall shape of the circles, as the radii are reduced, worsens in the square lattice compared to the hexagonal lattice. This is because of the improved angular resolution

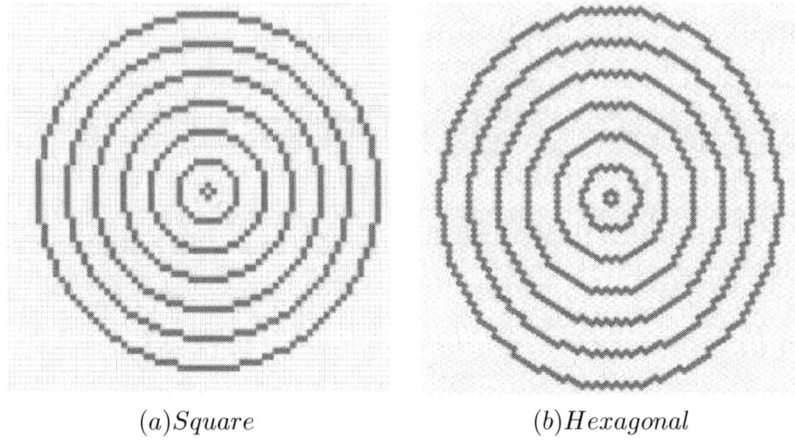

(a) Square   (b) Hexagonal

FIGURE 3.19: Scan-converted circles of radii R = 1, 5, 10, 15, 20, 25, and 30

afforded by having six equidistant neighboring pixels compared to only four in a square lattice. This is especially noticeable when the radius of the circle is 1. The oblique effect [45, 21] also plays a part in this perception.

We also compare the performance of hexagonal circle-drawing quantitatively by computing average error (AE), mean square error (MSE), and maximum error, as shown in Table 3.2. The values are comparable in both cases. In an attempt to compute a single statistical measure for quantitative comparison between quality of output on square and hexagonal grids, the overall average (mean) and standard deviation (SD) of the average error, maximum error, and mean square error for different radii of a circle (shown in Table 3.2) are computed and shown in the last two rows of the table. As seen from these results, the approximation error for circles on a hexagonal grid is slightly

TABLE 3.2: Deviation of circles in different radii in both square and hexagonal grids.

Radius	Circles on square grid			Circles on hexagonal grid		
	Average error	MSE	Maximum error	Average error	MSE	Maximum error
1	0.00	0.00	0.00	0.00	0.00	0.00
5	0.14	0.05	0.39	0.27	0.09	0.29
10	0.17	0.05	0.44	0.29	0.11	0.46
15	0.18	0.05	0.44	0.22	0.07	0.47
20	0.19	0.05	0.40	0.20	0.07	0.48
25	0.20	0.06	0.48	0.24	0.08	0.48
30	0.21	0.06	0.45	0.27	0.09	0.49
Mean	0.15	0.05	0.37	0.21	0.07	0.38
SD	0.07	0.02	0.17	0.10	0.04	0.18

higher than that on the square grid, and the overall variation of approximation error (last row of the table) across the radius of the circles is also slightly higher on the hexagonal grid than on the square grid [31].

$$AE = \sum_{x,y}(R - D_i)/n \quad (3.11)$$

$$MSE = \sum_{i=1}^{n}(R - D_i)^2/n \quad (3.12)$$

$$MaxE = Max_i(|R - D_i|) \quad (3.13)$$

where n is the number of pixels in a circle of given radius; R is the radius of the circle; R_a is radius of the circle obtained from the analog circle equation, and D is the distance between the center and the pixel locations.

3.5.3 Analysis of Deviations in Sampling Ellipses in both Square and Hexagonal Grids

The deviation of ellipses in both the hexagonal grid and the square grid is discussed as follows. Ellipses with different major and minor axis radii are drawn using the simulator. Qualitative and quantitative comparisons are performed in a way similar to that described earlier. The ellipses drawn are shown in Figure 3.20. The numerical results are shown in Table 3.3.

The comparison is performed quantitatively by computing average error (AE), mean square error (MSE), and maximum error (as shown in Table 3.3). The table values are comparable in both cases. The deviation of an ellipse on a hexagonal grid is slightly larger than that on the square grid, but

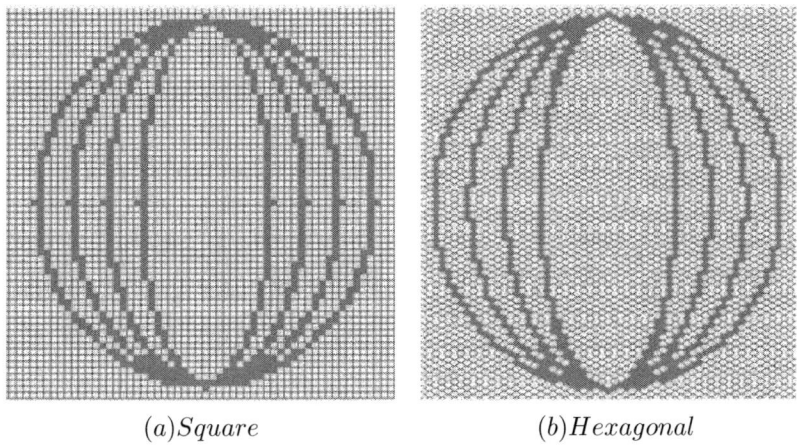

(a)Square   (b)Hexagonal

FIGURE 3.20: Scan-converted ellipses of a = 10 to 25 incremented by 5, b = 30

TABLE 3.3: Deviation of ellipse in both square and hexagonal grids.

a	Ellipse on a square grid			Ellipse on a hexagonal grid		
	Average error	MSE	Maximum error	Average error	MSE	Maximum error
10	0.50	0.13	0.49	0.53	0.65	0.66
15	0.47	0.26	0.48	0.60	0.47	0.69
20	0.48	0.14	0.50	0.56	0.44	0.68
25	0.41	0.12	0.47	0.43	0.27	0.46
Mean	0.46	0.16	0.49	0.53	0.46	0.62
SD	0.04	0.07	0.01	0.07	0.15	0.11

still the increment is too small. In an attempt to compute a single statistical measure for quantitative comparison between quality of output on square and hexagonal grids, the overall average (mean) and standard deviation (SD) of the average error, maximum error, and mean square error for different ellipses (shown in Table 3.3) are computed, and shown in the last two rows of the table. As seen from these results, the approximation error for an ellipse on the hexagonal grid is slightly higher than that on the square grid, and the overall variation of approximation error (last row of the table) across the radius of the ellipse is also slightly higher on the hexagonal grid than on the square grid.

$$AE = \sum_x |y - y_a|/n \tag{3.14}$$

$$MSE = \sum_x (y - y_a)^2/n \tag{3.15}$$

$$MaxE = Max_x(|y - y_a|) \tag{3.16}$$

Where y is the value of the y coordinate obtained from the analog ellipse equation for a different x value, y_a is the value of the y coordinate obtained from the algorithm, and n is the total number of pixels.

From Table 3.3 it is clear that the deviation of the analog curve from its digital version on a hexagonal grid is slightly higher than that on a square grid, but here also, the increment is not large enough to affect the visual perception. On the other hand, for representing the same ellipse on a hexagonal grid, more pixels per unit area are drawn than on a square grid, and this improves the quality and connectivity of the ellipse on a hexagonal grid, as shown in Figure 3.20. The facts in the case of a circle, such as better connectivity, equally distant neighbors, and better angular resolution, are true for an ellipse drawing, too, and hence, the quality of an ellipse produced on a hexagonal grid by the proposed algorithm is better than that of the existing ellipse-drawing algorithm on a square grid [31].

3.5.4 Character Generation

Any drawing and design needs to be labeled and dimensioned, which requires letters, numbers, and other special characters. It may be represented using a straight line and curve segments. The simulator has been used to draw characters from different languages, mathematical symbols, special characters, and some logos (Figure 3.21 to Figure 3.33).

This chapter has applied rasterization to some of the Chinese, Japanese, and English characters, and some special symbols and logos on a hexagonal grid, and compared the same with those on a square grid. The Chinese analog characters mostly consist of straight segments, whereas the Japanese characters mostly consist of high-curvature segments. The response of the hexagonal grid for high curvature regions is better than that of the square grid. The segments of the characters that have high curvature are better represented on a hexagonal grid than on a square grid. This is why the rasterization of Japanese characters produces visually more pleasing appearances than that of the Chinese characters.

When we consider the calligraphical information of Chinese and Japanese characters a hexagonal grid image preserves better way than a square grid image. The English characters consist of straight as well as curved segments and the curve segments are better represented on a hexagonal grid than the straight segments of the characters. The above conclusion is true for mathematical symbols too. As regard to the logos, the higher curvature regions of the logos are found to have less jaggies on a hexagonal grid than on the square grid. So we find that the segments of the characters that have higher curvature are better represented on a hexagonal grid than on a square grid because of the presence of less jaggies on these regions in a hexagonal grid.

The computer-aided design (CAD) tool uses basic primitive entities such as lines, circles, arcs, and text for designing more complex objects. From the experimental analysis, we may say that graphics primitives are better represented in the hexagonal grid than they are in the square grid. Owing to the hexagonal sampling approach, we may visualize design ideas through animations and photorealistic renderings, and simulate how a design will perform in the real world in the CAD tool with less aliasing artifact.

From the experimental results, it is clear that the deviation of hexagonal grids is slightly larger than that of square grids, but still the increment is too small to affect human visual judgment. On the other hand, for representing the same graphics primitives, the hexagonal grid takes a higher number of pixels per unit area than the square grid, and this results in increased quality and connectivity of the graphics primitives on the hexagonal grid. The facts of better connectivity, equally distant neighbors, and better angular resolution improve the quality of the primitives drawn on a hexagonal grid as compared to those drawn using the existing algorithm on a square grid.

(a) Square (b) Original (c) Hexagonal

FIGURE 3.21: Scan-converted character

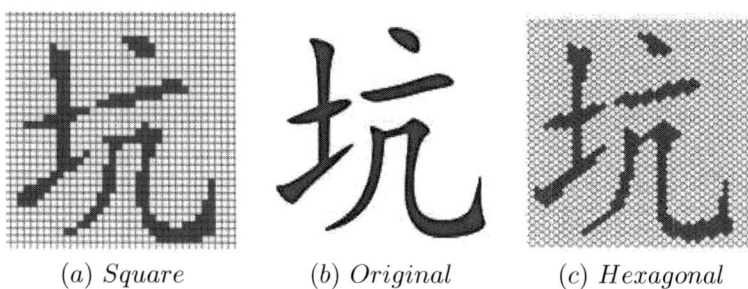

(a) Square (b) Original (c) Hexagonal

FIGURE 3.22: Scan-converted character

(a) Square (b) Original (c) Hexagonal

FIGURE 3.23: Scan-converted character

(a) Square (b) Original (c) Hexagonal

FIGURE 3.24: Scan-converted character

(a) *Square*   (b) *Original*   (c) *Hexagonal*

FIGURE 3.25: Scan-converted character

(a) *Square*   (b) *Original*   (c) *Hexagonal*

FIGURE 3.26: Scan-converted character

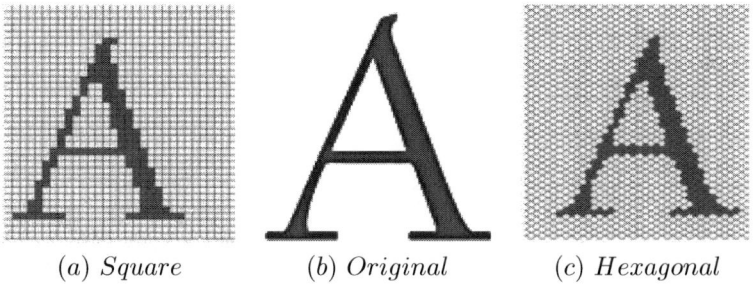

(a) *Square*   (b) *Original*   (c) *Hexagonal*

FIGURE 3.27: Scan-converted character

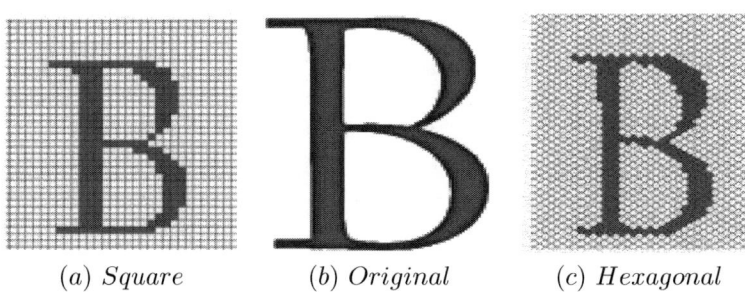

(a) *Square*   (b) *Original*   (c) *Hexagonal*

FIGURE 3.28: Scan-converted character

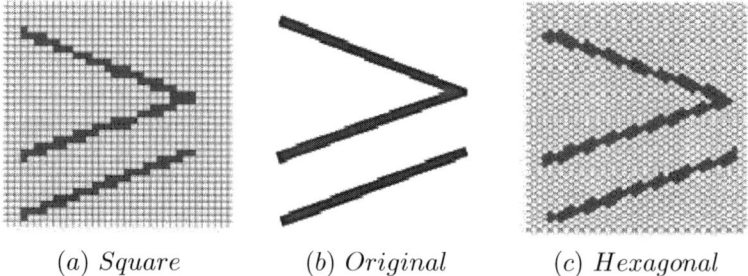

(a) Square (b) Original (c) Hexagonal

FIGURE 3.29: Scan-converted character

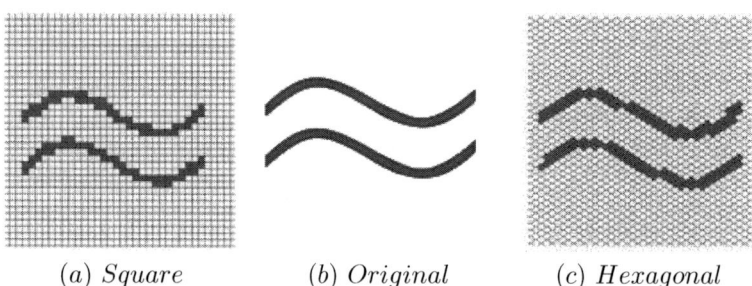

(a) Square (b) Original (c) Hexagonal

FIGURE 3.30: Scan-converted character

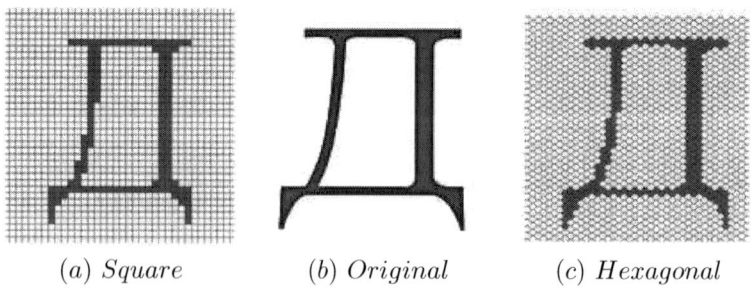

(a) Square (b) Original (c) Hexagonal

FIGURE 3.31: Scan-converted character

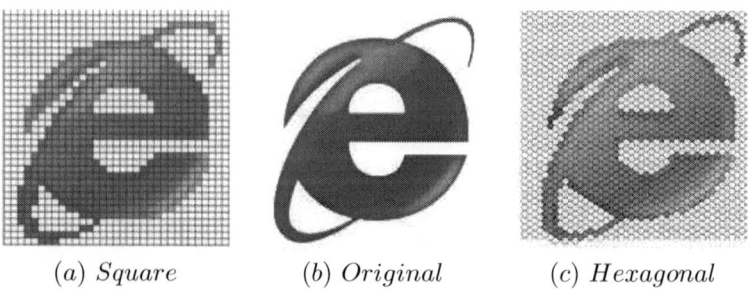

(a) Square (b) Original (c) Hexagonal

FIGURE 3.32: Scan-converted character

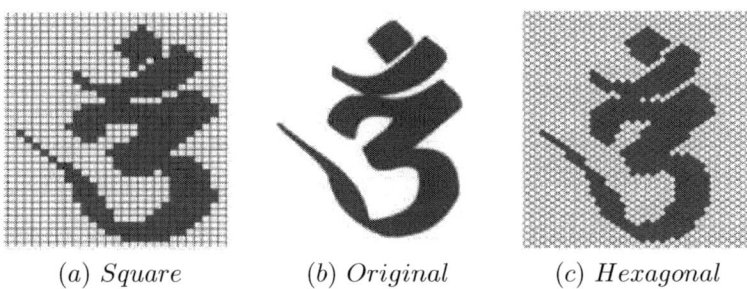

(a) *Square* (b) *Original* (c) *Hexagonal*

FIGURE 3.33: Scan-converted character

The simulator has also been used to draw characters from different languages, mathematical symbols, special characters, and some logos. The output has been compared with those on a square grid. It is found that, on a hexagonal grid, the high curvature regions produce less jaggies as compared to a square grid. The calligraphic information contained in characters is also better reflected on a hexagonal grid than on a square grid. Owing to this approach, we may model the objects with less aliasing artifact.

3.6 Conclusion

A survey on existing algorithms for drawing graphics primitives on hexagonal grids has been presented. Because of the non-existence of hexagonal grid hardware, a simulator has been developed for drawing hexagonal pixels. The simulator has been used for drawing lines, circles, and ellipses on hexagonal grids. The simulator has also been used to draw characters from different languages, mathematical symbols, special characters, and some logos. The output has been compared with that on a rectangular grid. It was found that on a hexagonal grid, the high curvature regions produce less jaggies as compared to those on a square grid. Moreover, the calligraphic information contained in characters is better reflected on hexagonal grids than on square grids.

Bibliography

[1] Andrés Almansa. Image resolution measure with applications to restoration and zoom. In *Geoscience and Remote Sensing Symposium, 2003. IGARSS'03. Proceedings. 2003 IEEE International*, volume 6, pages 3830–3832. IEEE, 2003.

[2] Paul G. Bao and Jon G. Rokne. Quadruple-step line generation. *Computers & Graphics*, 13(4):461–469, 1989.

[3] Jack Bresenham. A linear algorithm for incremental digital display of circular arcs. *Communications of the ACM*, 20(2):100–106, 1977.

[4] Peter J. Burt. Tree and pyramid structures for coding hexagonally sampled binary images. *Computer Graphics and Image Processing*, 14(3):271–280, 1980.

[5] Laurent Condat, Dimitri Van De Ville, and Brigitte Forster-Heinlein. Reversible, fast, and high-quality grid conversions. *IEEE Transactions on Image Processing*, 17(5):679–693, 2008.

[6] Per E. Danielsson. Incremental curve generation. *IEEE Transactions on Computers*, 100(9):783–793, 1970.

[7] Dan Field. Incremental linear interpolation. *ACM Transactions on Graphics (TOG)*, 4(1):1–11, 1985.

[8] Lijun He, Yongkui Liu, and Shuzhe Bao. 3d line generation algorithm on hexagonal prism grids. In *Congress on Image and Signal Processing, 2008. CISP'08*, volume 3, pages 741–745. IEEE, 2008.

[9] Innchyn Her. Geometric transformations on the hexagonal grid. *IEEE Transactions on Image Processing*, 4(9):1213–1222, 1995.

[10] Innchyn Her and Chi-Tseng Yuan. Resampling on a pseudohexagonal grid. *CVGIP: Graphical Models and Image Processing*, 56(4):336–347, 1994.

[11] Luis Ibáñez, Chafiaâ Hamitouche, and Christian Roux. A vectorial algorithm for tracing discrete straight lines in n-dimensional generalized grids. *IEEE Transactions on Visualization and Computer Graphics*, 7(2):97–108, 2001.

[12] Behzad Kamgar-Parsi and Behrooz Kamgar-Parsi. Quantization error in hexagonal sensory configurations. *IEEE transactions on pattern analysis and machine intelligence*, 14(6):665–671, 1992.

[13] Behzad Kamgar-Parsi and W.A. Sander. Quantization error in spatial sampling: comparison between square and hexagonal pixels. In *IEEE Computer Society Conference on Computer Vision and Pattern Recognition, 1989. Proceedings CVPR'8*, pages 604–611. IEEE, 1989.

[14] Yoshihiko Kimuro and Tadashi Nagata. Image processing on an omnidirectional view using a spherical hexagonal pyramid: vanishing points extraction and hexagonal chain coding. In *Intelligent Robots and Systems '95. Human Robot Interaction and Cooperative Robots, Proceedings. 1995 IEEE/RSJ International Conference on*, volume 3, pages 356–361. IEEE, 1995.

[15] Lewis N. Lester and John Sandor. Computer graphics on a hexagonal grid. *Computers & Graphics*, 8(4):401–409, 1984.

[16] Prabukumar Manoharan and Bimal Kumar Ray. An alternate line drawing algorithm on hexagonal grid. In *Proceedings of the 6th ACM India Computing Convention*, page 19. ACM, 2013.

[17] Erik G. Miller. Alternative tilings for improved surface area estimates by local counting algorithms. *Computer Vision and Image Understanding*, 74(3):193–211, 1999.

[18] B. Nagy. Nonmetrical distances on the hexagonal grid using neighborhood sequences. *Pattern Recognition and Image Analysis*, 17(2):183–190, 2007.

[19] Benedek Nagy. Characterization of digital circles in triangular grid. *Pattern Recognition Letters*, 25(11):1231–1242, 2004.

[20] Benedek Nagy. Distances with neighbourhood sequences in cubic and triangular grids. *Pattern Recognition Letters*, 28(1):99–109, 2007.

[21] Ian Overington. *Computer vision: a unified, biologically-inspired approach*. Elsevier Science Inc., 1992.

[22] Daniel P. Petersen and David Middleton. Sampling and reconstruction of wave-number-limited functions in n-dimensional euclidean spaces. *Information and Control*, 5(4):279–323, 1962.

[23] Michael L.V. Pitteway and Dereck J. Watkinson. Bresenham's algorithm with grey scale. *Communications of the ACM*, 23(11):625–626, 1980.

[24] Michael L.V. Pitteway. Algorithm for drawing ellipses or hyperbolae with a digital plotter. *The Computer Journal*, 10(3):282–289, 1967.

[25] Michael LV Pitteway. Drawing conics on a hexagonal grid. In *Information Visualisation, 2001. Proceedings. Fifth International Conference on*, pages 489–491. IEEE, 2001.

[26] M. Prabukumar. An improved mid-point circle drawing algorithm on a hexagonal grid. *Journal of Advanced Research in Computer Science*, 3(1):41–57, 2011.

[27] M. Prabukumar and Bimal Kumar Ray. A mid-point ellipse drawing algorithm on a hexagonal grid. *International Journal of Computer Graphics*, 3(1):17–24, 2012.

[28] Manoharan Prabukumar and Bimal Kumar Ray. Best approximate hyperbola drawing algorithm on hexagonal grid. *International Journal of Applied Engineering Research*, 8(8), 2013.

[29] Manoharan Prabukumar and Bimal Kumar Ray. An alternative conics drawing algorithm on a hexagonal grid using method of deviation. *International Journal of Computer Applications in Technology*, 51(1):69–79, 2015.

[30] Manoharan Prabukumar and Bimal Kumar Ray. Semi-circular angle-based one bit circle generation algorithm on a hexagonal grid. *International Journal of Computer Aided Engineering and Technology*, 8(3):199–216, 2016.

[31] K.S. Ray and B.K. Ray. Drawing lines and conic sections using method of deviation. *International Journal of Computer Graphics*, 3(2):1–26, 2012.

[32] Jon G. Rokne, Brian Wyvill, and Xiaolin Wu. Fast line scan-conversion. *ACM Transactions on Graphics (TOG)*, 9(4):376–388, 1990.

[33] Azriel Rosenfeld and John L. Pfaltz. Distance functions on digital pictures. *Pattern Recognition*, 1(1):33–61, 1968.

[34] William L, Sansom. Fiber optic display device and method for producing images for same, April 19, 1988. U.S. Patent 4,738,510.

[35] Jean Serra. Introduction to mathematical morphology. *Computer vision, graphics, and image processing*, 35(3):283–305, 1986.

[36] Phil Sheridan, Tom Hintz, and David Alexander. Pseudo-invariant image transformations on a hexagonal lattice. *Image and Vision Computing*, 18(11):907–917, 2000.

[37] Robert F. Sproull. Using program transformations to derive line-drawing algorithms. *ACM Transactions on Graphics (TOG)*, 1(4):259–273, 1982.

[38] Richard C. Staunton. The design of hexagonal sampling structures for image digitization and their use with local operators. *Image and vision computing*, 7(3):162–166, 1989.

[39] Richard C. Staunton and Neil Storey. A comparison between square and hexagonal sampling methods for pipeline image processing. In *1989 Symposium on Visual Communications, Image Processing, and Intelligent Robotics Systems*, pages 142–151. International Society for Optics and Photonics, 1990.

[40] Tran Thong. A symmetric linear algorithm for line segment generation. *Computers & Graphics*, 6(1):15–17, 1982.

[41] Theodore R. Touw. Hexagonal display array having close-packed cells, January 7, 1975. U.S. Patent 3,859,553.

[42] Krzysztof T. Tytkowski. Hexagonal raster for computer graphics. In *Information Visualization, 2000. Proceedings. IEEE International Conference on*, pages 69–73. IEEE, 2000.

[43] Jerry R. Van Aken. An efficient ellipse-drawing algorithm. *IEEE Computer Graphics and Applications*, 4(9):24–35, 1984.

[44] Dimitri Van De Ville, Thierry Blu, Michael Unser, Wilfried Philips, Ignace Lemahieu, and Rik Van de Walle. Hex-splines: A novel spline family for hexagonal lattices. *IEEE Transactions on Image Processing*, 13(6):758–772, 2004.

[45] Andrew B. Watson and Cynthia H. Null. *Digital images and human vision*. 1997.

[46] H-S Wu. Hexagonal discrete cosine transform for image coding. *Electronics Letters*, 27(9):781–783, 1991.

[47] Xiaolin Wu and Jon G. Rokne. Double-step incremental generation of lines and circles. *Computer Vision, Graphics, and Image Processing*, 37(3):331–344, 1987.

[48] Xiaolin Wu and Jon G. Rokne. Double-step generation of ellipses. *IEEE Computer Graphics and Applications*, 9(3):56–69, 1989.

[49] Charles Albert Wüthrich and Peter Stucki. An algorithmic comparison between square and hexagonal-based grids. *CVGIP: Graphical Models and Image Processing*, 53(4):324–339, 1991.

[50] Liu Yong-Kui. The generation of circular arcs on hexagonal grids. In *Computer Graphics Forum*, volume 12, pages 21–26. Wiley Online Library, 1993.

[51] Liu Yong-Kui. The generation of straight lines on hexagonal grids. In *Computer Graphics Forum*, volume 12, pages 27–31. Wiley Online Library, 1993.

[52] Whitehouse, D. J., & Phillips, M. J. (1985). Sampling in a two-dimensional plane. Journal of Physics A: *Mathematical and General*, 18(13), 2465–2477.

[53] Watson, A. B., & Ahumada, A. J. (1989). A hexagonal orthogonal-oriented pyramid as a model of image representation in visual cortex. *IEEE Transactions on Biomedical Engineering*, 36(1), 97–106.

[54] McIlroy, M. D. (1983). Best approximate circles on integer grids. *ACM Transactions on Graphics*, 2(4), 237–263.

[55] Kappel, M. R. (1985). An ellipse-drawing algorithm for raster displays. In *Fundamental algorithms for computer graphics* (pp. 257–280). Springer, Berlin, Heidelberg.

[56] Surany, A. P. (1985). An Algorithm for Determining the Draw Start Point of a Hyperbola given the System Direction of Draw and the Coordinates of the Video Window. In *Fundamental Algorithms for Computer Graphics*, (pp. 281–285). Springer, Berlin, Heidelberg.

[57] Jordan, B. W., Lennon, W. J., & Holm, B. D. (1973). An improved algorithm for the generation of nonparametric curves. *IEEE Transactions on Computers*, 100(12), 1052–1060.

[58] Golay, M. J. (1969). Hexagonal parallel pattern transformations. *IEEE Transactions on computers*, 100(8), 733–740.

[59] Mersereau, R. M. (1979). The processing of hexagonally sampled two-dimensional signals. *Proceedings of the IEEE*, 67(6), 930–949.

[60] Gibson, L., & Lucas, D. (1982). Vectorization of raster images using hierarchical methods. *Computer Graphics and Image Processing*, 20(1), 82–89.

Chapter 4

Electrical Impedance Tomography Using Evolutionary Computing: A Review

Wellington Pinheiro dos Santos
Universidade Federal de Pernambuco, Recife, Brazil

Ricardo Emmanuel de Souza
Universidade Federal de Pernambuco, Recife, Brazil

Reiga Ramalho Ribeiro
Universidade Federal de Pernambuco, Recife, Brazil

Allan Rivalles Souza Feitosa
Universidade Federal de Pernambuco, Recife, Brazil

Valter Augusto de Freitas Barbosa
Universidade Federal de Pernambuco, Recife, Brazil

Victor Luiz Bezerra Arajo da Silva
Universidade de Pernambuco, Recife, Brazil

David Edson Ribeiro
Universidade Federal de Pernambuco, Recife, Brazil

Rafaela Covello de Freitas
Universidade Federal de Pernambuco, Recife, Brazil

4.1	Introduction		95
	4.1.1	Advantages and Disadvantages of EIT	95
	4.1.2	Mathematical Formulation	96
	4.1.3	Reconstructing Images in EIT	96

4.2	EIT Computing Tools		97
	4.2.1	Fitness Function	99
4.3	Evolutionary Computation		100
4.4	Genetic Algorithms		100
	4.4.1	Approaches to Genetic Algorithms	101
	4.4.2	Method of Searching and Optimization of Genetic Algorithms	103
	4.4.3	General Characteristics of Experiments	103
	4.4.4	Application Results of AG	105
	4.4.5	Discussion of the Results for AG	105
4.5	Simulated Annealing		106
	4.5.1	Simulated Annealing Approaches	106
	4.5.2	Proposed Implementation	107
4.6	Particle Swarm Optimization		108
	4.6.1	The Particle Optimization Swarm Algorithm	110
4.7	Differential Evolution		113
	4.7.1	Differential Evolution Approach	114
	4.7.2	Implementation Purpose	117
4.8	Fish School Search		118
	4.8.1	Fish School Search Approaches	119
	4.8.2	Implementation Purpose	121
	Bibliography		123

Electrical impedance tomography (EIT) is an image technique based on the application of an alternating electrical current through electrodes placed around the region to be imaged and on the measurement of the resulting electrical potentials in the remaining electrodes. Therefore, the EIT images represent a map of the estimation of the electrical properties of the medium. EIT has the advantage of being non-invasive, free of ionizing radiation, and low cost compared to other image techniques, having applications in geophysics, industrial, biological, and medical areas. However, the EIT image reconstruction process is an inverse, ill-posed, and ill-conditioned problem that results in high computational cost and images with low spatial resolution. One way to reconstruct an EIT image is by treating this problem as an optimization problem that can be solved by methods of evolutionary computing. The aim of this approach is to minimize the root mean square relative error between the electrical potentials measured and simulated. This chapter addresses the basic principles of the EIT and the image reconstruction as an optimization problem solved by the following search and optimization algorithms: genetic algorithms, simulated annealing, particle swarm optimization, and fish school search. A brief explanation of these algorithms is given as well.

4.1 Introduction

Electrical impedance tomography (EIT) is a non-invasive set of techniques of image reconstruction, which can be used to obtain estimated images of electrical conductivity or permittivity in the inside of a section of any body/object through electrical quantity measured in its surface [11, 18, 51]. Getting images through EIT techniques consists of solving an inverse, ill-posed, and ill-conditioned problem, where it is observed that the distribution of electrical conductivity inside a body/object is estimated, given the electrical current of excitement and the electrical potentials measured in its surface [4, 53]. Currently, EIT has applications in several areas, e.g., geophysics, medicine, and industry.

EIT has largely been used in geophysics to support finding underground storage of minerals and different geological formations [34]. Furthermore, EIT is used in numerous industrial processes, e.g., monitoring multiphase flow in pipes; non-destructive control of quality tests of corrosion and defects, as fissures or empty spaces in metal pieces; and leak monitoring and detection in underground tanks [34]. Since 1980, EIT has been used in biological tissues monitoring through medical image reconstruction, being motivated by the fact that different biological tissues have different electrical conductivities. Some values of conductivity (in microsiemens per centimeter mS/cm) of body tissues are given in Table 4.1. Therefore, determining the electrical conductivity inside a body makes it possible to rebuild an organ's image in order to monitor it [27, 6].

From the perspective of medical applications, EIT can be employed in breast cancer detection and cerebrovascular accidents, and to monitor the lung ventilation imposed by mechanical ventilation [33, 14]. Other medical applications of EIT are: evaluation of lung perfusion [43], redistributing pulmonary pressure detection [39], detection of pleural inclusions [9], detection of subdural hematomas [17], evaluation of changes in stroke volume [32], support in the diagnosis of prostate cancer [55], and breast cancer detection [37].

4.1.1 Advantages and Disadvantages of EIT

EIT is especially feasible for clinical applications due to its portability, given its relatively reduced size in comparison with other non-invasive tomographic techniques (e.g., magnetic resonance imaging and free-contrast x-ray computerized tomography), relative low cost, and high safety, because EIT does not employ ionizing radiation [4]. Nevertheless, when compared to other tomographic imaging techniques, EIT images present low spatial resolution [57]. Furthermore, image reconstruction algorithms are computationally intensive [34].

TABLE 4.1: Typical values of tissue's electrical conductivity.

Tissue	Conductivity (mS/cm)
Blood	6.79
Liver	2.8
Muscle (longitudinal)	8
Muscle (transversal)	0.6
Cardiac muscle (longitudinal)	6.3
Cardiac muscle (transversal)	2.3
Neural tissue	1.7
Lung (exhalation)	1
Lung (inhalation)	0.4
Fat	0.36
Bone	0.06

4.1.2 Mathematical Formulation

With EIT, the estimation of the distribution of conductivity inside a heterogeneous body/object is performed by solving a partial differential equation known as the Poisson equation, as will be shown in the following vector expressions:

$$\nabla \cdot [\sigma(\vec{u})\nabla\phi(\vec{u})] = 0, \quad \forall \vec{u} \in \Omega \quad (4.1)$$

with boundary conditions [12]:

$$\phi_{ext}(\vec{u}) = \phi(\vec{u}), \quad \forall \vec{u} \in \partial\Omega \quad (4.2)$$

$$I(\vec{u}) = -\sigma(\vec{u})\nabla\phi(\vec{u})\hat{n}(\vec{u}), \quad \forall \vec{u} \in \partial\Omega \quad (4.3)$$

Where $u = (x, y, z)$ is the position of a particular object, $\phi(\vec{u})$ is the overall distribution of potentials, $\phi_{ext}(\vec{u})$ is the distribution of electric potentials in the surface electrodes, $(\vec{u})$ is the electrical current applied to the surface, $\sigma(\vec{u})$ is the distribution of electric conductivity (image of interest), Ω is the volume of interest or domain, $\partial\Omega$ is the edge of this area, and $\hat{n}(\vec{u})$ is the normal vector to the edge into position $\vec{u} \in \partial\Omega$. According to Cheney et al. (1999) [14], Equation 4.1 can be obtained by Maxwell's equations.

4.1.3 Reconstructing Images in EIT

EIT is based on injecting an alternating electrical current of low amplitude and high frequency through electrodes arranged and fixed on the surface of the body section/object, where it is desired to obtain the image, and measuring the resulting electrical potentials in these surface electrodes [18, 51]. This is schematized in Figure 4.1. Possible forms of excitation of the body by injecting electric current can be classified into two types: adjacent and diametrical. The pattern of the excitation adjacent electric current is applied

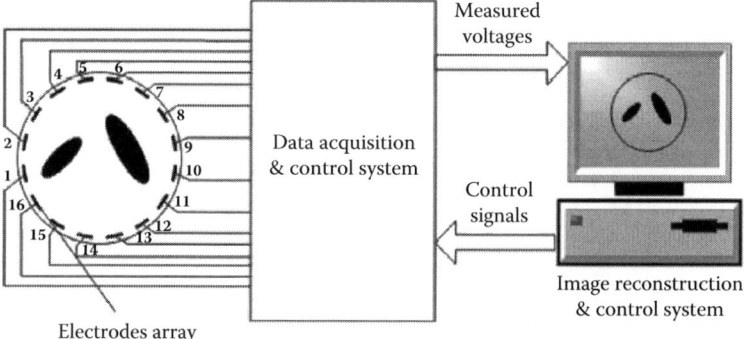

FIGURE 4.1: EIT schematization [56]

to an electrode and the nearest electrode (neighbor) is taken as a reference point [34, 10]. As for the diametric pattern, the electric current is applied to an electrode and the diametric opposite electrode is taken as a reference point [34, 10]. Thus, for both excitement patterns of the pair of electrodes (injection and the reference electrode), current is successively alternated around the body section until it obtains a set of linearly independent data determined by the number of electrodes distributed around the body [34, 10].

EIT image is generated by hardware and software. The hardware is used to implement the excitation and measurement of responses to excitation and the software is used to define the pattern of excitation, while reconstruct the image by obtaining an approximate solution to the inverse problem of EIT [53].

4.2 EIT Computing Tools

The experiments realized in this work were made using the tool EIDORS (Electrical Impedance Tomography and Diffuse Optical Tomography Reconstruction Software), a computational tool open sourced and used in EIT. It was developed starting in 1999 for MatLab (versions $\geq$ 2008a) and Octave (versions $\geq$ 3.6). EIDORS is characterized by providing several free, inverse problem resolution algorithms of EIT, for example: Gauss-Newton [2], GREIT [1], NOSER [15] non-trivial and Backprojection [45]. This tool can realize some, tasks of EIT such as the resolution of direct and inverse problems for any domain, or the graphical representation of a candidate vector to the solution in a mesh of triangular finite elements in order to visually evaluate the distribution of electrical conductivity.

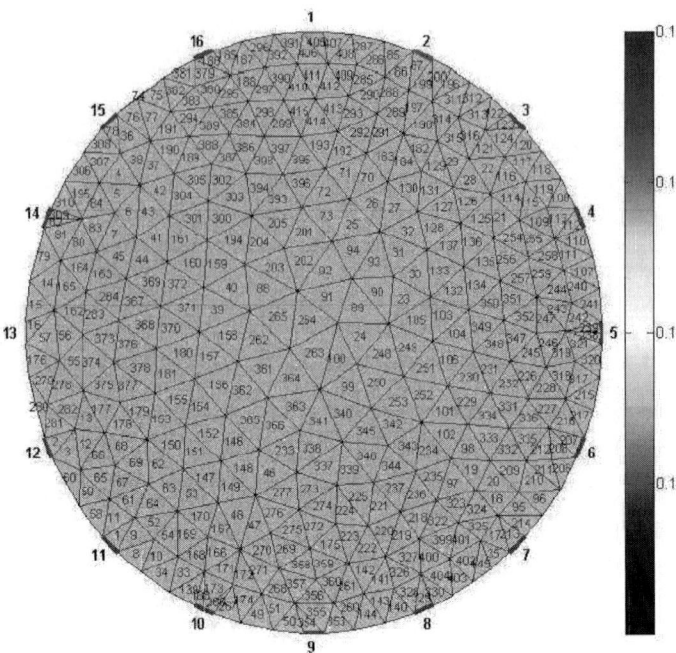

FIGURE 4.2: Mesh of triangular finite elements generated by EIDORS

With EIDORS, the candidate vector to the solution contains the electric conductivity of each of the triangular finite elements of the field under study. Figure 4.2 show a mesh generated by EIDORS. In this mesh, each triangle graphically represent a finite element, and its number corresponds to the candidate vector index to the solution containing the electrical conductivity value of this element. The EIDORS configurations used by the authors of this chapter (see Table 4.2), obtained empirically to represent a numerical phantom, generated a circular mesh containing 415 triangular finite elements. Each candidate vector to the solution has 415 positions (dimension vector), and each contains the electrical conductivity of the respective mesh element.

For all experiments, these items were used as comparison objects (ground-truth images): three electrical conductivity distribution settings in the circular area, with objects positioned in the center, on the border and between the

TABLE 4.2: EIDORS parameter configurations.

Parameter	Value
Number of electrodes	16
Mesh density	b
Mesh refinement	2
Size of solution candidates	415 elements

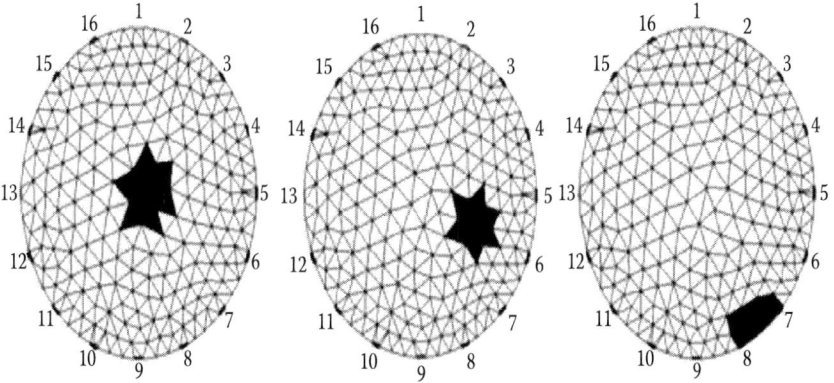

FIGURE 4.3: Ground-truth images used as in the experiments

center and the border (intermediate). Figure 4.3 shows the ground-truth images created by the authors of this chapter to be reconstructed by bioinspired methods.

In Figure 4.3, the darker area is the higher conductivity, defined as 5 S/m, and the white area is a lower conductivity, defined as 0 S/m. The figure represents objects with high conductivity located in the center; between the center and the border; and near the border of a circular area, respectively. These settings were used to simulate computationally the presence of a tumor characterized by high electrical conductivity, inside a human body.

4.2.1 Fitness Function

For the realization of computational experiments, the authors of this chapter used the root mean square relative error between the electric potentials on surface electrodes of the ground-truth image and the candidate to the solution X as fitness function $f_o(X)$ to be minimized by bioinspired optimization methods. This relative error is calculated by the follow expression [41]:

$$f_o(X) = \sqrt{\frac{\sum_{i=1}^{n_p}(U_i(X) - V_i)^2}{\sum_{i=1}^{n_p}(V_i)^2}} \quad (4.4)$$

where n_p is the number of electrodes, V_i are the edge potentials in the ground-truth image, and $U_i(X)$ are the electric potentials on surface electrodes of the candidate to the solution obtained by the reconstruction algorithm.

4.3 Evolutionary Computation

Living organisms are consummate problem solvers. They exhibit a versatility that puts the best computer programs to shame. This observation is especially galling for computer scientists, who may spend months or years of intellectual effort on an algorithm, whereas organisms come by their abilities through the apparently undirected mechanism of evolution and natural selection. The state of the art of the resolution of the problem of EIT walks toward the use of computational intelligence techniques, where the following methods of evolutionary computation can be cited: genetic algorithms and differential evolution [29, 3, 38].

> According to Bäck et al. [5], evolutionary computation can be considered a branch of the computational approach known as natural computation. It is a line of research in computer science that uses the principles of Darwin's evolution and Mendel's genetics, aiming to inspire the development of algorithms able to solve several complex problems of engineering.

The following sections will discuss some bioinspired methods used for reconstruction of EIT images published by the authors of this work.

4.4 Genetic Algorithms

> Computer programs that "evolve" in ways that resemble natural selection can solve complex problems even their creators do not fully understand.
>
> by John H. Holland

According to Cintra (2007) [16], genetic algorithms (GA) were created around 1960 and published by Holland in 1975. Such algorithms are inspired by natural phenomena associated with the adaptation of species and natural selection. According to Lopes (2006) [30], these algorithms are iterative processes that can solve a general problem at one time considered acceptable, although that response may not be the optimal solution to the problem under study. Genetic algorithms strongly follow the principles of Mendelian theory, implementing the functions recombination and gene mutation to generate new individuals with better adaptation to the environment [30]. The following outlines what happens in each basic GA generation [44, 13]. A genetic algorithm (GA) is a generalized, computer-executable version of Fisher's formulation [21]. The generalizations consist of:

- Concern with interaction of genes on a chromosome, rather than assuming alleles act independently of each other;

- Enlargement of the set of genetic operators to include other well-known genetic operators, such as crossing-over (recombination) and inversion.

4.4.1 Approaches to Genetic Algorithms

The genetic algorithm, following Fisher's formulation, uses the differing fitness of variants in the current generation to propagate improvements to the next generation, but the GA places strong emphasis on the variants produced by crossover. Outline of the algorithm is depicted in Figure 4.4. The basic GA subroutine, which produces the next generation from the current one, executes the following steps (where, for simplicity, it is assumed that each individual is specified by a single chromosome):

1. Start with a population of N individual strings of alleles (perhaps generated at random).

2. Select two individuals at random from the current population, biasing selection toward individuals with higher fitness.

3. Use crossover (with occasional use of mutation) to produce two new individuals, which are assigned to the next generation.

4. Repeat steps (1) and (2) N/2 times to produce a new generation of N individuals.

5. Return to step (1) to produce the next generation.

Of course, there are several ways to modify these steps, but most characteristics of a GA are already exhibited by this basic program [19]. The basic concepts of genetic algorithms are:

- **Generation of the initial population:** From the literature, we have chosen the simplest way to generate an initial population of solution candidates, as a random model based on the randomness of the genes composing the individual. Such a model is constructed to avoid repeated representation of the same value, to combat ambiguous solutions [35].

- **Selection:** The selection processes of the individuals to compose an intermediate population of solution candidates can be classified into the following methods [35]. Selection is simple: If an organism fails some test

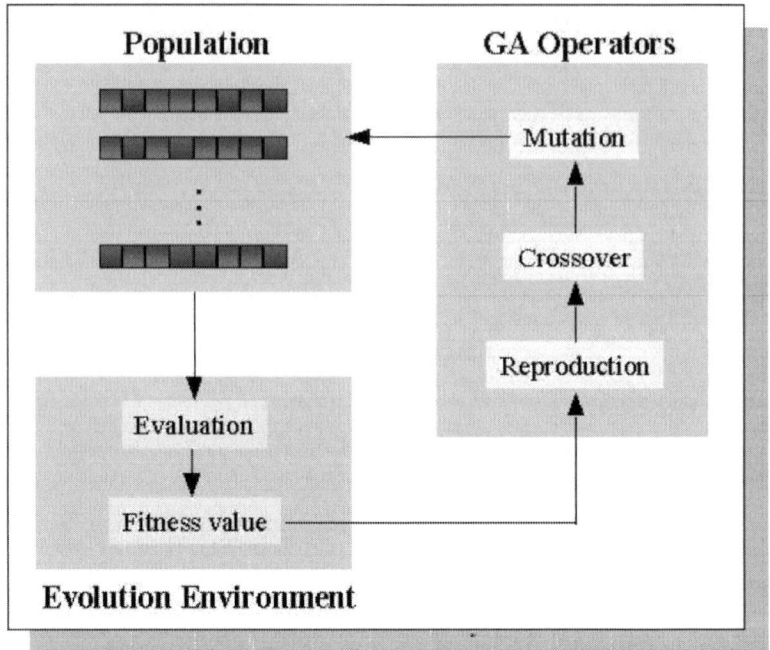

FIGURE 4.4: Genetic algorithm scheme

of fitness, such as recognizing a predator and fleeing, it dies. Similarly, computer scientists have little trouble weeding out poorly performing algorithms. If a program is supposed to sort numbers in ascending order, for example, one need merely check whether each entry of the program's output is larger than the preceding one [19].

1. **Roulette:** This method is characterized by individuals chosen to be part of the next generation through a lottery where each individual of the population is represented in roulette as a slice that has size proportional to his own fitness index. This index is obtained from the evaluation of the objective function (or fitness function) set to the problem. Thus, the greater the aptitude of the individual to the environment, the greater the chance of being selected to compose the intermediate population;

2. **Tournament:** This method is characterized by randomly choosing NP individuals in a population (usually NP = 3), then selecting from among these NP individuals, make the intermediate population, that individual which has the highest fitness index. This index corresponds to the value obtained by the objective function (or fitness function) set to the problem. The process is repeated until all the intermediate population of solution candidates is obtained.

TABLE 4.3: Parameters used for genetic algorithms.

Parameter	Value
Initial population	100 chromosomes
Selection	10 chromosomes top rated
Crossing Probability	100%
Mutation Probability	100% of the offspring
Stop criteria	500 generations

- **Crossover:** This process performs exchange of genetic information between two individuals of the intermediate population, selected through their fitness. Thus, this process can result in the generation of one or two child individuals, depending on the genetic information of the parent individuals [35].

- **Mutation:** The mutation operation seeks to change one or more genes of an individual of the intermediate population taken as a basis, thus seeking to obtain a new individual with diversified genetic material [35].

- **Stop criteria:** This operator is used as an algorithm termination condition and can be: running time, number of generations, or fitness value [35].

4.4.2 Method of Searching and Optimization of Genetic Algorithms

To implement the method of searching and optimization of genetic algorithms (GAs) in the reconstruction of EIT, images were necessary to propose some analogies based on Mendel's genetic theory. Chromosomes in the AG technique were represented in EIT by electrical conductivity vectors; that is, images of candidates for the EIT solution. The genes were represented by such vector positions, i.e., the triangular finite elements of EIT images. In applying this technique, the parameters mentioned in Table 4.3 are empirically used.

It is important to emphasize that the chromosomes of the initial population were defined with a random distribution of internal conductivity in the range $[0, 1]$.

4.4.3 General Characteristics of Experiments

Quantitative: This describes the error calculation made on the objective function to the solution of the best candidate found by the algorithm at the end of each generation, in order to evaluate the cost and computation time for the reconstruction of EIT images. As a search technique, it has a smaller relative error value in a smaller number of evaluations of the objective function and/or

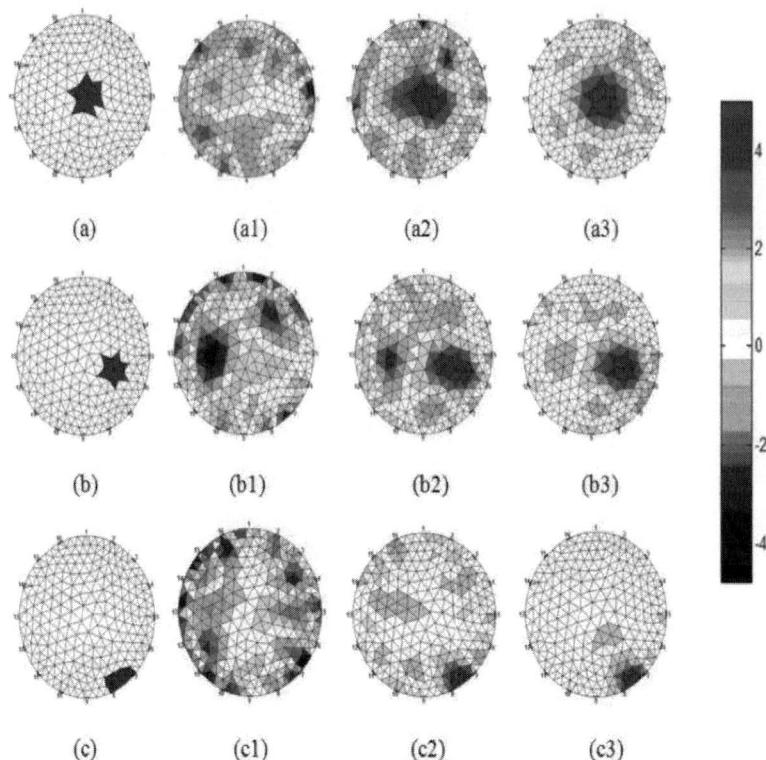

FIGURE 4.5: Results using GA for an object placed in the center (a1, b1, and a3), between the center and the border (b1, b2, and b3), and near the border (c1, c2, and c3) of the circular domain for 50, 300, and 500 iterations

a smaller reconstruction time. The **evaluation of the objective function** is the most computationally expensive step among all the techniques due to the cost of solving the problem of direct EIT for each candidate solution.

Qualitative: This evaluation was performed by visual comparison of the reconstructed images with the ground-truth images, to assess how similar the best candidate solution found by the algorithm at the end of each generation was to the ground-truth image.

To allow comparison of results, the parameters used were, also the same for all experiments, available in EIDORS software. The number of solution candidates was also a common factor for all the algorithms implemented: 100 initial solution candidates. These candidates had their conductivity distributions filled randomly.

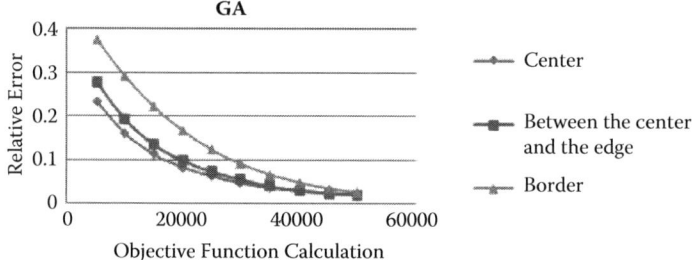

FIGURE 4.6: Relative error along the iterations for GA

4.4.4 Application Results of AG

The following shows the results for an AG canonical for an isolated object located in the center, between the center and the border, and near the border, respectively: Where (a), (b), and (c) represent the individual objects positioned in the center between the center and the edge, and near the edge, respectively. Figure 4.5 (a1), (a2), and (a3) represents the image of the best candidate solution for the object in the center at 50, 300, and 500 iterations, respectively. Figure 4.5 (b1), (b2), and (b3) represents the image of the best candidate solution for the object between the center and the edge at 50, 300, and 500 iterations; respectively. Figures 4.5 (c1), (c2), and (c3) represents the image of the best candidate solution for the object near the edge at 50, 300, and 500 iterations, respectively.

The following quantitative results depicted in Figure 4.6 are shown in the relative error behavior by this individual in the population, with respect to the lower cost and computational time. The lowest and medium gray line represents the evolution of the objective function for the object located in the center; the darkest gray line for the object located between the center and the edge of the area; and the highest and lightest gray line for the object located close to the edge.

4.4.5 Discussion of the Results for AG

Qualitatively, from Figure 4.5 you can see that the AG could generate anatomically consistent images; that is, it found the individual objects shown in just 300 iterations, and was also conclusive, given that it could delimit the edges of isolated objects. The technique was shown to be able to reconstruct images with few artifacts in a reasonable number of iterations.

Quantitatively, from Figure 4.6 it can be seen that the computational costs for the reconstruction of EIT images in three settings are equal. Also, the AG showed a slow decline of the relative error for the reconstruction of isolated objects positioned on the field edge; for such a configuration, the AG needed more computational time for EIT reconstruction. Finally, the AG proved to be a technique that made good reconstructions of EIT images.

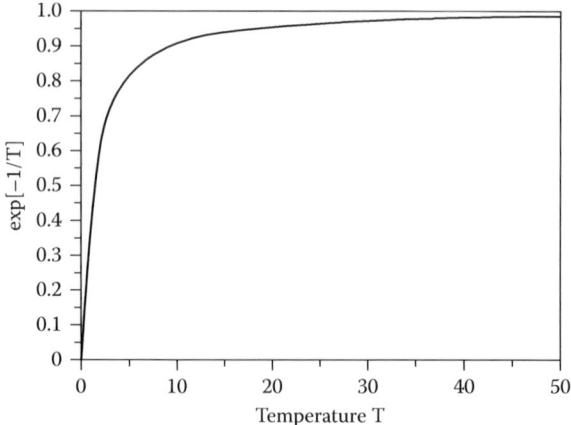

FIGURE 4.7: Temperature during the cooling time

4.5 Simulated Annealing

The simulated annealing method is used to search the optimal solutions based on the variation of the temperature of the metals annealing method. It was proposed by Scott Kirkpatrick et al. (1983) [26], for a physical process of solidifying and melting metals by controlling the temperature variation. Metal structures alter their shape and arrangement depending on the time in which their physical state changes: in this context, the cooling of a metal in a gradual and controlled manner, subjected to a high temperature, leads to the minimum energy states that contribute to its perfection and quality, avoiding structural disorders. The temperature during cooling time is depicted in Figure 4.7.

4.5.1 Simulated Annealing Approaches

The application algorithm should abstract the metal annealing problem, in order to execute a heuristic search for the optimal solution.

- Search space: set of possible states of a metal subjected to cooling.
- Objective function: the measured power in each state.
- Energy minimum or maximum: value corresponding to an optimal solution site that can be at best an overall large value.

The Metropolis algorithm [20] implements the simulation through a sequence of decreasing temperature (T) to generate optimization problem solutions, starting with a high T value and reducing it gradually to the equilibrium. With each iteration of the algorithm, the system power (E) is changed, enabling the conclusions given below. Parameters used to simulate annealing are

TABLE 4.4: Parameters used to simulate annealing.

Parameter	Value
Start temperature	200.000
Process parameter	1
Maximum iteration for temperature stability	100
Stopping criterion	500 iterations

presented in Table 4.4. The procedure is repeated upto the possible nearest values of balance (Metropolis step).

- If $\Delta E < 0$, the new state has less power than the current happens to have.

- If $\Delta E > 0$, the new state has more energy than the current state, and then the probability to change the current state to another through the expression $P = e^{-\frac{\Delta E}{(kT)}}$, where k is the Boltzman constant and T is the actual temperature.

4.5.2 Proposed Implementation

The problem of image reconstruction of EIT can be abstracted for the use of simulated annealing (SA), defined in accordance with the previous section of analogies. The body will be cooled and represented by an electrical conductivity vector that ultimately results in the image being reconstructed.

Pseudocode is described in Ribeiro (2016) [42], who implemented a function called *generateAnyNeighbor*(S) that generates a new candidate solution in a search space of the current solution (S), a result of the insertion of a random noise, at a maximum of 5%, of a randomly chosen element. By the time the change in the value of a conductivity element results in the generation of a new candidate for the worst solution, this can result in an increase of the objective function, and then the algorithm prevents modification of this conductivity element for all other iterations until there is convergence. In the following algorithm, the pseudocode of simulated annealing is presented.

The candidate solution that has high value in the associated objective function can also be maintained with a probability obtained from Equation (6d) from Simulated Annealing Algorithm. This fact enables the SA to avoid points of local minima. In this context, it can be concluded that the higher the temperature associated with the candidate, the more likely it is to avoid local minima, thus contributing to a more rapid convergence of the algorithm. So, the value of 200,000 was assigned as the initial temperature. Figure 4.8 shows the results of using SA for an object placed in the center, between the center and the border, and near the border of the circular domain for 50, 300, and 500 iterations whereas, Figure 4.9 shows the relative error along the iterations.

Algorithm [Simulated Annealing]
1. $S^* \longleftarrow S$; (Initial solution)
2. $T \longleftarrow T_0$; (Initial temperature)
3. k; (Process parameter)
4. γ; (Random constant $\in [0,1]$)
5. $IterT_{max}$; (Iteration maximum for temperature stability)
6. While (stop criterion not satisfied) do

    1. $IterT \longleftarrow 0$; (Iteration for temperature stability)

    2. While ($IterT < IterT_{max}$) do

        (a) $IterT \longleftarrow IterT + 1$

        (b) $S' \longleftarrow genereateAnyNeighbor(S)$

        (c) $\Delta E \longleftarrow f(S') - f(S)$; (Objective function changing f)

        (d) If ($\Delta E < 0$) Then $S \longleftarrow S'$;

        (e) If ($f(S') < f(S^*)$) Then $S^* \longleftarrow S'$;

        (f) Else

            i. Take random $P \in [0,1]$

            ii. If ($P < e^{-\frac{\Delta E}{kT}}$) Then $S \longleftarrow S'$

    3. End-While

    4. $T \longleftarrow \frac{T}{1+\gamma\sqrt{T}}$

    5. $IterT \longleftarrow 0$

7. End-While
8. $S \longleftarrow S^*$
9. Return S^*
10. End SimulatedAnnealing

4.6 Particle Swarm Optimization

Particle swarm optimization was developed initially for continuous nonlinear functions optimization by the social psychologist James Kennedy and the electrical engineer Russell Eberhart, in 1995. This technique is a metaheuristic based on the motion of birds flock searching for food. The method was discovered through the simulation of this particular social behavior, and the observance of the optimization of the food searching and flying organization of the birds [24].

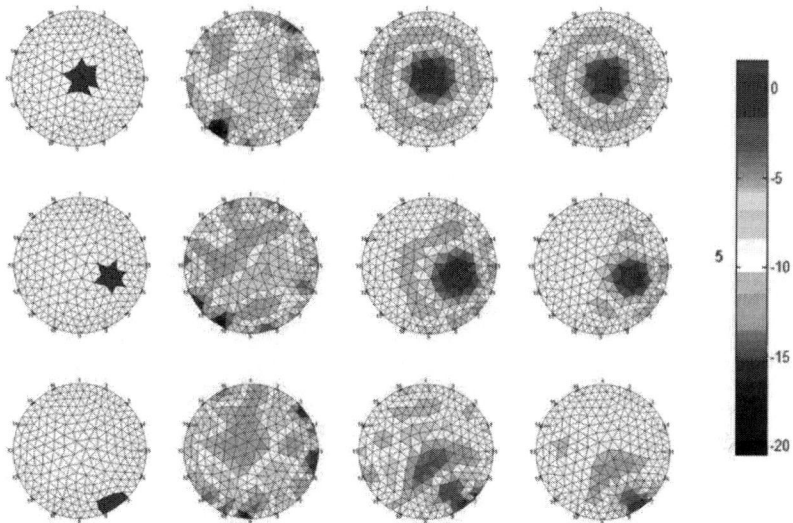

FIGURE 4.8: Results of using SA for an object placed in the center, between the center and the border, and near the border of the circular domain

The movement of the whole flock is based on the overall intelligence of the group, as shown in Figure 4.10. The most experienced particle (bird) leads the others in the way to food sources. But the others have their self-knowledge of places where there is food as well, and their trajectory is influenced by these two knowledge sources.

The PSO technique has been applied in a diverse range of fields, such as weights adjustment in neural networks [47], job scheduling [54], vehicle routing [48], simulations for optimal design of hybrid renewable energy systems [25], estimating the parameters for solar cell components [25] and building a rules system for extraction for medical diagnosis [22]. Since its creation in 1995, PSO has received several contributions. These efforts have since added factors, like

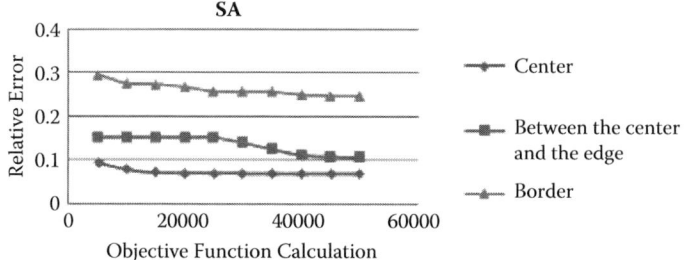

FIGURE 4.9: Relative error along the iterations for SA

FIGURE 4.10: Birds flock flight (adapted from: media.salon.com)

the insertion of the inertia factor [49], the dynamic changing of the leader distribution for maintaining diversity [23], and several other alterations in the structure and operation of the algorithm, that are intended to improve the convergence and provide efficiency to the PSO technique.

4.6.1 The Particle Optimization Swarm Algorithm

The particle in the PSO algorithm is represented by two vectors: velocity (v) and position (x), and they are calculated by the expressions shown below. Here, (w) is an inertia factor, as created by Shi and Eberhart in 1998 [49], to reduce the velocity along the iterations and try to avoid local minima places within the objective function, while (c_1 and c_2) are constants that ponder the individual influence and social influence factors, respectively. (P_i and P_g) are the best position of the i-th particle and the overall fittest particle, respectively, and (r_1 and r_2) are random numbers between 0 and 1 that provide stochasticity to the search process.

$$x_i(t+1) = x_i(t) + v_i(t+1) \tag{4.5}$$

$$v_i(t+1) = wv_i(t) + c_1 r_1 (P_i - x_i(t)) + c_2 r_2 (P_g - x_i(t)) \tag{4.6}$$

Thus, in the PSO version used in EIT experiments, the particle movement is influenced by the individual's knowledge from their self-best fitness, along with the social knowledge from the best fitness of all the history, and its velocity is pondered by the inertia factor as well. The following figure shows a representation of the vectorial forces acting on the particle. The PSO steps are represented by the flowchart shown in Figure 4.11.

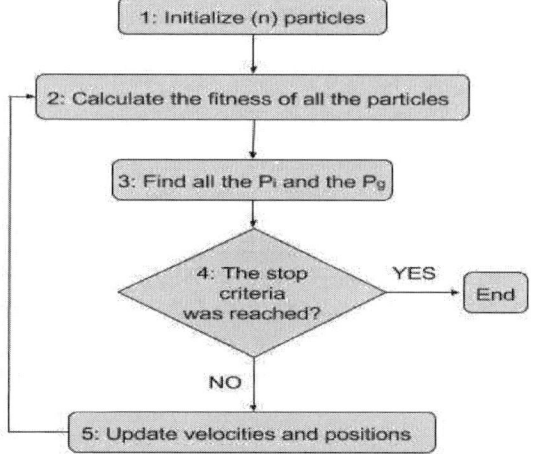

FIGURE 4.11: Flowchart of PSO algorithm

1. This step initializes (n) d-dimensional vectors, where d is the number of dimensions of the problem domain, and n is the number of particles to be initialized.

2. According to the given objective function, the fitness of all the particles in the system is calculated and stored.

3. Looking at the fitness previously calculated in step 2, the best fitness of each particle is found.

4. The stop criteria, which can be a number of iterations or a fitness value from the best particle, is verified. If reached, the program ends and the best solution found is the result of the PSO execution. If not, the algorithm execution goes on.

5. According to Equations 4.5 and 4.6, the velocities and positions of all the particles are updated.

To be implemented to electrical impedance tomography, the PSO steps were modelled as follows:

1. The particles' initial set was initialized with $n = 100$ particles of 415 dimensions each. These numbers come from the experience with the best convergence number of initial candidates of solution for the other techniques used, and the dimension of an EIT candidate image, respectively.

2. For each particle, the function objective was calculated through the euclidian distance between the border potentials generated by the candidate distribution and the border potentials from the ground-truth image.

3. The best distribution so far and the best distribution for each particle are found.

4. The experiment was made with 50, 300, and 500 iterations, in order to find the best image in a smaller number of iterations, because of the computational time spent for each iteration.

5. The particles' distributions were updated by the PSO velocity and position in Equations 4.5 and 4.6.

Figure 4.12 shows the resulting images obtained by PSO, where (a), (b), and (c) are the ground-truth images placed in center, between the center and the border and in the border, (a1), (b1), and (c1) are the images from 50 iterations; (a2), (b2), and (c2) are the images from 300 iterations; and (a3), (b3), and (c3) are the images from 500 iterations. The relative error obtained by PSO is presented in Figure 4.13.

For all the techniques, the hardest image reconstruction was the border image. That happened because the potentials of a border object is heterogeneous and it makes the image reconstruction difficult. The best images were generated at 500 iterations; however, they still are blurred with a lot of noise (darker gray elements) in the image distribution. In the qualitative analysis, the PSO was shown to be a technique that reaches the images; however generates noise images with very low quality.

For the image placed in the center, the relative error, it did not fall after 20,000 objective function calculations; this happened because of the tendency of PSO to fall in local minimum and stagnate the increase of the particles' fitness after some time. In the center-border localized object, the error falls after 20,000 function evaluations, but at the end of the execution, it has still not reached a very low error rate. In PSO implementation, the object placed in the border generated the worst result; despite the error fall not being stopped in a local minimum, the end result was not a low error rate. This can be seen in the image generated, where is not very easy to identify the target image.

Electrical Impedance Tomography Using Evolutionary Computing

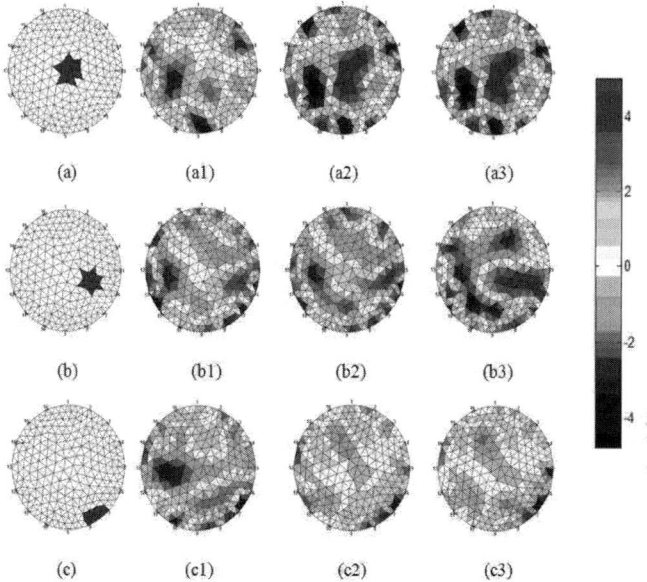

FIGURE 4.12: Results of using PSO for an object placed in the center, between the center and the border, and in the border of the circular domain for 50, 300, and 500 iterations

4.7 Differential Evolution

The differential evolution (DE) metaheuristic is one of the several techniques of global optimization methods used to find the best solution for a problem that cannot be modeled. It basically uses an objective function, which defines

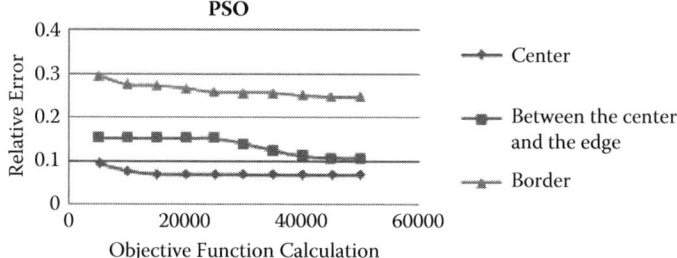

FIGURE 4.13: Relative error along the objective function calculation number for an object placed at the border, between the center and the border, and in the border

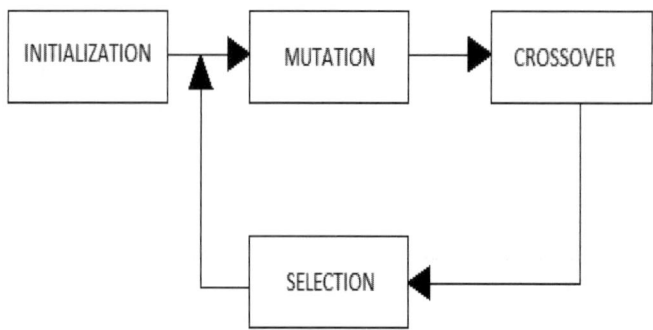

FIGURE 4.14: Differential evolution algorithm's scheme

the problem as a minimization (or maximization) task, and has the following main advantages over other techniques: simplicity, and the ability to deal with non-differentiable, nonlinear, and multimodal functions, variables' robustness to reduce computational cost while minimizing the objective function [50]. An overall view of differential evolution algorithm is depicted in Figure 4.14.

DE began with genetic annealing, created by Kenneth Price. This method, published in 1994, consisted of a population of possible solutions created as the result of a combinatorial algorithm from populations of previous iterations, using, at the same time, an annealing criterion that has a threshold based on the average performance of the population. Interested in solving the Chebychev polynomial problem, Rainer Storn contacted Kenneth Price in order to use GA as a possible solution method for the situation. After a few experiments, Price noticed that substituting logical operations for arithmetic ones and using floating points instead, bit-string resulted in a method based on differential mutation. They also realized that discrete recombination and pairwise selection of the algorithm recently discovered resulted in an algorithm that did not need an annealing factor. Therefore, in doing these alterations, the algorithm obtained they was the first definition of differential evolution [52].

Nowadays, differential evolution has applications in several fields of study, such as gas transmission network design, radio network design, neural network learning, optimization of nonlinear chemical processes, mineral exploration, reconstruction of images obtained from electric tomography impedance (EIT), and so on.

4.7.1 Differential Evolution Approach

Differential evolution is considered simple and efficient when its purpose is to optimize a function described in a continuous space [46]. Its main idea is

TABLE 4.5: Mutation equations.

Equation ($V_{i,j,G} =$)	Notation
$X_{\alpha,G} + F_i(X_{\beta,G} - X_{\gamma,G})$	ED/rand/1/bin
$X_{\alpha,G} + F_i(X_{\rho,G} - X_{\beta,G}) + F_i(X_{\gamma,G} - X_{\delta,G})$	ED/rand/2/bin
$X_{best,G} + F_i(X_{\beta,G} - X_{\gamma,G})$	ED/best/1/bin
$X_{best,G} + F_i(X_{\rho,G} - X_{\beta,G}) + F_i(X_{\gamma,G} - X_{\delta,G})$	ED/best/2/bin
$X_{i,G} + F_i(X_{\rho,G} - X_{i,G}) + F_i(X_{\gamma,G} - X_{\delta,G})$	ED/rand-to-best/2/bin
$X_{\alpha,G} + F_i(X_{\beta,G} - X_{\gamma,G})$	ED/rand/1/exp
$X_{\alpha,G} + F_i(X_{\rho,G} - X_{\beta,G}) + F_i(X_{\gamma,G} - X_{\delta,G})$	ED/rand/2/exp
$X_{best,G} + F_i(X_{\beta,G} - X_{\gamma,G})$	ED/best/1/exp
$X_{best,G} + F_i(X_{\rho,G} - X_{\beta,G}) + F_i(X_{\gamma,G} - X_{\delta,G})$	ED/best/2/exp
$X_{i,G} + F_i(X_{\rho,G} - X_{i,G}) + F_i(X_{\gamma,G} - X_{\delta,G})$	ED/rand-to-best/2/exp

to generate new parametrization vectors using mutation and crossover (based on Mendel's theory) and to use natural selection (Darwinism) as the operator that will define which vector will survive until the next generation.

DE population is peculiar, since it has a constant size, with NP vectors $X_{j,i,G}$ with real values and D dimensions, where $i = 1, 2, 3, \ldots, NP$ (i means the population index), $j = 1, 2, 3, \ldots, D$ (j means the dimension index), and G is the number of the current generation [36]. The optimization process by DE is composed of the following steps, shown by the fluxogram below and explained in more detail in the following paragraphs.

Initialization: This first step consists of creating an initial population of vectors, whose components are chosen randomly and with uniform distribution, respecting the search space limit's. Therefore, they are usually limited by two border values, one being superior ($X_j^{(U)}$) and one inferior ($X_j^{(L)}$), meaning that:

$$X_j^{(L)} < X_j < X_j^{(U)}, j = 1, 2, \ldots, D \qquad (4.7)$$

and

$$X_{j,i,G} = X_j^{(L)} + rand_j[0,1](X_j^{(U)} - X_j^{(L)}) \qquad (4.8)$$

where $rand_j[0,1]$ implies a random number between 0 and 1, chosen every time j changes its value.

Mutation: The generation of new parameters' vectors is important to help the algorithm's process escaping of from local maximums/minimums. DE can do it in several ways, as shown in Table 4.5. The $\alpha, \beta, \gamma, \delta$, and ρ indexes are integers chosen between 1 and NP and different from i, selected once randomly for each donating vector. The amplification vector F_i is a positive parameter's vector that expands the different vectors, enlarging the diversity of the candidates to the solution of the problem. The vector X_{best} is one of the current generation in which the objective function presented the best value (minor for a minimization problem and bigger for a maximization problem).

It is important to add that those equations were tested by Storn and Price by being combined with binomial or exponential mutation, making them 10 equations instead of 5. Another mutation equation was proposed by Ribeiro et al. (2014a) [40], based on Equation ED/rand-to-best/2/bin of the previous table.

$$V_{i,j,G} = X_{best,G} + F_i(X_{i,G} - X_{best,G}) + F_i(X_{\gamma,G} - X_{\delta,G}) \quad (4.9)$$

This mutation method, according to Ribeiro et al. (2014a) [40], made the algorithm converge faster than the other ways, when compared to the classic process of mutation presented in Equation ED/rand-to-best/2/bin.

Crossover: Here the parameters of the $V_{i,j,G}$ vector are mixed with the parameters of another predetermined vector, known as the judgment vector. In the case of binomial crossover, this vector is created as $W_{i,j,G+1} = (W_{i,1,G+1}, W_{i,2,G+1}, \ldots, W_{i,D,G+1})$, where:

$$W_{i,j,G+1} = \begin{cases} V_{i,j,G} & \text{if} rand_{(z)} \leq P_{CR} \text{ or } z = rnbr(i) \\ X_{i,j,G} & \text{if} rand_{(z)} > P_{CR} \text{ or } z \neq rnbr(i) \end{cases} \quad (4.10)$$

It is important to say that z is the number of evaluations, $rand_{(z)}$ is a random number between $[0,1]$, $rnbr(i) \in 1, 2, \ldots, D$ is a randomly chosen index that guarantees $W_{i,1,G+1}$ has at least one parameter from $V_{i,1,G}$, and $P_{CR} \in [0, 1]$ is the input parameter influencing the number of elements to be changed by crossover. In binomial crossover, the relation between the probability of crossover and P_{CR} is linear.

For exponential crossover, its star position is chosen between $1, \ldots, D$, and L consecutive elements are taken from the mutant vector $V_{i,j,G}$, meaning that L adjacent elements are changed in exponential variant. Also, the relation between the probability of crossover and P_{CR} is nonlinear and the deviation from linearity is bigger with the increasing dimension of the problem.

$$W_{i,j,G+1} = \begin{cases} V_{i,j,G} & \text{for } (j = \langle n \rangle_D, \langle n+1 \rangle_D \ldots \langle n+L-1 \rangle_D) \\ X_{i,j,G} & \text{for all other } j \in [1, D] \end{cases} \quad (4.11)$$

and, to change L:

$$L = \begin{cases} L+1, & \text{while } (rand_{(z)} \leq P_{CR}) \text{ and } (L \leq D) \\ 0, & \text{otherwise} \end{cases} \quad (4.12)$$

Selection: in a minimization problem, if the objective function of judgment vector $W_{i,j,G+1}$ produces a value lower than the target vector $X_{i,j,G}$, $W_{i,j,G+1}$ replaces $X_{i,j,G}$ in the given generation. If the opposite occurs, $X_{i,j,G}$ is maintained. For maximization problems, the situation described above is reversed.

Stop criterion: the algorithm stops running when the objective function of the best vector reaches a predetermined value or the number of iterations also reaches a predefined value, usually specified by the algorithm's operator.

Control parameter selection: Usually, the size of the population, NP, is chosen as $5D$ or $10D$, and F_i and P_{CR} are predetermined inputs, as mentioned before. Those values must be discovered by trial and error, since they change depending on the application.

4.7.2 Implementation Purpose

As described before, DE has several applications, one of them being image reconstruction based on EIT. Since EIT is the main study subject of this chapter, here the applications of DE are basically turned to this goal, meaning that differential evolution is applied as a minimization problem and the mutation is binomial instead of exponential. In the following paragraph, we present the pseudocode used to implement the technique for EIT.

1. Generate the population of n agents, each one represented by a vector $X_{i,j,G}$ where $i = 1, 2, \ldots, n$
2. Repeat until the maximum number of iterations is reached
   (a) For $i = 1, 2, \cdots, n$ do
      i. Get a random number r within $[0,1]$
      ii. If $r \leq P_{CR}$ then
         A. Generate the crossed mutated new agent $V_{i,j,G}$;
         B. If $f_0(V_{i,j,G}) < f_0(X_{i,j,G})$, then $X_{i,j,G} \leftarrow V_{i,j,G}$ (minimization)

As mentioned in Section 4.7.1, in the Mutation paragraph, there are 10 types of methods used to mix the vectors in order to create new ones. Since the DE applied in this research uses binomial distribution, Equation ED/rand-to-best/2/bin was tested in order to check their efficiency in the matter of images reconstruction for EIT. The initial parameters were set up as $NP = 100$, $P_{CR} = 0.90$. Also, it is important to mention that the algorithm was tested to 50, 300, and 500 iterations, in order to observe its development across iterations.

As the qualitative results showed, the 5th version of the differential evolution mutation algorithm exhibited a consistent result to EIT images reconstruction for all of the three configurations studied, as presented in Figure 4.15. It is also possible to show that 300 iterations were enough to anatomically find reliable and conclusive results for all experiments.

Quantitatively, in Figure 4.16, it is possible to notice that the DE algorithm's (5th version) computational cost is almost the same for all configurations, being, however, more efficient to rebuild the situation where the object is placed in the center of the area of interest. This fact could be noticed since the relative error is lower than that of the other configurations (object on the border and between the center and the border), when analyzed along iterations.

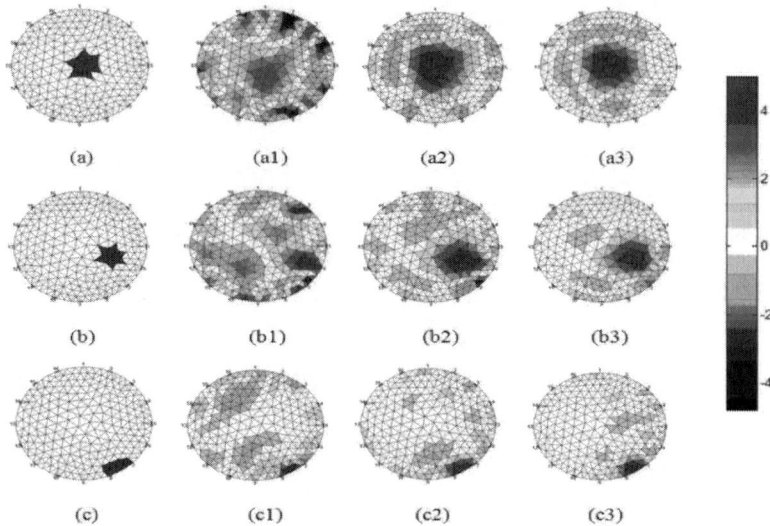

FIGURE 4.15: Results using DE for an object placed in the center (a1, b1, and a3), between the center and the border (b1, b2, and b3), and in the border (c1, c2, and c3) of the circular domain for 50, 300, and 500 iterations

---

4.8 Fish School Search

The fish school search (FSS) algorithm is a metaheuristic based on fish behavior in the search for food, developed by Bastos Filho e Lima Neto in 2008 [8]. The technique is indicated for search and optimization problems of high dimension. A population of limited-memory individuals (the fish) carries out the search process in FSS [28]. Each fish in the school represents one possible

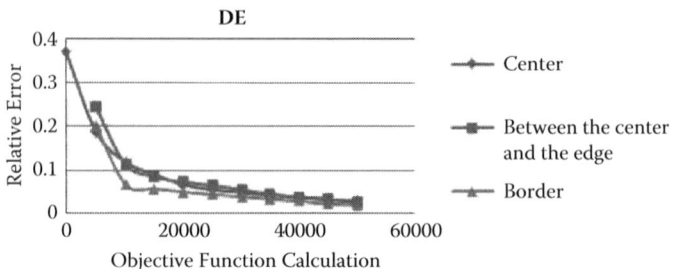

FIGURE 4.16: Relative error along iterations to an isolated object placed at the center, between the center and the border, and at the border, for DE, 5th version

solution of the optimization problem [8]. In this metaphor, an aquarium is the search space of the optimization problem.

4.8.1 Fish School Search Approaches

The technique has four operators, classified as feeding and swimming operators. These operators are detailed as follows.

Individual Movement Operator: The individual movement is the movement responsible for making each fish in the school move to some random point in his neighborhood. In this movement, each fish moves independently with respect to other fish in the school. The displacement is executed only if the new position randomly determined by a fish is better than the previous one; in other words, the movement will only be executed if the new calculated point has a better value in the fitness function; otherwise, this fish will not execute this movement.

The individual displacement for each fish i is given in 4.13, where $rand[-1,1]$ is a vector with values within the interval $[-1,1]$ and $step_{ind}$ is an algorithm parameter that represents the fish displacement capacity in the individual movement. After the calculation of the individual displacement, the new fish position is updated as in 4.14 and thus the condition of the movement described above is satisfied.

$$\Delta x_{ind_i}(t+1) = step_{ind} \cdot rand(-1,1) \quad (4.13)$$
$$x_{ind_i}(t+1) = x_{ind_i}(t) + \Delta x_{ind_i}(t+1) \quad (4.14)$$

Analyzing Equations 4.13 and 4.14, it is possible conclude that the individual movement is a perturbation in the fish's position in the search for promising regions. To guarantee the final convergence of the search process and one larger exploitation of the search space in the last iterations, the value of the individual movement parameter decays linearly, making the search process in this movement more refined. Equation 4.15 gives the decrease in value of the $step_{ind}$, where $step_{indInit}$ and $step_{indEnd}$ are the initial and final values of the parameter, and $iterations$ is the maximum possible value of iterations in the algorithm.

$$step_{ind}(t+1) = step_{ind}(t+1) - \frac{step_{indInit} - step_{indEnd}}{iterations} \quad (4.15)$$

Feeding Operator: The feeding operator is given by the fish weight. Where the weight is the success indicator of the fish, the heavier a fish is, the better the solution it represents is [31]. Therefore, the objective of the fish school search is to maximize the fish weights. If the problem to be resolved by the FSS is a minimization problem, the amount of food in a region is inversely proportional to the function evaluation in this region [8].

The fish weight update depends on its fitness variation after the individual movement and the larger fitness variation occurred in the school. Equation 4.16 gives the update weight expression, where $W_i(t)$ and $W_i(t+1)$ represent the weight of fish i before and after the update, Δf_i is the fitness variation of the fish i after the realization of the individual movement, and $max(\Delta f)$ is the larger variation that occurred in the school in the current iteration.

$$W_i(t+1) = W_i(t) + \frac{\Delta f_i}{max(\Delta f)} \quad (4.16)$$

Collective Instinctive Operator: The second movement in the fish school search is a collective movement where the fish that are more successful in their individual movements guide the school to a promising region. This movement considers the resulting direction (I) calculated by the weighted average of the individual displacement of each fish having like weight. Equation 4.17 gives a calculation of the vector's resulting direction, where N is the number of fish in the school. The position of all fish is updated in Equation 4.18.

$$I(t) = \frac{\sum_{i=1}^{n} \Delta x_{ind_i} \Delta f_i}{\sum_{i=1}^{n} \Delta f_i} \quad (4.17)$$

$$x_i(t+1) = x_i(t) + I(t) \quad (4.18)$$

Collective Volitive Operator: The last movement of the algorithm is based on the global performance of the school [28]. The collective volitive movement is the tool that provides the capacity for the algorithm to adjust the search radius of the fish, being responsible for controlling the search and making a balance between the explorative and exploitative searches. The capability to escape from local optima is due to this radius control of the search [31].

The movement takes place in the following way: If the global weight of the school increases, the search is most successful and the search radius must decrease; otherwise, the search radius must increase to increase the exploration of the aquarium, in attempting to find better results. These expand and contract movements of the search radius of the school are made with respect to the school barycenter given in 4.19.

$$Bary(t) = \frac{\sum_{i=1}^{N} x_i(t) W_i(t)}{\sum_{i=1}^{N} W_i(t)} \quad (4.19)$$

Thus, the movement of each fish is made following Equation 4.20 in the case of the school increasing in weight, or is made by Equation 4.21 if the school is decreasing in weight. In Equations 4.20 and 4.21, $rand[0,1]$ is a vector with values within the interval $[0,1]$ and $step_{vol}$ is an adjust parameter of the movement. The $step_{vol}$ value needs to be at the same order of magnitude as the $step_{ind}$. As in the individual step, it is considered to be a random interval $[-1,1]$ twice as bigg as the interval considered in this movement, and it usually utilizes to be a value of the stepvol twice as bigg as the stepind [31].

$$x(t+1) = x(t) - step_{vol} \cdot rand(0,1)(x(t) - Bary(t)) \quad (4.20)$$
$$x(t+1) = x(t) + step_{vol} \cdot rand(0,1)(x(t) - Bary(t)) \quad (4.21)$$

4.8.2 Implementation Purpose

The pseudocode of the FSS is presented in the following algorithm. The parameter's values relative to FSS utilized in the resolution of the EIT problem were: number of fish $N = 100$, $step_{indInit} = 0.01$ e $step_{indEnd} = 0.0001$ e $step_{vol} = 2step_{ind}$ [7]. The stop criterion utilized was the max number of iterations equal to 500. Figure 4.17 gives the images obtained by the method for an object placed in the center (a), in the border (c), and between the center and the border (c) of the circular domain. The images (a1, b1, and c1) and (a2, b2, and c2) are the partial results for 50 and 300 iterations, respectively, and (a3, b3 and c3) are the final results of the method after 500 iterations.

Algorithm [Fish School Search]

1. Initialize all fish in random positions
2. Repeat the steps from (a) to (f) until some stop criterion is satisfied
    (a) For each fish:
        i. Execute the individual movement.
        ii. Evaluate the fitness of the fish.
        iii. Execute the feeding operator.
    (b) Calculate the resultant direction vector - $I(t)$.
    (c) For each fish:
        i. Execute the collective-instinctive movement.
    (d) Calculate the barycenter
    (e) For each fish:
        i. Execute the collective-volitive movement.
    (f) Update the individual and collective-volitive step
3. Select the fish in the final school that has the best fitness

---

Qualitatively, the performance of the method can be evaluated by the obtained images. In the beginning of the iterative process in 50 iterations, the obtained images are still neither conclusive nor anatomically consistent compared with the ground-truth images, whereas in 300 iterations the method was able to identify the object with reasonably good resolution and was anatomi-

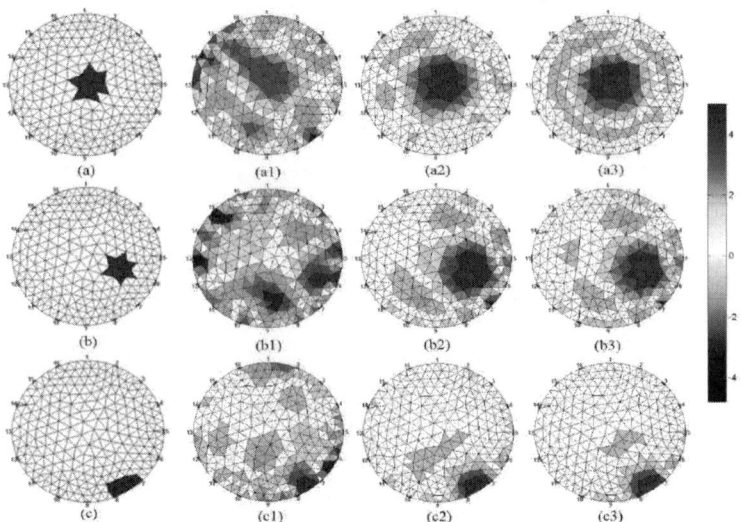

FIGURE 4.17: Results using FSS for an object placed in the center (a1, b1, and a3), in the border (b1, b2, and b3), and between the center and the border (c1, c2, and c3) of the circular domain for 50, 300, and 500 iterations

cally consistent. Yet with 500 iterations the obtained images were soft modifications of the images obtained with 300 iterations.

Figure 4.18 gives a graph of the decrease of the square error relative in function to the number of calculations of the fitness functions for each image

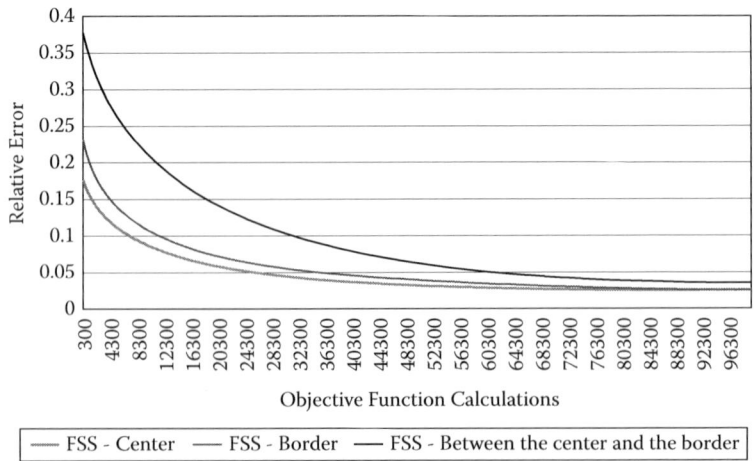

FIGURE 4.18: Error decreasing according to the number of calculations of the fitness functions, considering an object placed in the center, in the border, and between the center and the border of the domain

standard. The quantitative analysis of the results and the algorithm convergence can be made with this graph where it is possible observe that the decay of the relative error has a behavior similar to an exponential decreasing. Therefore, at the initial search process, the method easily finds a good region in the search space, but at the end the method convergence is very slow, and that is why the images obtained in 300 and 500 iterations are too similar. In addition, one can say that the three images accommodating the relative error of the reconstructed images are inferior by 5%.

Bibliography

[1] Andy Adler, John H. Arnold, Richard Bayford, Andrea Borsic, Brian Brown, Paul Dixon, Theo J.C. Faes, Inéz Frerichs, Hervé Gagnon, Yvo Gärber, et al. GREIT: a unified approach to 2D linear eit reconstruction of lung images. *Physiological Measurement*, 30(6):S35, 2009.

[2] Andy Adler, Tao Dai, and William R.B. Lionheart. Temporal image reconstruction in electrical impedance tomography. *Physiological Measurement*, 28(7):S1, 2007.

[3] Andy Adler and William R.B. Lionheart. Uses and abuses of EIDORS: an extensible software base for eit. *Physiological Measurement*, 27(5):S25, 2006.

[4] Juan Carlos Zavaleta Aguilar. *Estudos numéricos para o problema da tomografia por impedância elétrica*. Ph.D. thesis, Universidade de São Paulo, 2009.

[5] Thomas Bäck, David B. Fogel, and Zbigniew Michalewicz. *Handbook of evolutionary computation. New York: Oxford*, 1997.

[6] Lee E. Baker. Principles of the impedance technique. *IEEE Engineering In Medicine and Biology Magazine*, 8(1):11–15, 1989.

[7] Valter A.F. Barbosa, Reiga R. Ribeiro, Allan R.S. Feitosa, Victor L. B.A. da Silva, Arthur D.D. Rocha, Rafaela C. Freitas, Ricardo E. de Souza, and Wellington P. dos Santos. Reconstrução de imagens de tomografia por impedância elétrica usando cardume de peixes, busca não-cega e algoritmo genético. In C.J.A. Bastos Filho, A.R. Pozo, and H.S. Lopes, editors, *Anais do 12 Congresso Brasileiro de Inteligência Computacional*, pages 1–6, Curitiba, PR, 2015. ABRICOM.

[8] Carmelo J.A. Bastos Filho, Fernando B. de Lima Neto, Anthony J.C.C. Lins, Antonio IS Nascimento, and Marilia P. Lima. A novel search algorithm based on fish school behavior. In *IEEE International Conference*

on *Systems, Man and Cybernetics, 2008. SMC 2008*, pages 2646–2651. IEEE, 2008.

[9] Risha Bhatia, Georg M. Schmölzer, Peter G. Davis, and David G. Tingay. Electrical impedance tomography can rapidly detect small pneumothoraces in surfactant-depleted piglets. *Intensive Care Medicine*, 38(2):308–315, 2012.

[10] Liliana Borcea. Electrical impedance tomography. *Inverse Problems*, 18(6):R99, 2002.

[11] B.H. Brown, D.C. Barber, and A.D. Seagar. Applied potential tomography: possible clinical applications. *Clinical Physics and Physiological Measurement*, 6(2):109, 1985.

[12] Timothy C. Brown, Barbara Gail Ivanoff, and N.C. Weber. Poisson convergence in two dimensions with application to row and column exchangeable arrays. *Stochastic Processes and Their Applications*, 23(2):307–318, 1986.

[13] Grazieli L.C. Carosio, Vanessa Rolnik, and Paulo Seleghim, Jr. Hybrid parallel genetic algorithm in electrical impedance tomography. In *Proceedings of the 19th International Congress of Mechanical Engineering, Brasilia (Brazil)*, 2007.

[14] Margaret Cheney, David Isaacson, and Jonathan C. Newell. Electrical impedance tomography. *SIAM Review*, 41(1):85–101, 1999.

[15] Margaret Cheney, David Isaacson, Jonathan C. Newell, S. Simske, and J. Goble. NOSER: An algorithm for solving the inverse conductivity problem. *International Journal of Imaging Systems and Technology*, 2(2):66–75, 1990.

[16] Marcos Evandro Cintra and Heloisa de Arruda Camargo. Fuzzy rules generation using genetic algorithms with self-adaptive selection. In *Information Reuse and Integration, 2007. IRI 2007. IEEE International Conference on*, pages 261–266. IEEE, 2007.

[17] Meng Dai, Bing Li, Shijie Hu, Canhua Xu, Bin Yang, Jianbo Li, Feng Fu, Zhou Fei, and Xiuzhen Dong. In vivo imaging of twist drill drainage for subdural hematoma: a clinical feasibility study on electrical impedance tomography for measuring intracranial bleeding in humans. *PloS one*, 8(1):e55020, 2013.

[18] Felipe Dalvi-Garcia, Marcio Nogueira de Souza, and Alexandre Visintainer Pino. Algoritmo de reconstrução de imagens para um sistema de tomografia por impedância elétrica (TIE) baseado em configuração multiterminais. *Rev. Bras. Eng. Bioméd., Rio de Janeiro*, 29(2):133–143, 2013.

[19] Ronald Aylmer Fisher. *The genetical theory of natural selection: a complete variorum edition.* Oxford University Press, 1930.

[20] E.M. Gibbs. Metropolis, missing data. *TECHNOMETRICS*, 45(3), 2003.

[21] John H. Holland. Genetic algorithms. *Scientific American*, 267(1):66–72, 1992.

[22] Yi-Zeng Hsieh, Mu-Chun Su, and Pa-Chun Wang. A PSO-based rule extractor for medical diagnosis. *Journal of Biomedical Informatics*, 49:53–60, 2014.

[23] A. Rezaee Jordehi. Enhanced leader PSO (ELPSO): A new PSO variant for solving global optimisation problems. *Applied Soft Computing*, 26:401–417, 2015.

[24] J. Kennedy and R. Eberhart. Particle swarm optimization. In *IEEE International Conference on Neural Networks, 1995. Proceedings*, volume 4, pages 1942–1948, 1995.

[25] Vandana Khanna, B.K. Das, Dinesh Bisht, P.K. Singh, et al. A three diode model for industrial solar cells and estimation of solar cell parameters using PSO algorithm. *Renewable Energy*, 78:105–113, 2015.

[26] Scott Kirkpatrick, C. Daniel Gelatt, Mario P. Vecchi, et al. Optimization by simulated annealing. *Science*, 220(4598):671–680, 1983.

[27] Ying Li, Guizhi Xu, Lei Guo, Lei Wang, Shuo Yang, Liyun Rao, Renjie He, and Weili Yan. Resistivity parameters estimation based on 2D real head model using improved differential evolution algorithm. In *Proceedings of the 28th IEEE Annual International Conference EBMS, page*, 2006.

[28] A.J.C.C. Lins, Carmelo J.A. Bastos-Filho, Débora N.O. Nascimento, Marcos A.C. Oliveira Junior, and Fernando B. de Lima-Neto. Analysis of the performance of the fish school search algorithm running in graphic processing units. *Theory and New Applications of Swarm Intelligence*, pages 17–32, 2012.

[29] Yang Liu and Fan Sun. A fast differential evolution algorithm using k-nearest neighbour predictor. *Expert Systems with Applications*, 38(4):4254–4258, 2011.

[30] Heitor S. Lopes and Guilherme L. Moritz. A graph-based genetic algorithm for the multiple sequence alignment problem. In *International Conference on Artificial Intelligence and Soft Computing*, pages 420–429. Springer, 2006.

[31] Salomão Sampaio Madeiro, Fernando Buarque de Lima-Neto, Carmelo José Albanez Bastos-Filho, and Elliackin Messias do Nascimento Figueiredo. Density as the segregation mechanism in fish school search for multimodal optimization problems. In *Advances in Swarm Intelligence*, pages 563–572. Springer, 2011.

[32] Stefan Maisch, Stephan H. Bohm, Josep Solà, Matthias S. Goepfert, Jens C. Kubitz, Hans Peter Richter, Jan Ridder, Alwin E. Goetz, and Daniel A. Reuter. Heart-lung interactions measured by electrical impedance tomography. *Critical Care Medicine*, 39(9):2173–2176, 2011.

[33] Olavo H. Menin and Vanessa Rolnik Artioli. Tomografia de impedância elétrica: uma nova técnica de imageamento em medicina. *Revista Eletrônica-Iluminart*, 1(5), 2010.

[34] Olavo Henrique Menin. Método dos elementos de contorno para tomografia de impedância elétrica. Master's thesis, Faculdade de Filosofia, Ciências e Letras de Ribeirão Preto da Universidade de São Paulo, 2009.

[35] Zbigniew Michalewicz and Cezary Z. Janikow. GENOCOP: a genetic algorithm for numerical optimization problems with linear constraints. *Communications of the ACM*, 39(12es):175, 1996.

[36] Giovana Trindade da Silva Oliveira. *Estudo e aplicações da evolução diferencial*. Ph.D. thesis, 2006.

[37] D.D. Pak, N.I. Rozhkova, M.N. Kireeva, M.V. Ermoshchenkova, A.A. Nazarov, D.K. Fomin, and N.A. Rubtsova. Diagnosis of breast cancer using electrical impedance tomography. *Biomedical Engineering*, 46(4):154–157, 2012.

[38] Kenneth Price, Rainer M. Storn, and Jouni A. Lampinen. *Differential evolution: a practical approach to global optimization*. Springer Science & Business Media, 2006.

[39] Oliver C. Radke, Thomas Schneider, Axel R. Heller, and Thea Koch. Spontaneous breathing during general anesthesia prevents the ventral redistribution of ventilation as detected by electrical impedance tomography a randomized trial. *The Journal of the American Society of Anesthesiologists*, 116(6):1227–1234, 2012.

[40] Reiga R. Ribeiro, Allan R.S. Feitosa, Ricardo E. de Souza, and Wellington P. dos Santos. A modified differential evolution algorithm for the reconstruction of electrical impedance tomography images. In *5th ISSNIP-IEEE Biosignals and Biorobotics Conference (2014): Biosignals and Robotics for Better and Safer Living (BRC)*, pages 1–6. IEEE, 2014.

[41] Reiga R. Ribeiro, Allan R.S. Feitosa, Ricardo E. de Souza, and Wellington P. dos Santos. Reconstruction of electrical impedance tomography

images using chaotic self-adaptive ring-topology differential evolution and genetic algorithms. In *2014 IEEE International Conference on Systems, Man, and Cybernetics (SMC)*, pages 2605–2610. IEEE, 2014.

[42] Reiga Ramalho Ribeiro. Reconstrução de imagens de tomografia por impedância elétrica usando evolução diferencial. Master's thesis, Universidade Federal de Pernambuco, 2016.

[43] J. Riera, P.J. Riu, P. Casan, and J.R. Masclans. Tomografía de impedancia eléctrica en la lesión pulmonar aguda. *Medicina Intensiva*, 35(8):509–517, 2011.

[44] Vanessa P, Rolnik and Paulo Seleghim, Jr. A specialized genetic algorithm for the electrical impedance tomography of two-phase flows. *Journal of the Brazilian Society of Mechanical Sciences and Engineering*, 28(4):378–389, 2006.

[45] Fadil Santosa and Michael Vogelius. A backprojection algorithm for electrical impedance imaging. *SIAM Journal on Applied Mathematics*, 50(1):216–243, 1990.

[46] João Guilherme Sauer. Abordagem de evolução diferencial híbrida com busca local aplicada ao problema do caixeiro viajante. Master's thesis, Pontifícia Universidade Católica do Paraná, 2007.

[47] D.Y. Sha and Hsing-Hung Lin. A multi-objective PSO for job-shop scheduling problems. *Expert Systems with Applications*, 37(2):1065–1070, 2010.

[48] Masoud Sharafi and Tarek Y. ELMekkawy. Multi-objective optimal design of hybrid renewable energy systems using PSO-simulation based approach. *Renewable Energy*, 68:67–79, 2014.

[49] Yuhui Shi and Russell Eberhart. A modified particle swarm optimizer. In *Evolutionary Computation Proceedings, 1998. IEEE World Congress on Computational Intelligence, The 1998 IEEE International Conference on*, pages 69–73. IEEE, 1998.

[50] Rainer Storn and Kenneth Price. Differential evolution–a simple and efficient heuristic for global optimization over continuous spaces. *Journal of Global Optimization*, 11(4):341–359, 1997.

[51] Joubin Nasehi Tehrani, Craig Jin, Alistair McEwan, and André van Schaik. A comparison between compressed sensing algorithms in electrical impedance tomography. In *Engineering in Medicine and Biology Society (EMBC), 2010 Annual International Conference of the IEEE*, pages 3109–3112. IEEE, 2010.

[52] Josef Tvrdik. Adaptive differential evolution and exponential crossover. In *International Multiconference on Computer Science and Information Technology, 2008. IMCSIT 2008*, pages 927–931. IEEE, 2008.

[53] Miguel Fernando Montoya Vallejo. Algoritmo de tomografia por impedância elétrica utilizando programação linear como método de busca da imagem. Master's thesis, Escola Politécnica da Universidade de São Paulo, 2007.

[54] S.R. Venkatesan, D. Logendran, and D. Chandramohan. Optimization of capacitated vehicle routing problem using PSO. *International Journal of Engineering Science and Technology (IJEST)*, 3(1):7469–7477, 2011.

[55] Yuqing Wan, Andrea Borsic, John Heaney, John Seigne, Alan Schned, Michael Baker, Shaun Wason, Alex Hartov, and Ryan Halter. Transrectal electrical impedance tomography of the prostate: Spatially coregistered pathological findings for prostate cancer detection. *Medical Physics*, 40(6), 2013.

[56] Qi Wang, Huaxiang Wang, Ronghua Zhang, Jinhai Wang, Yu Zheng, Ziqiang Cui, and Chengyi Yang. Image reconstruction based on l1 regularization and projection methods for electrical impedance tomography. *Review of Scientific Instruments*, 83(10):104707, 2012.

[57] Alvaro Yuh Yojo. Construção do hardware de um tomógrafo por impedância elétrica, 2008.

Part II
Bio-Inspired Optimization Techniques

Chapter 5

An Optimized False Positive Free Video Watermarking System in Dual Transform Domain

L. Agilandeeswari
VIT University, Vellore, India

K. Ganesan
VIT University, Vellore, India

5.1	Introduction	132
5.2	Related Works	134
5.3	Basic Concepts	137
	5.3.1 Firefly Algorithm	137
	5.3.2 Contourlet Transform	138
	5.3.3 Hilbert Transform	139
5.4	Proposed Watermarking Algorithm	140
	5.4.1 Firefly-Based Embedding Algorithm	140
	5.4.2 Extraction Algorithm	143
	5.4.3 Firefly Algorithm for Optimal Selection of Embedding Strength	145
5.5	Experimental Results	146
	5.5.1 Performance Evaluation Metrics	147
	5.5.2 Attacks	147
	5.5.3 No Attacks	149
	5.5.4 Image Processing Attacks or Spatial Attacks	149
	5.5.5 Video Processing or Temporal Attacks	152
5.6	Comparative Analysis	153
5.7	Conclusions and Future Work	156
	Bibliography	157

In this chapter, a novel optimization-based video watermarking technique is proposed, to increase robustness and imperceptibility. This is an extension work of [2], where the optimized image watermarking technique was introduced. The high level of imperceptibility is achieved by embedding the watermark in the dual transform domain of color video frames, namely, contourlet transform (CT) and Hilbert transforms. The robustness level of the proposed system is improved by determining the optimal scaling values among the multiple scaling factors, using the modified firefly optimization algorithm. In addition, the authors also concentrate on the false positive problem where it is very common in the existing transform domain techniques, mainly due to embedding in the singular values of the SVD. Since the embedding of a watermark onto the cover content happens in the Hilbert domain, the above-mentioned problem is completely removed. The experimental results prove that the proposed method is more robust against different kinds of spatial attacks, such as noise addition, filtering attack, contrast adjustment, histogram equalization, row-column deletion, geometrical attack, JPEG, and video attacks such as frame dropping, frame averaging, frame swapping or exchange attacks, and MPEG compression, when compared to the similar existing algorithms.

5.1 Introduction

The advancement in technologies of the Internet and digital multimedia processing allows the easy transmission and accessibility of digital data by everyone in the world. In parallel to the growing diversity in multimedia applications, technology has also facilitated unauthorized copying, tampering, and distribution of digital video. The ease of such manipulations emphasizes the need for data authentication schemes. Hence, for verifying authenticity of the multimedia content, various authentication techniques have been proposed. The authentication techniques can be classified into two broad categories, namely, digital signature-based and digital watermark-based schemes. A digital signature-based technique mainly deals with either an encrypted or a signed hash value of image contents or image characteristics [24]. The main drawback of this digital signature scheme is that it can detect only the modification of data, not the location of modified regions of an image [23]. This problem can be solved by many researchers through the second method, digital watermarking techniques [24]. Digital watermarking is a technique that mainly involves two steps: (i) watermark embedding, which is an algorithm to embed small pieces of authentication information called watermark content on the host content; (ii) watermark extraction, which is an algorithm to retrieve or extract the embedded watermark with less or acceptable distortion. Based on the domain of embedding, watermarking techniques can be broadly categorized into two groups: spatial domain methods and transform or frequency domain methods. The spatial domain methods embed by modifying directly on the pixels of an image [22, 29]. The spatial domain techniques require only low operation costs and the implementation is also easy, but it is not robust against transformations and image processing attacks. On the other hand, the transform domain or frequency domain method involves modifying the transform domain coefficients for embedding the watermark [10, 30].

The discrete fourier transform (DFT) [36], discrete cosine transforms (DCT) [6], singular value decomposition (SVD) [37], and discrete wavelet transform (DWT) [39] are the single or the fusion techniques. DFT-Radon [25], DCT-SVD [41], DWT-SVD [39], DWT-Hilbert [4], and CT-DWT-SVD [2] are the transform techniques used for embedding the single watermark or embedding multiple watermarks [3] on the cover content. Each technique has its own advantages and disadvantages, and thus the researchers are motivated towards introducing new transform techniques. For example, DFT-based frequency domain algorithms produce phase and magnitude representations when they are applied to images. The main advantage of DFT is the rotation, scaling and translation (RST) invariant property, which makes it suitable against geometric distortions [13]. When we use the DCT, the watermark bits are embedded into the selected DCT coefficients [7, 22, 16]. The DCT-based transform domain watermarking algorithms are more robust against various image processing attacks, such as noise addition, brightness modification, contrast adjustment, smoothing, lossy compression [14], etc. But they fail to provide robustness against geometrical attacks such as cropping, rotation, etc. The discrete wavelet transform (DWT) decomposes the image into three spatial directions. Due to this, DWT mimics the HVS better than DCT. DWT has the property of spatial frequency locality, which means it affects the image locally based on the embedding location. This characteristic of wavelets helps us to achieve both frequency and spatial description for an image [39] and also robustness against additions of noise [8]. Another new transform technique used for watermarking is SVD. It is possible to take the SVD of the cover image and the watermark image, and replace the singular values of the watermark image with the singular values of the cover image. This can result in embedding [34]. As discussed in [3], the big problem here is the false positive detection, and it can be overcome for the video watermarking technique, as discussed in this research work.

The embedding strength normally reveals the effect of watermarking on the cover content. The use of a static embedding strength parameter or scaling factor is very common in traditional image watermarking techniques, which may lead to some visual artifacts in the watermarked image [11]. These distortions mostly affect the smooth regions of an image. By reducing the embedding strength, the above-mentioned distortions can be reduced, which in turn can affect the robustness. However, the robustness can be improved by increasing the embedding strength, which consecutively affects the fidelity or visual quality by causing more distortions. Hence, the trade-off between the fidelity (visual quality) and robustness can be achieved by selecting the suitable scaling factor [29]. Cox et al. [11] also argued that a single scaling factor may not be applicable for perturbing all the coefficients of the cover image, as different spectral components may exhibit more or less tolerance to modification. As an alternative, the user can use multiple embedding strength values. Another important issue in this approach is to choose the best optimal value among the multiple embedding strength values.

Thus, the evolutionary algorithms can be used to determine the optimal scaling factor. Some of the existing optimization algorithms are particle swarm optimization (PSO) [40, 4], bacterial foraging [2], genetic algorithm [21, 26, 33, 20], ant-colony optimization [26], and firefly algorithm [32, 27]. These evolutionary algorithms are either used to find the suitable location for embedding [40, 4, 2] or to determine the optimal embedding strengths [21, 27]. These algorithms also suffer from the problem of false positive detection [26, 5].

In this chapter, the authors would like to propose an optimal embedding strength-based robust and imperceptible video watermarking algorithm, which is the extension work of [2], where it concentrates on avoiding false positive detection problems in the image watermarking technique. In addition, the authors use the hybrid transform domain of contourlet transform [18], and the Hilbert transform [1] for embedding the gray scale watermark image within the video frame covers. The rest of the chapter is organized as follows. In Section 2 the related works are discussed. The basic concepts are discussed in Section 3. A detailed explanation of the proposed watermarking scheme is presented in Section 4. The experimental results are shown in Section 5. Finally, the conclusions and future works are drawn in Section 6.

5.2 Related Works

According to the need for visual quality or robustness of the cover content, the domain for embedding the watermark might be chosen as either a spatial or a transform domain. Even though the selection of a domain is feasible, optimal selection of suitable location for embedding or the embedding strength parameter are very big issues. This section discusses some of the existing watermarking techniques in the spatial and transform domain with a constant scaling factor, and the optimization-based watermarking algorithm for optimal selection of suitable location and scaling factor.

The spatial domain watermarking on compressed videos proposed by Mobasseri [28] showed the possibility of embedding a watermark in the raw video and also the possibility of recovering it from the MPEG decoder by exploiting the inherent processing gain of DSSS (direct sequence spread spectrum). Tsai and Chang [38] proposed a compressed video sequence via VLC decoding and VLC code substitution. They used Watson's DCT-based video watermarking to achieve better imperceptibility. Novel adaptive approaches to video watermarking have been proposed [25]. In order to guarantee the robustness and perceptual invisibility of the watermark, they used both intra-frame and inter-frame information of video content. The main advantage of this method is that the extraction of the watermark can be done without using the original video, since the embedding was done adaptively based on the signal characteristics and human visual system.

Doerr and Dugelay [13] have proposed video watermarking based on spread spectrum techniques, in order to improve robustness. Here each watermark bit is spread over a large number of chip rate (CR) and then modulated by a pseudorandom sequence of binary data. This algorithm increases the robustness with the increase of the variance of the pseudo-noise sequence. As a result, the increase of CR will reduce the embedding rate of watermark information, whereas the increase of variance may result in the perceptibility of the watermark. A new type of watermarking scheme was proposed by Niu et al. [31], using two-dimensional and three-dimensional multi resolution signal decomposing. The watermark image, which is decomposed with different resolution, is embedded in the corresponding resolution of the decomposed video. The robustness of watermarking is enhanced by coding the watermark information using the Hamming error correction code. This approach is robust against attacks such as frame dropping, averaging, and lossy compression.

Haneih et al. [19] have proposed a multiplicative video watermarking scheme with semi-blind maximum likelihood decoding for copyright protection. They first divided the video signal into non-overlapping pixel cubes. Then, the 2D wavelet transform is applied on each plane of the selected cubes. For extraction, a semi-blind likelihood decoder is employed. This method was robust against linear collusion, frame swapping, dropping, noise insertion, and median filtering. Gaurav et al. [9] have proposed a novel digital wavelet packet transform-based robust video watermarking technique. The authors used a meaningful binary image as a watermark. Once the WPT is applied on the frames of a video, the robust sub-band is selected based on the block-mean intensity value. The experimental results proved that the proposed algorithm achieves better robustness when compared to the existing algorithms.

In [15], the author proposes a technique for efficient video watermarking using the DWT and SVD. Here, the author embedded the watermark image in three sub-bands namely, HL, LH, and HH bands, to improve the visual perception and robustness. A highly secured watermarking scheme has been introduced by Rupachandra et al. in [35], where the authors used the concept of visual cryptography and scene change detection to identify the motion frames, which helps to avoid collusion attack. Hasnaoui and Mitrea [18] introduce the theoretical framework allowing for the binary quantization index modulation (QIM) embedding techniques to be extended towards multiple symbol. The underlying technique is optimized. Nisreen et al. [43] proposed a blind scheme for digital video watermarking, where a secret key is used in the retrieval of the watermark image. Each video frame is decomposed into sub-images using 2-level discrete wavelet transform, and then the principal component analysis (PCA) transformation is applied on each block in the two bands, LL and HH. Then, the quantization index modulation (QIM) is used for quantizing the maximum coefficient of the PCA blocks, followed by embedding the watermarks on such selected quantizer values. The authors proved that the proposed system achieves a maximum imperceptibility in terms of PSNR (exceeding 45dB) and is also robust against various attacks in both regular and medical videos.

A detailed study of security analysis of the recent image watermarking schemes based on redundant discrete wavelet transform and singular value decomposition was done by Jing-Ming Guo et al. [17]. They have presented three vulnerable attacks to show the weakness of the redundant discrete wavelet transform and singular value decomposition techniques. From the experimental results, they concluded that these schemes cannot provide trustworthy evidence in rightful ownership protection. Wu et al. [40] have presented an intelligent watermarking method using particle swarm optimization. This technique is used to identify the suitable transform coefficients for embedding the watermark in the transform domain. It improves a high level of visual quality but fails to optimize the embedding strength parameter. Huang et al. [33] have proposed an optimized watermarking technique using swarm based bacterial foraging, which is mainly used to determine the optimal embedding location for embedding the watermark. Another existing optimization system for determining the suitable transform coefficients is proposed by Shieh et al. [33].

A new approach for optimization in image watermarking using genetic algorithms is proposed by Kumsawat et al. in [20]. Lai [21] proposed a tiny genetic algorithm-based watermarking technique to determine the embedding strength optimally. This approach suffered from the problem of false positive detection, which is commented on by Loukhaoukha [26]. An ant colony optimization-based watermarking technique in the LWT-SVD domain by Loukhaoukha [26] was used to find the optimal values of multiple scaling factors. In [27], Mishra et al. have proposed a gray scale image watermarking using DWTSVD and the firefly algorithm system for determining the optimal scaling factor value among the multiple scaling factors. This technique achieves good visual quality and robustness, but suffers from the problem of false positive detection, which is commented on by Musrat et al. in [5]. This literature review makes it clear that the hybrid approach improves the performance of the watermarking scheme and one of the metaheuristic evolutionary techniques, namely the firefly algorithm (FA), can be used for various applications; its performance is also better than that of other metaheuristic techniques such as GA and PSO. Thus, this FA will greatly help us in selecting the suitable embedding strength for embedding the watermark on the cover image. But it suffers from the problem of false positive detection. Hence, there is a requirement to find a watermarking system that uses an optimization algorithm for selecting the best scaling factor without the false positive detection problem. From the review, the authors found that a lot of work has been done for optimizing the image watermarking systems, but not much research work is done in optimizing video watermarking systems.

The present work proposes an efficient contourlet and Hilbert transform technique for video watermarking, using the nature-inspired firefly evolutionary algorithm. The proposed video watermarking scheme provides improved fidelity and robustness constraints in the watermarking system. To achieve this objective, the firefly optimization algorithm is a more suitable option, one that uses a linear combination of PSNR and NCC, where a PSNR metric determines the fidelity level of the watermarked image, and the NCC metric helps us to verify the robustness level.

5.3 Basic Concepts

This section is devoted to explaining the basic concepts used in the proposed watermarking algorithm.

5.3.1 Firefly Algorithm

The firefly algorithm (FA) is a nature-inspired metaheuristic optimization algorithm inspired by the fireflies' flashing behavior. The crucial purpose for a firefly's flash is to act as a signal system to attract other fireflies. Xin-She Yang [42] formulated this firefly algorithm using the following rules. The two important points in the firefly algorithm are: the variation in the light intensity and formulation of the attractiveness. In general, the attractiveness of a firefly is determined by its brightness, which in turn is proportional to the encoded objective function.

1. All fireflies are unisexual, so that any individual firefly will be attracted to all other fireflies.

2. The brightness of a firefly is determined by the encoded objective function.

3. Attractiveness is proportional to their brightness, and for any two flashing fireflies, the less bright one will be attracted by the brighter one; however, the intensity decreases as their mutual distance increases. If there are no fireflies brighter than a given firefly, it will move randomly.

Now the general firefly algorithm can be stated [42] as below.

Algorithm [Firefly Algorithm]

Input: State the objective function $f(x), where x = (x1, ..., xd)T$; Generate the initial population of fireflies x_i, where $i = 1, 2, ..., n$; Determine the light intensity of G_i at x_i via $f(x_i)$; Define absorption coefficient γ, α, β_o and Maximum generation Gmax

Output: Obtain the maximum objective function value $x_i{}^1$. This determines the suitable scaling factor α for various attacks in this case.

1. While $(t < G_{max})$
2. For $i = 1$ to n (all n fireflies)
3. For $j = 1$ to n (all n fireflies)
4. If $(G_j > G_i)$
5. Move firefly i towards j by using equation (7)
6. Attractiveness varies with distance d via $exp(-\gamma d^2)$
7. Evaluate new solutions and update light intensity
8. End for loop
9. End for loop
10. Rank the fireflies and find the current best xi^{max}
11. End while loop
12. Post process results and visualization

5.3.2 Contourlet Transform

An efficient multi-scale directional transform developed by Minh N. Do and Martin Vetterli [12] is the contourlet transform. It uses a double filter bank structure, which can be constructed by combining two distinct and successive decomposition stages: a Laplacian pyramid (LP) and a directional filter bank (DFB). The Laplacian pyramid is used to perform multi-scale decomposition, i.e., it decomposes an image into a number of detailed (high frequency) sub-bands and an approximation of (low frequency) sub-bands. Then a DFB is used to perform directional decomposi-tion on the detail sub-bands. The discrete contourlet transform can capture the directional edges in a better way when compared to wavelets [1] There are various options for pyramid and directional filters. Here, the authors have used a 9-7 pyramid filter and pkva

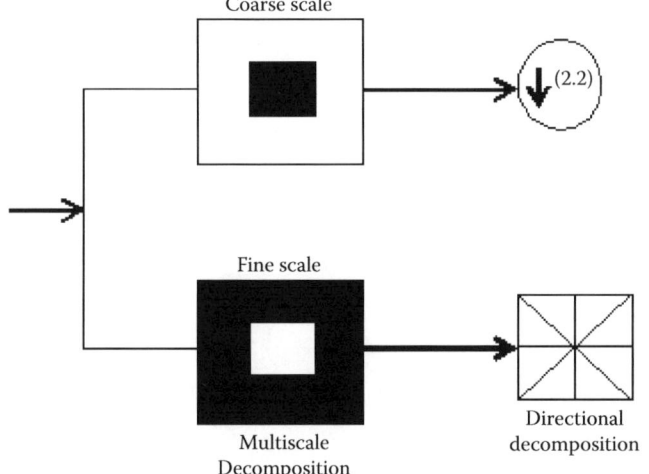

FIGURE 5.1: Schematic diagram of contourlet transform

directional filter. The schematic diagram of contourlet transform is given in Figure 5.1, and its application to one of the image frames of a video sample is shown in Figure 5.2.

5.3.3 Hilbert Transform

Any analytic signal $X'(t)$ that is associated to a signal X(t) is defined as below, where, $X_H(t)$ is the Hilbert transform of $X(t)$. The Hilbert transform

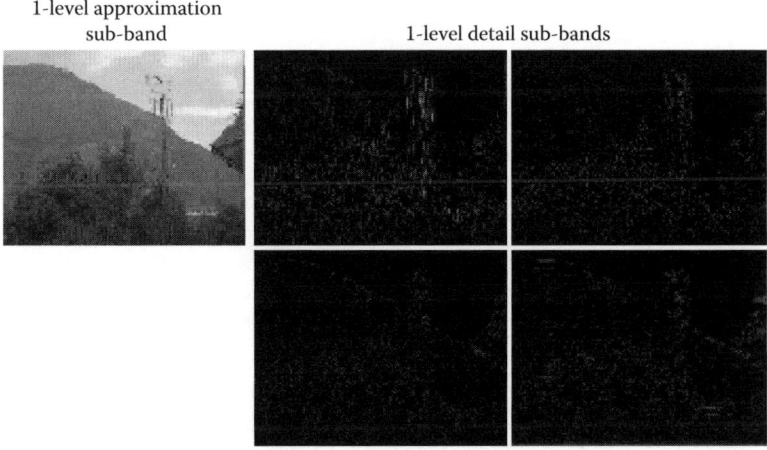

FIGURE 5.2: CT on mountain.avi

of a signal $X_H(t)$ is defined as the transform in which the phase angle of all components of the signal is shifted by $\pm 90°$. The $X'(t)$ can be written in the form of amplitude and phase as $A(t)e^{i\theta(t)}$.

$$X'(t) = X(t) + iX_H(t) \tag{5.1}$$

5.4 Proposed Watermarking Algorithm

In this section, the proposed false positive free dual contourlet and Hilbert transform-based video watermarking using the firefly nature-inspired algorithm is discussed in detail. The tradeoff between imperceptibility and robustness can be balanced by determining the suitable location for embedding the watermark, and by choosing the optimal scaling factor among the multiple scaling values. The former is achieved using a PCA-based sub-band selection procedure, and the latter is achieved by using the firefly optimization algorithm. The proposed technique involves three phases: (i) firefly-based embedding algorithm, (ii) firefly-based extraction algorithm, and (iii) firefly-based optimal embedding strength selection. They are explained below.

5.4.1 Firefly-Based Embedding Algorithm

The various steps involved in the firefly-based embedding algorithm are discussed here. A diagramatic representation of the proposed nature-inspired embedding algorithm is depicted in Figure 5.3.

1. Consider the color cover video CV of size $M \times M$ and the gray watermark image W of size $M/4 \times M/4$.

2. Perform video-to-frame conversion as CV to CV^f, where, $f = 1, 2, 3, \cdots$, the number of frames in a video.

3. Apply RGB to YC_bC_r conversion on the cover video frame CV^f and name it C^f_{YCbCr}.

4. Perform 2-level CT on the Y-component of C^f_{YCbCr}, namely C^f_Y, as

$$\begin{aligned}(LL1, (D11, D12, D13, D14)) &= CT(C^f_Y) \\ (LL2, (D21, D22, D23, D24)) &= CT(LL1)\end{aligned} \tag{5.2}$$

5. Let's represent the CT frame D^f with its elements as D^f_{ij} and the watermark image W with its coefficients as W_{ij}.

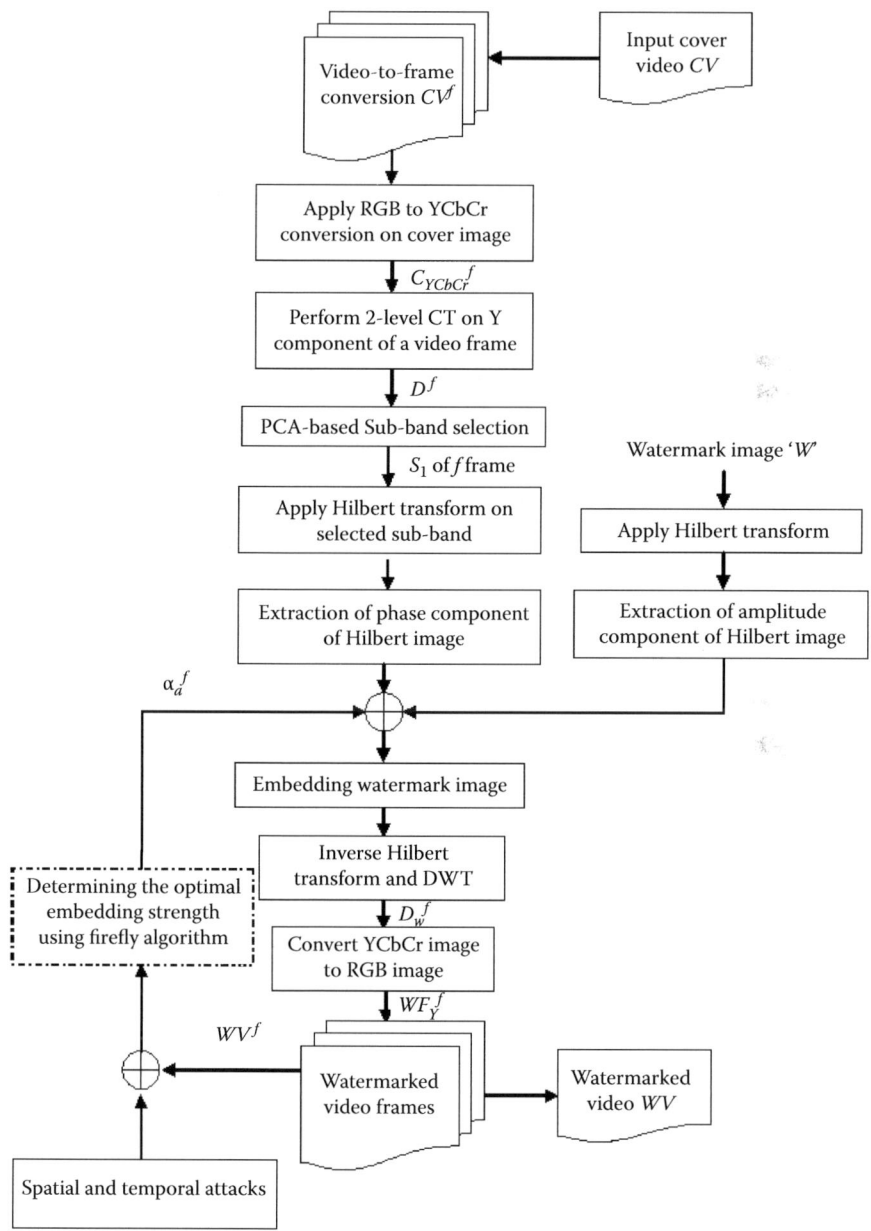

FIGURE 5.3: Proposed nature-inspired embedding algorithm

6. Apply the PCA-based sub-band selection algorithm on D^f as follows:

    (a) Find zero mean A_i for each block
    $$A_i^f = E(B_i^f - m_i^f) \tag{5.3}$$

    (b) Calculate covariance matrix
    $$Cm_i^f = A_i^f (A_i^f)^T \tag{5.4}$$

    (c) Transform each block into PCA components by calculating the eigenvectors corresponding to the eigenvalues of the covariance matrix, as
    $$Cm_i^f \phi = \gamma_i^f \phi \tag{5.5}$$

    (d) Compute PCA transformation of each block to get a block of uncorrelated coefficients using the following equation, where Pc_i^f is the principal component of the ith block of the fth frame.
    $$Pc_i^f = \phi^T A_i^f \tag{5.6}$$

7. Choose the sub-band with the highest score and name it the sub-band S_1 of $'f'$ frame.

8. Perform Hilbert transform on sub-band S_1 of frame D^f and W as below, where p_{ij}^f and q_{ij} are amplitude components of D and W, repectively; θ_{ij}^f and φ_{ij} are phase components of D and W, respectively.
    $$D_{ij}^f = p_{ij}^f \cos \theta_{ij}^f \tag{5.7}$$
    $$W_{ij} = q_{ij} \cos \varphi_{ij} \tag{5.8}$$

9. Embed the watermark image in the sub-band that has the highest score, as follows:

    (a) Divide the selected sub-bands into non-overlapping blocks of size equal to the size of the watermark frame.

    (b) Compute the score for each block using Step 6.

    (c) Extract the phase component and the amplitude component of the Hilbert cover frame and watermark image, respectively.

    (d) Modify the coefficients of the selected Hilbert block of the sub-bands with the watermark image as follows, where, α_a^f represents the optimal robustness factor chosen using the firefly algorithm, q_{ij} represents the amplitude component of the watermark image, and θ_{ij}^f represents the phase component of the cover frame f.
    $$\psi_{ij}^f = \theta_{ij}^f + \alpha_a^f q_{ij} \tag{5.9}$$

10. Reconstruct the modified sub-band CT coefficient D^w using the inverse Hilbert transform as follows:

$$D_w^f = p_{ij}^f \cos \psi_{ij}^f \tag{5.10}$$

11. Obtain the luminance watermarked frame WI_Y^f using inverse DWT.

12. Convert the resultant YC_bC_r frame WF_Y^f to RGB watermarked frame WF^f.

13. Repeat Step 3 to Step 12 for various frames $f = 1, 2, 3, \cdots$ to the total number of frames of the cover video.

14. The resultant modified frames are combined to form the watermarked video WV.

5.4.2 Extraction Algorithm

The various steps involved in the extraction algorithm are discussed below. A diagramatic representation of the proposed nature-inspired embedding algorithm is depicted in Figure 5.4.

1. From the received attacked video WV_A, perform a video-to-frame conversion as WV_A^f. The users can extract the watermark W' from the computed frames if $p_{ij}, \theta_{ij}, \phi_{ij}$ is known for all the values of i and j and the robustness factor α_a^f. This is computed optimally using the firefly algorithm for optimal embedding strength selection.

2. Apply RGB to YC_bC_r conversion on WV_A and name it WV_A^Y.

3. Perform 2-level contourlet transform on the Y component of WV_A^Y.

4. Apply the PCA-based sub-band selection procedure.

5. Perform Hilbert transform on Dw' and extract its phase component ψ_{ij} from Equation (5.10) as

$$D_w^{f'} = p_{ij}^f \cos \psi_{ij}^f \tag{5.11}$$

Now divide both sides of Equation (5.10) by p_{ij}^f and also replace it with Equation (5.10), then

$$\begin{aligned} D_w^{f'}/p_{ij}^f &= \cos \psi_{ij}^f \\ \psi_{ij}^f &= \cos^{-1}(D_w^{f'}/p_{ij}^f) \\ \theta_{ij}^f + \alpha_a^f q_{ij} &= \cos^{-1}(D_w^{f'}/p_{ij}^f) \\ q_{ij}' &= [\cos^{-1}(D_w^{f'}/p_{ij}^f) - \theta_{ij}^f]\alpha_a^f \end{aligned} \tag{5.12}$$

6. Repeat Step 2 to Step 5 for every frame of the received watermarked video, by changing the values of $'f'$.

7. Thus, the extracted watermark W_e^f can be obtained using the equation below, where $f = 1, 2, 3, \cdots$, the number of frames in a watermarked video.

$$(W_{ij})_e^f = q'_{ij} \cos \varphi_{ij}^f \tag{5.13}$$

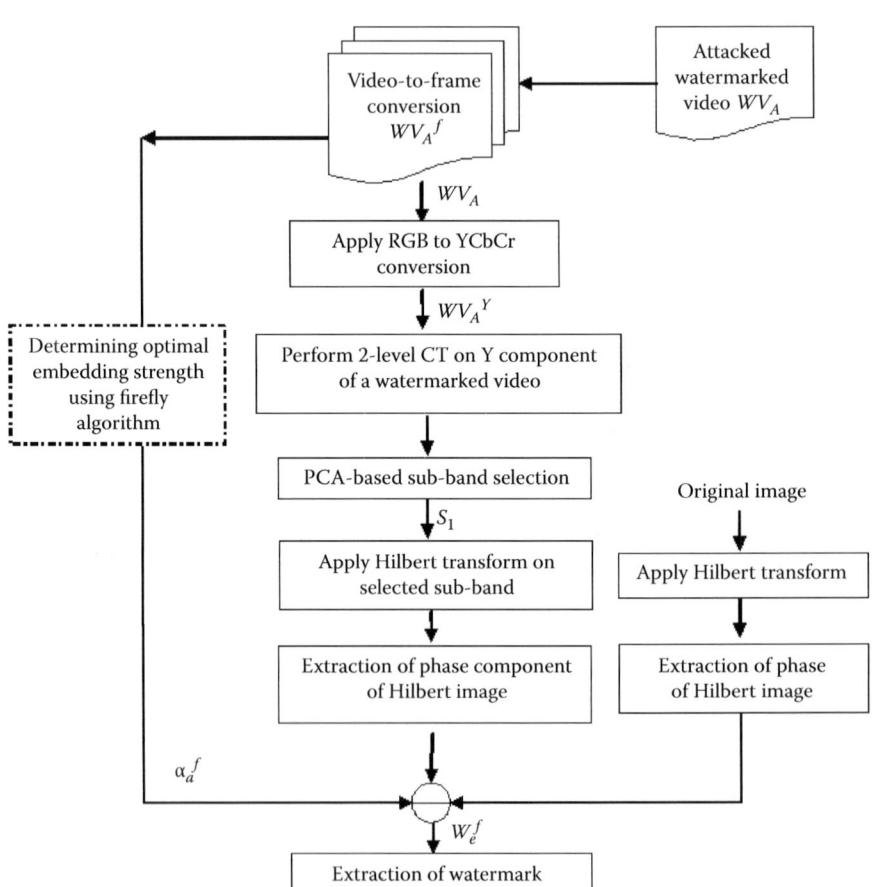

FIGURE 5.4: Proposed firefly-based extraction algorithm

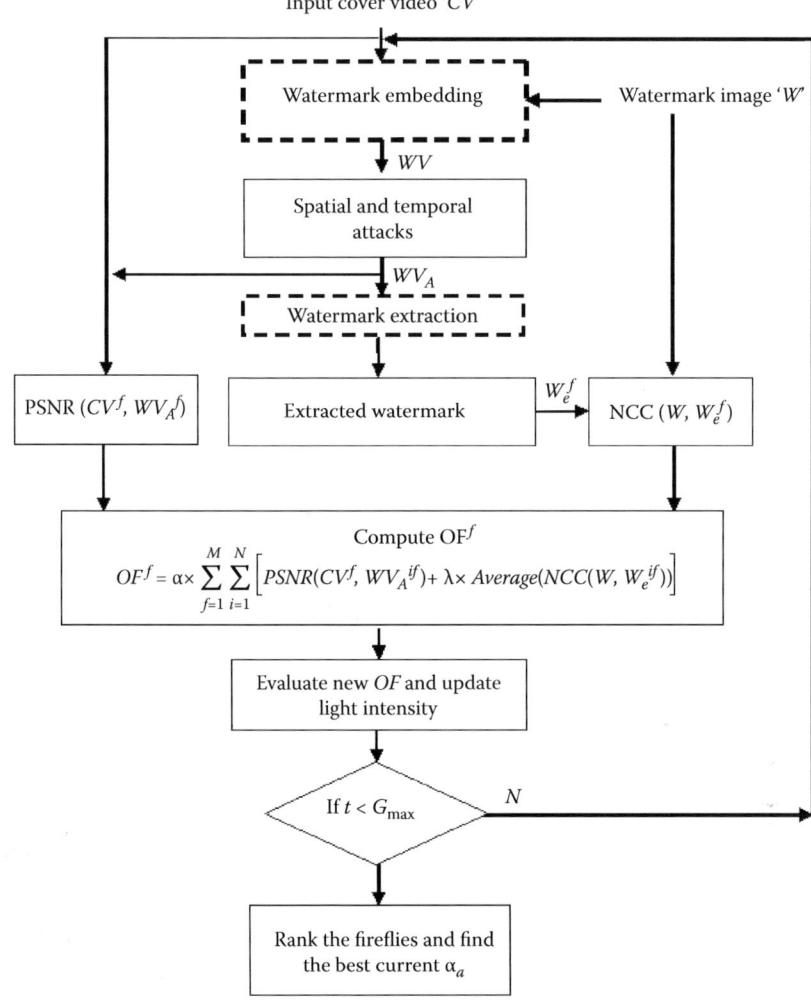

FIGURE 5.5: Firefly-based optimal embedding strength selection

5.4.3 Firefly Algorithm for Optimal Selection of Embedding Strength

The various steps involved in the firefly algorithm for optimal selection of embedding strength α_a^f is discussed below. A diagramatic representation of the proposed nature-inspired embedding algorithm is depicted in Figure 5.5.

1. Initialization:

    (a) State the objective function (OF) to maximize the imperceptibility and robustness as below, where CV is cover video, WV_A^f is attacked watermarked video, W is watermark image, W_e^{if} is extracted watermark image, M is number of frames, N is number of image and video processing attacks, λ is adjustment factor between PSNR and NCC, and f is frame number.

$$OF^f = \alpha \times M\Sigma_{f=1} N\Sigma_{i=1}[PSNR(CV^f, WV_A^{if}) + \lambda] \times [Average(NCC(W, W_e^{if}))] \quad (5.14)$$

    (b) Create initial population of n fireflies

    (c) Initialize parameters $t = 0$, $\beta = 1$, $\alpha = 0.01$, $\lambda = 20$, $\gamma = 1.0$, $G_{max} = 10$, n=10 (as per [27, 42])

2. Output: Choose the suitable af value for every frame while on embedding and extraction under various attacks, using the objective function (OF^f) as given in Equation (5.13).

3. Process:

    (a) For each firefly F_n, perform embedding using Algorithm 2 and assume the various image and video processing attacks on watermarked images: finally, perform extraction using Algorithm 3.

    (b) While $t < G_{max}$

    (c) Evaluate the objective function OF_n^f for various values of α, using Equation (5.13)

    (d) Repeat (a) to (c) for various fireflies.

    (e) Rank the objective function value for various n fireflies, $R(OF_n^f)$.

    (f) Find the best OF with its value used and name it α_a^f.

5.5 Experimental Results

The proposed video watermarking technique has been validated in terms of its imperceptibility and robustness against possible image processing attacks, such as noise addition, filtering, contrast adjustment, histogram equalization, JPEG compression, and row-column deletion; geometric attacks, namely, cropping and rotation; and video processing or temporal attacks such as MPEG compression, frame dropping, frame swapping, and frame averaging. The test videos chosen to evaluate the proposed optimized video watermarking algorithm are of two categories (i) 15 standard videos and (ii) 4 natural videos. The standard videos are akiyo, rhinos, bus, carphone, clarie, coastguard, container, football, foreman, garden, grandma, hallmonitor, missam, mobile and motherdaughter; the natural videos are mountain, manwalk, deer, baby all were of standard pixel dimension 512 × 512. The watermark image is VIT-logo.jpg, of dimension 128 × 64. The cover videos are shown in Figure 5.6, and the binary watermark image is shown in Figure 5.7. For embedding the watermark, the optimal scaling factor is determined using the nature-inspired firefly algorithm. For better comparison, the initialization parameters of firefly algorithm listed in the firefly algorithm for optimal selection of embedding strength are chosen as used in the existing systems [27, 42].

5.5.1 Performance Evaluation Metrics

The proposed watermarking algorithms are evaluated in terms of the watermarking characteristics, namely, imperceptibility and robustness. The imperceptibility level of the algorithm is measured in terms of peak signal-to-noise ratio (PSNR) and structural similarity index measure (SSIM) value, while the robustness of the algorithm is measured in terms of normalized correlation coefficient (NCC) and bit error rate (BER).

Imperceptibility Measure: The peak signal-to-noise ratio (PSNR) and structural similarity index measure (SSIM) are used as common measures to evaluate the degradation caused by various attacks. Low PSNR values indicate higher degradation, and high PSNR values indicate lower degradation. Hence, high PSNR values indicate that the watermarking technique is more robust to that type of attack. On the other hand, SSIM's acceptable value varies between 0 (no match) to 1 (exact match).

Robustness Measure: The robustness of the embedded watermark to various attacks is measured in terms of normalized correlation coefficient (NCC), and bit error rate (BER). Both metric's acceptable value is between 0 and 1 with a peak value of 1. If two images are identical, then the correlation coefficient value will be 1. If two images are uncorrelated, then the correlation, coefficient value will be 0.

5.5.2 Attacks

Most of the transmission is happening over an insecure communication channel, which results in the introduction of degradation in the cover video contents. This in turn affects the watermark, which is hidden. Therefore, in order to test the performance of the proposed video watermarking system and also to compute the optimal embedding strength parameter using the firefly algorithm, the following attacks were introduced, namely: (i) No attacks, (ii) 11 image processing or spatial attacks, and (iii) 4 video processing or temporal attacks. Spatial attacks used in this investigation were Gaussian noise, Poisson noise, salt and pepper noise, Gaussian low-pass filtering, median filtering, rotation (90), cropping 20%, contrast adjustment, JPEG compression (QF= 90), row-col deletion ($10R, 10C$), and histogram equalization. Temporal attacks used in this investigation were MPEG compression, frame dropping, frame averaging, and frame swapping.

(a) akiyo (b) rhinos (c) bus (d) carphone

(e) clarie (f) coastguard (g) container (h) football

(i) foreman (j) garden (k) grandma (l) hallmonitor

(m) missam (n) mobile (o) motherdaughter (p) mountain

(q) manwalk (r) baby (s) deer

FIGURE 5.6: Video frame covers

FIGURE 5.7: Original watermark image

After performing all these attacks on the watermarked video, the PSNR metric is computed between the original cover video and the attacked watermarked videos, respectively, for knowing the level of degradation. These two values are used to compute the optimal embedding strength using the firefly algorithm.

5.5.3 No Attacks

The effect of embedding on the cover video is measured effectively under a "no attack" scenario in terms of visual quality and robustness. The watermarked videos without any attack are shown in Figure 5.8. Figure 5.9 represents the watermark images extracted from the watermarked video, that is assumed to be uncorrupted.

Table 5.1 shows the computed PSNR and correlation coefficient on various sample videos after embedding and extraction of the watermark under the no-attack scenario, based on the nature-inspired optimal embedding strength selection algorithm.

5.5.4 Image Processing Attacks or Spatial Attacks

To know the level of degradation on the watermarked video after various image processing attacks, the visual quality metrics, namely PSNR and SSIM, are used. Similarly, the robustness measure is evaluated using the correlation coefficients (NCC) and the bit error rate (BER). Figure 9 shows the effect of image processing attacks on the watermarked videos, namely, akiyo, container, hallmonitor, baby, deer, rhinos. The parameters of the various image processing attacks are Gaussian noise with variance 0.5, Poisson noise, salt and pepper noise with density 0.05, median filtering of size 5×5, Gaussian low-pass filtering of size 5×5, rotation with an angle of about $90°$, cropping at the rate of 20%, contrast adjustment, JPEG compression with quality factor of 90, row-column deletion by deleting 10 rows and 10 columns randomly from the watermarked frames, and histogram equalization attacks. For simplicity, these spatial attacks are represented with an index as given in Table 5.2. Figure 5.10 shows the effect of various image processing attacks on the watermarked video and its corresponding watermark extraction.

From Figure 5.10, it is clear that the proposed optimized watermarking system is able to extract 99%, 99%, 98%, 99%, 98%, 98%, 99%, 99%, 99%, 99%, and 98% of the watermark, when the frames are attacked by attack index 1, 2, 3, 4, 5, 6, 7, 8, 9, 10, and 11, respectively, for the test video akiyo. Similarly, it can extract 99%, 99%, 98%, 99%, 98%, 95%, 98%, 99%, 99%, 97%, 98% (container), 99%, 99%, 98%, 99%, 99%, 96%, 97%, 99%, 98%, 99%, 97% (hallmonitor), 99%, 97%, 97%, 99%, 97%, 97%, 98%, 97%, 98%, 99%, 99% (baby), 98%, 98%, 96%, 99%, 98%, 97%, 97%, 96%, 98%, 99%, 99% (deer), and 99%, 98%, 97%, 97%, 97%, 98%, 96%, 98%, 99%, 96%, 98% (rhinos) of the

(a) akiyo (b) rhinos (c) bus (d) carphone
(e) clarie (f) coastguard (g) container (h) football
(i) foreman (j) garden (k) grandma (l) hallmonitor
(m) missam (n) mobile (o) motherdaughter (p) mountain
(q) manwalk (r) baby (s) deer

FIGURE 5.8: Watermarked video frames

(a) rhinos (b) foreman (c) akiyo

FIGURE 5.9: Extracted watermark from watermarked video

TABLE 5.1: No attack vs. imperceptibility and robustness values.

Test Videos	Imperceptibility and robustness metrics			
	Avg. PSNR	Avg. SSIM	NCC	BER
akiyo	57.53	0.9875	0.9998	0.0000
rhinos	56.87	0.9510	0.9996	0.0001
bus	56.25	0.9541	0.9997	0.0001
carphone	56.63	0.9574	0.9995	0.0002
clarie	56.75	0.9661	0.9998	0.0001
coastguard	57.72	0.9845	0.9996	0.0000
container	56.95	0.9523	0.9995	0.0001
football	55.38	0.9641	0.9992	0.0002
foreman	56.84	0.9516	0.9996	0.0001
garden	56.23	0.9489	0.9995	0.0001
grandma	56.75	0.9490	0.9996	0.0001
hallmonitor	57.12	0.9863	0.9994	0.0000
missam	55.74	0.9412	0.9997	0.0002
mobile	56.91	0.9889	0.9993	0.0001
motherdaughter	55.53	0.9506	0.9996	0.0002
mountain	57.85	0.9889	0.9998	0.0000
manwalk	56.21	0.9619	0.9992	0.0001
baby	56.98	0.9745	0.9991	0.0001
deer	56.87	0.9728	0.9992	0.0001

watermark for these test videos. From the above results, the authors conclude that the proposed watermarking algorithm is able to withstand spatial attacks for the test videos akiyo and hallmonitor in a better way than the other test cases. In particular, when a frame is attacked by the index 4, the 99% watermark extraction is achieved for all the test videos.

TABLE 5.2: Spatial attacks and their attack index.

Attack index	Spatial Attacks
1	Gaussian noise (0.5)
2	Poisson noise
3	Salt and pepper noise (0.05)
4	Median filter (5 × 5)
5	Gaussian low-pass filter (5 × 5)
6	Rotation (90°)
7	Cropping (20%)
8	Contrast adjustment
9	JPEG compression (QF=90)
10	Row-col deletion (10R, 10C)
11	Histogram equalization

Attack Index	Attacked Video akiyo	Attacked Video container	Attacked Video hall_monitor	Attacked Video baby	Attacked Video deer	Attacked Video rhinos
1	PSNR=55.75 SSIM=0.9875 NC=0.9999 BER=0.0001	PSNR=53.52 SSIM=0.9523 NC=0.9999 BER=0.0001	PSNR=54.18 SSIM=0.9897 NC=0.9999 BER=0.001	PSNR=55.23 SSIM=0.9749 NC=0.9989 BER=0.001	PSNR=55.78 SSIM=0.9743 NC=0.9889 BER=0.002	PSNR=55.75 SSIM=0.9793 NC=0.9999 BER=0.0001
	a) Gaussian Noise (0.05)					
2	PSNR=55.67 SSIM=0.9846 NC=0.9998 BER=0.001	PSNR=54.26 SSIM=0.9516 NC=0.9964 BER=0.001	PSNR=54.87 SSIM=0.9852 NC=0.9987 BER=0.002	PSNR=54.32 SSIM=0.9812 NC=0.9705 BER=0.003	PSNR=53.85 SSIM=0.9689 NC=0.9823 BER=0.002	PSNR=53.63 SSIM=0.9592 NC=0.9851 BER=0.002
	b) Poisson Noise					
3	PSNR=53.86 SSIM=0.9743 NC=0.9891 BER=0.002	PSNR=50.45 SSIM=0.9438 NC=0.9887 BER=0.002	PSNR=50.54 SSIM=0.9617 NC=0.9894 BER=0.001	PSNR=52.25 SSIM=0.9623 NC=0.9784 BER=0.003	PSNR=53.99 SSIM=0.9671 NC=0.9686 BER=0.004	PSNR=53.43 SSIM=0.9513 NC=0.9755 BER=0.004
	c) Salt and Pepper Noise (0.5)					
4	PSNR=57.13 SSIM=0.9879 NC=0.9999 BER=0.0001	PSNR=56.59 SSIM=0.9589 NC=0.9995 BER=0.001	PSNR=57.25 SSIM=0.9899 NC=0.9999 BER=0.0001	PSNR=55.43 SSIM=0.9628 NC=0.9989 BER=0.002	PSNR=54.58 SSIM=0.9752 NC=0.9954 BER=0.002	PSNR=54.65 SSIM=0.9871 NC=0.9972 BER=0.003
	Median filtering (3*3)					
5	PSNR=55.75 SSIM=0.9986 NC=0.9899 BER=0.002	PSNR=53.52 SSIM=0.9587 NC=0.9895 BER=0.002	PSNR=54.18 SSIM=0.9934 NC=0.9983 BER=0.003	PSNR=55.23 SSIM=0.9985 NC=0.9748 BER=0.003	PSNR=55.78 SSIM=0.9952 NC=0.9856 BER=0.002	PSNR=55.75 SSIM=0.9688 NC=0.9796 BER=0.003
	Gaussian low pass filtering (3*3)					

FIGURE 5.10: Imperceptibility and robustness measure for various spatial attacks on watermarked videos and extracted watermark

5.5.5 Video Processing or Temporal Attacks

Normally, the temporal attacks are used to validate the efficiency of the video watermarking algorithm in addition to spatial attacks. Prominent among them are: (i) MPEG compression, (ii) frame dropping, (iii) frame averaging, and (iv) frame swapping. In general, for better storage and distribution of digital video, it is compressed with the MPEG standard. The robustness of the extracted watermark to MPEG coding at a bit rate of 2 Mbps is evaluated here.

TABLE 5.3: Spatial attacks and their attack index.

Attack index	Spatial Attacks
1	MPEG compression (2Mbps)
2	Frame dropping (20%)
3	Frame averaging (5 frames)
4	Frame swapping (20%)

Frame dropping involves random dropping of frames, whereas frame averaging and swapping involve averaging of successive frames and random swapping of frames, respectively. Here, the proposed approach is tested against a frame dropping rate of 20%, frame averaging of 5 successive frames, and a frame swapping rate of 20%. This is shown in Figure 5.11 and its indexing is as mentioned in Table 5.3.

From Figure 5.11, the authors infer that the proposed optimized approach is robust against all kinds of temporal attacks of index 1, 2, 3, and 4, with a correlation value of about 99%, 99%, 99%, 98%; 99%, 97%, 98%, 98%; 98%, 98%, 96%, 97%; 99%, 98%, 99%, 99%; 98%, 99%, 99%, 99%; and 99%, 99%, 98%, 99% for the test videos akiyo, conainer, hallmonitor, baby, deer and rhinos respectively. The average correlation coefficient of various video processing attacks across all test videos is about 98%.

5.6 Comparative Analysis

To justify the efficiency of the proposed approach, some of the similar existing watermarking algorithms were considered for comparison and implemented. These includes (i) Video watermarking system with no optimization; that is, a single scaling factor is used for embedding across the video frames a namely V-SSF proposed by Nisreen et al. [43]; (ii) optimized multiple scaling; factor-based image watermarking system (I-MSF) by Mishra et al. [27]; (iii) optimal embedding location but with a single-scaling-factor-based video watermarking system (V-OL1-SSF) by Wu et al. [40]; and (iv) another optimal embedding location, but with a single-scaling-factor-based video watermarking system (V-OL2-SSF) by De Li et al. [4].

The authors tested the visual quality or imperceptibility and robustness of the proposed optimized video watermarking algorithm against common spatial and temporal distortions, and also compared it with the existing watermarking algorithms. These observations are plotted in Figure 5.12 and Figure 5.13. In the first test, watermarks were decoded from the watermarked video, which is affected by some of the image processing or spatial attacks, such as Gaussian

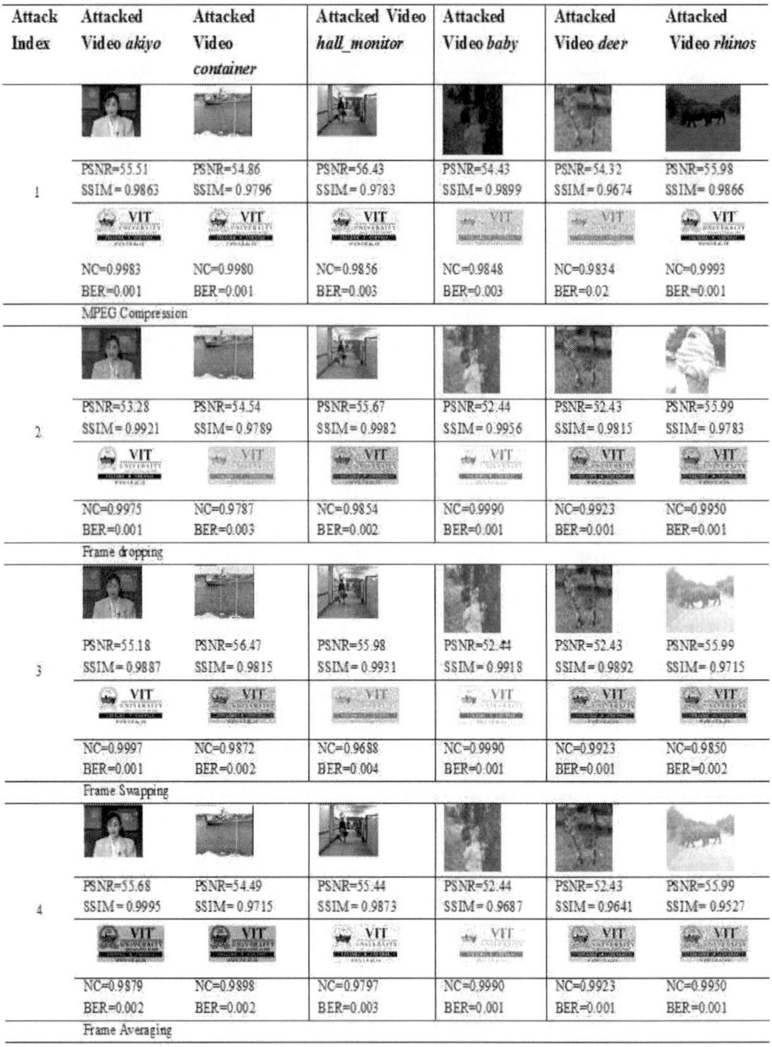

Attack Index	Attacked Video *akiyo*	Attacked Video *container*	Attacked Video *hall_monitor*	Attacked Video *baby*	Attacked Video *deer*	Attacked Video *rhinos*
1	PSNR=55.51 SSIM=0.9863 VIT NC=0.9983 BER=0.001 MPEG Compression	PSNR=54.86 SSIM=0.9796 VIT NC=0.9980 BER=0.001	PSNR=56.43 SSIM=0.9783 VIT NC=0.9856 BER=0.003	PSNR=54.43 SSIM=0.9899 VIT NC=0.9848 BER=0.003	PSNR=54.32 SSIM=0.9674 VIT NC=0.9834 BER=0.02	PSNR=55.98 SSIM=0.9866 VIT NC=0.9993 BER=0.001
2	PSNR=53.28 SSIM=0.9921 VIT NC=0.9975 BER=0.001 Frame dropping	PSNR=54.54 SSIM=0.9789 VIT NC=0.9787 BER=0.003	PSNR=55.67 SSIM=0.9982 VIT NC=0.9854 BER=0.002	PSNR=52.44 SSIM=0.9956 VIT NC=0.9990 BER=0.001	PSNR=52.43 SSIM=0.9815 VIT NC=0.9923 BER=0.001	PSNR=55.99 SSIM=0.9783 VIT NC=0.9950 BER=0.001
3	PSNR=55.18 SSIM=0.9887 VIT NC=0.9997 BER=0.001 Frame Swapping	PSNR=56.47 SSIM=0.9815 VIT NC=0.9872 BER=0.002	PSNR=55.98 SSIM=0.9931 VIT NC=0.9688 BER=0.004	PSNR=52.44 SSIM=0.9918 VIT NC=0.9990 BER=0.001	PSNR=52.43 SSIM=0.9892 VIT NC=0.9923 BER=0.001	PSNR=55.99 SSIM=0.9715 VIT NC=0.9850 BER=0.002
4	PSNR=55.68 SSIM=0.9995 VIT NC=0.9879 BER=0.002 Frame Averaging	PSNR=54.49 SSIM=0.9715 VIT NC=0.9898 BER=0.002	PSNR=55.44 SSIM=0.9873 VIT NC=0.9797 BER=0.003	PSNR=52.44 SSIM=0.9687 VIT NC=0.9990 BER=0.001	PSNR=52.43 SSIM=0.9641 VIT NC=0.9923 BER=0.001	PSNR=55.99 SSIM=0.9527 VIT NC=0.9950 BER=0.001

FIGURE 5.11: Imperceptibility and robustness measure for various temporal attacks on watermarked videos and extracted watermark VITlogo

noise with variance 0.5, Poisson noise, salt and pepper noise with noise density 0.05, Gaussian low-pass filtering of size 5×5, median filtering of size 5×5, rotation with 90°, cropping of 20%, contrast adjustment, JPEG, row-column deletion of 10R and 10C, and histogram equalization; the results are shown in Figure 5.12.

From Figure 5.12, the authors infer that the proposed approach is able to extract 99.99%, 99.98%, 98.91%, 99.99%, 98.99%, 98.81%, 99.98%, 99.97%,

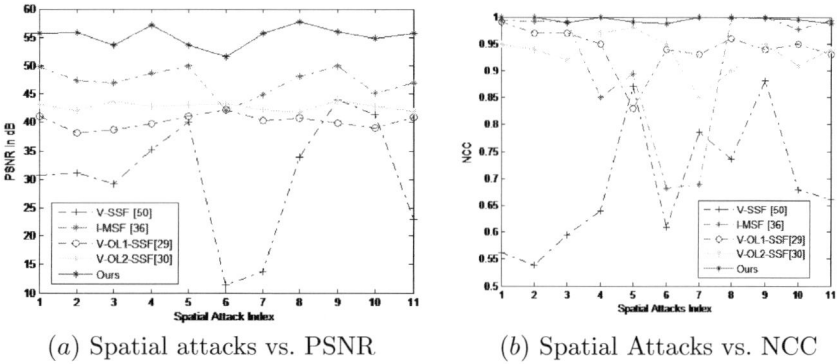

(a) Spatial attacks vs. PSNR (b) Spatial Attacks vs. NCC

FIGURE 5.12: Spatial attacks vs. PSNR and NCC

99.83%, and 99.50% of the watermark when the frames are affected by index 1, 2, 3, 4, 5, 6, 7, 8, 9, 10, and 11, respectively. Similarly, it can extract 99%, 97%, 97%, 95%, 83%, 94%, 93%, 96%, 94%, 95%, 93%; 95%, 94%, 92%, 97%, 98%, 95%, 85%, 90%, 95%, 91%, 94%; 56%, 53%, 59%, 63%, 87%, 60%, 78%, 73%, 88%, 67%, 66%; and 99%, 99%, 99%, 85%, 89%, 68%, 69%, 99%, 100%, 97%, 97%, 99% of the watermark for the existing watermarking algorithms V-SSF-OL1 [40], V-SSF-OL2 [4], V-SSF[43], and I-MSF [27], respectively. From the above results, the authors infer that the proposed watermarking algorithm is able to withstand image processing attacks in a better way when compared to the existing watermarking algorithms in terms of imperceptibility level, as shown in Figure 5.12. From the figure, it is also clear that the proposed approach is highly imperceptible against attack indices 4 and 8, with an average PSNR value of 57.13 and 57.41, respectively, when compared to other spatial attacks. In particular, when frames are corrupted by attack index 1, 2, 7, and 8, one can extract 99.99%, 99.98%, 99.98%, and 99.97% of the watermark, respectively, which is superior when compared to the existing system.

In the temporal test, the authors compared the effect of video processing attacks, such as MPEG compression, frame dropping, frame averaging and frame swapping, with the similar conventional algorithms [40, 4]. The proposed approach is able to extract 99%, 99%, 99% and 98% of the watermark when the frames are attacked by the index 1, 2, 3 and 4 respectively. From Figure 5.13, the authors found that the robustness measure of the watermark that is extracted is superior for all MPEG, frame dropping, and frame averaging attacks, except for the frame swapping attack. These values are found to be better when compared with the similar optimization algorithms.

From the overall inspection, the authors infer that the proposed nature-inspired optimal watermarking algorithm is best suited for video watermarking-based copyright protection applications. The efficiency of the proposed video watermarking algorithm has been measured and evaluated

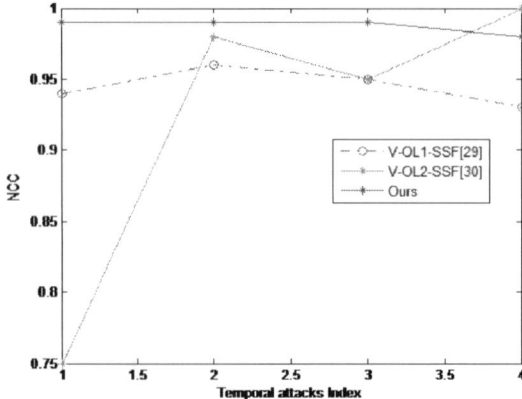

FIGURE 5.13: Temporal attacks vs. NCC

in terms of imperceptibility metrics (PSNR, SSIM) and robustness metrics (NCC, BER) against various spatial and temporal attacks. The obtained value in Figure 5.13 shows that the proposed watermarking scheme produced better imperceptibility (57.99) and robustness (0.9999), which is superior when compared to the existing system. The obtained results are due to the optimal selection of scaling factor.

5.7 Conclusions and Future Work

In this chapter, a new nature-inspired video watermarking algorithm, which is highly imperceptible and robust, and based on the firefly algorithm and dual transforms (like contourlet and Hilbert, using a PCA-based region selection) has been presented. It is imperceptible because a PCA-based sub-band selection is applied on each contourlet transform coefficient in order to obtain the uncorrelated coefficients for embedding the watermark. The experimental analysis shows that the proposed approach is robust against various spatial attacks, such as Gaussian noise, Poisson noise, salt and pepper attack, median filtering, Gaussian low-pass filtering, rotation, cropping, contrast adjustment, JPEG compression, row column deletion, and histogram equalization. It is robust because a firefly optimization algorithm is used to find the optimized scaling factor among the multiple scaling values obtained Figure 12, and it is also appropriate for the frame. The comparison plots in and Figure 13 show that the proposed approach is superior when compared to the similar existing systems, except for frame swapping attack (index 4 of temporal attack). In addition, the proposed approach is also free from the false positive detection problem, which is most common in the existing system. As a future work, the authors can extend this optimization for video in the transform domain.

Bibliography

[1] Rashmi Agarwal, R. Krishnan, M.S. Santhanam, K. Srinivas, and K Venugopalan. Digital watermarking: An approach based on Hilbert transform. *arXiv preprint arXiv:1012.2965*, 2010.

[2] L. Agilandeeswari and K. Ganesan. A robust color video watermarking scheme based on hybrid embedding techniques. *Multimedia Tools and Applications*, 75(14):8745–8780, 2016.

[3] Loganathan Agilandeeswari and K. Ganesan. A bi-directional associative memory based multiple image watermarking on cover video. *Multimedia Tools and Applications*, 75(12):7211–7256, 2016.

[4] Loganathan Agilandeeswari, K. Ganesan, and K. Muralibabu. A side view based video in video watermarking using dwt and hilbert transform. In *International Symposium on Security in Computing and Communication*, pages 366–377. Springer, 2013.

[5] Musrrat Ali and Chang Wook Ahn. Comments on optimized gray-scale image watermarking using dwt-svd and firefly algorithm. *Expert Systems with Applications*, 42(5):2392–2394, 2015.

[6] Mauro Barni, Franco Bartolini, Vito Cappellini, and Alessandro Piva. A dct-domain system for robust image watermarking. *Signal Processing*, 66(3):357–372, 1998.

[7] Mauro Barni, Franco Bartolini, Vito Cappellini, and Alessandro Piva. A dct-domain system for robust image watermarking. *Signal Processing*, 66(3):357–372, 1998.

[8] Gaurav Bhatnagar, Balasubramanian Raman, and Q.M. Jonathan Wu. Robust watermarking using fractional wavelet packet transform. *IET Image Processing*, 6(4):386–397, 2012.

[9] Gaurav Bhatnagar and Balasubrmanian Raman. Wavelet packet transform-based robust video watermarking technique. *Sadhana*, 37(3):371–388, 2012.

[10] Ingemar Cox, Matthew Miller, Jeffrey Bloom, Jessica Fridrich, and Ton Kalker. *Digital watermarking and steganography*. Morgan Kaufmann, 2007.

[11] Ingemar J. Cox, Joe Kilian, F. Thomson Leighton, and Talal Shamoon. Secure spread spectrum watermarking for multimedia. *IEEE Transactions on Image Processing*, 6(12):1673–1687, 1997.

[12] Minh N. Do and Martin Vetterli. The contourlet transform: an efficient directional multiresolution image representation. *IEEE Transactions on Image Processing*, 14(12):2091–2106, 2005.

[13] Gwenael Doerr and Jean-Luc Dugelay. A guide tour of video watermarking. *Signal Processing: Image Communication*, 18(4):263–282, 2003.

[14] F.Y. Duan, Irwin King, Lai-Wan Chan, and Lei Xu. Intra-block algorithm for digital watermarking. In *Fourteenth International Conference on Pattern Recognition, 1998. Proceedings*, volume 2, pages 1589–1591. IEEE, 1998.

[15] Osama S. Faragallah. Efficient video watermarking based on singular value decomposition in the discrete wavelet transform domain. *AEU-International Journal of Electronics and Communications*, 67(3):189–196, 2013.

[16] Yonggang Fu. Robust oblivious image watermarking scheme based on coefficient relation. *Optik-International Journal for Light and Electron Optics*, 124(6):517–521, 2013.

[17] Jing-Ming Guo and Heri Prasetyo. Security analyses of the watermarking scheme based on redundant discrete wavelet transform and singular value decomposition. *AEU-International Journal of Electronics and Communications*, 68(9):816–834, 2014.

[18] Marwen Hasnaoui and Mihai Mitrea. Multi-symbol qim video watermarking. *Signal Processing: Image Communication*, 29(1):107–127, 2014.

[19] Hanieh Khalilian and Ivan V. Bajić. Multiplicative video watermarking with semi-blind maximum likelihood decoding for copyright protection. In *IEEE Pacific Rim Conference on Communications, Computers and Signal Processing (PacRim)*, 2011 , pages 125–130. IEEE, 2011.

[20] Prayoth Kumsawat, Kitti Attakitmongcol, and Arthit Srikaew. A new approach for optimization in image watermarking by using genetic algorithms. *IEEE Transactions on Signal Processing*, 53(12):4707–4719, 2005.

[21] Chih-Chin Lai. A digital watermarking scheme based on singular value decomposition and tiny genetic algorithm. *Digital Signal Processing*, 21(4):522–527, 2011.

[22] Chang-Hsing Lee and Yeuan-Kuen Lee. An adaptive digital image watermarking technique for copyright protection. *IEEE Transactions on Consumer Electronics*, 45(4):1005–1015, 1999.

[23] Ching-Yung Lin and Shih-Fu Chang. Robust image authentication method surviving jpeg lossy compression. In *Photonics West'98 Electronic Imaging*, pages 296–307. International Society for Optics and Photonics, 1997.

[24] Eugene T. Lin, Christine I. Podilchuk, and Edward J. Delp III. Detection of image alterations using semifragile watermarks. In *Electronic Imaging*, pages 152–163. International Society for Optics and Photonics, 2000.

[25] Yan Liu and Jiying Zhao. A new video watermarking algorithm based on 1d dft and radon transform. *Signal Processing*, 90(2):626–639, 2010.

[26] Khaled Loukhaoukha. Comments on a digital watermarking scheme based on singular value decomposition and tiny genetic algorithm. *Digital Signal Processing*, 23(4):1334, 2013.

[27] Anurag Mishra, Charu Agarwal, Arpita Sharma, and Punam Bedi. Optimized gray-scale image watermarking using dwt–svd and firefly algorithm. *Expert Systems with Applications*, 41(17):7858–7867, 2014.

[28] Bijan G. Mobasseri. A spatial digital video watermark that survives mpeg. In *International Conference on Information Technology: Coding and Computing, 2000. Proceedings*, pages 68–73. IEEE, 2000.

[29] Nikos Nikolaidis and Ioannis Pitas. Robust image watermarking in the spatial domain. *Signal Processing*, 66(3):385–403, 1998.

[30] Nikos Nikolaidis and Ioannis Pitas. Robust image watermarking in the spatial domain. *Signal Processing*, 66(3):385–403, 1998.

[31] Xiamu Niu, Shenghe Sun, and Wenjun Xiang. Multiresolution watermarking for video based on gray-level digital watermark. *IEEE Transactions on Consumer Electronics*, 46(2):375–384, 2000.

[32] J. Senthilnath, S.N. Omkar, and V. Mani. Clustering using firefly algorithm: performance study. *Swarm and Evolutionary Computation*, 1(3):164–171, 2011.

[33] Chin-Shiuh Shieh, Hsiang-Cheh Huang, Feng-Hsing Wang, and Jeng-Shyang Pan. Genetic watermarking based on transform-domain techniques. *Pattern Recognition*, 37(3):555–565, 2004.

[34] Jieh-Ming Shieh, Der-Chyuan Lou, and Ming-Chang Chang. A semi-blind digital watermarking scheme based on singular value decomposition. *Computer Standards & Interfaces*, 28(4):428–440, 2006.

[35] Th. Rupachandra Singh, Kh. Manglem Singh, and Sudipta Roy. Video watermarking scheme based on visual cryptography and scene change detection. *AEU-International Journal of Electronics and Communications*, 67(8):645–651, 2013.

[36] Vassilios Solachidis and Loannis Pitas. Circularly symmetric watermark embedding in 2-d dft domain. *IEEE transactions on image processing*, 10(11):1741–1753, 2001.

[37] Qingtang Su, Yugang Niu, Hailin Zou, and Xianxi Liu. A blind dual color images watermarking based on singular value decomposition. *Applied Mathematics and Computation*, 219(16):8455–8466, 2013.

[38] Han-Min Tsai and Long-Wen Chang. Highly imperceptible video watermarking with the Watson's dct-based visual model. In *2004 IEEE International Conference on Multimedia and Expo, 2004. ICME'04*, volume 3, pages 1927–1930. IEEE, 2004.

[39] Chuntao Wang, Jiangqun Ni, and Jiwu Huang. An informed watermarking scheme using hidden Markov model in the wavelet domain. *IEEE Transactions on Information Forensics and Security*, 7(3):853–867, 2012.

[40] C.H. Wu, Y. Zheng, W.H. Ip, C.Y. Chan, K.L. Yung, and Z.M. Lu. A flexible h. 264/avc compressed video watermarking scheme using particle swarm optimization based dither modulation. *AEU-International Journal of Electronics and Communications*, 65(1):27–36, 2011.

[41] Xiaotian Wu and Wei Sun. Robust copyright protection scheme for digital images using overlapping dct and svd. *Applied Soft Computing*, 13(2):1170–1182, 2013.

[42] Xin-She Yang. *Nature-inspired metaheuristic algorithms*. Luniver Press, 2010.

[43] Nisreen I. Yassin, Nancy M. Salem, and Mohamed I. El Adawy. Qim blind video watermarking scheme based on wavelet transform and principal component analysis. *Alexandria Engineering Journal*, 53(4):833–842, 2014.

Chapter 6

Bone Tissue Segmentation Using Spiral Optimization and Gaussian Thresholding

Hugo Aguirre-Ramos

Universidad de Guanajuato, Salamanca, GTO, Mexico

Juan-Gabriel Avina-Cervantes

Universidad de Guanajuato, Salamanca, GTO, Mexico

Ivan Cruz-Aceves

CONACYT - Centro de Investigacion en Matematicas A.C. (CIMAT), Guanajuato, GTO, Mexico

6.1	Introduction	162
6.2	Background	164
	6.2.1 Spiral Optimization	164
	6.2.2 Implementation Details	167
	6.2.3 Example: Optimization of Sphere Function	170
	6.2.4 Gabor Filters	178
	6.2.5 Optimal Parameter Selection Using Gabor Filter	181
	6.2.6 Filter Bank Definition	183
6.3	The SO Initialized Gaussian Threshold Algorithm	185
	6.3.1 Image Preprocessing	186
	6.3.2 The Gabor Filter Bank	187
	6.3.3 SO Seeds	188
	6.3.4 Seed-Based Region Growing	189
	6.3.5 Gaussian Modeling	190
	6.3.6 Image Post-Processing	191
6.4	Computational Experiments	192
	6.4.1 Enhancement of CT Images	193
	6.4.2 Segmentation of Bone Tissue	194
6.5	Conclusions	196
	Acknowledgments	199
	Bibliography	199

In this work, a novel procedure for automatic bone tissue segmentation using the spiral optimization (SO) strategy and a Gaussian modeled thresholding is presented. As a pre-processing stage, a set of directional filters are applied over the computer tomography (CT) images in order to enhance and remove differently oriented features. In the segmentation stage, the SO strategy is used to initialize region-growing operators according to its intensity over the Gabor response image, and two Gaussian models are created from its result. Then, based on the intersection of these models, a thresholding strategy is introduced to classify bone and non-bone tissue pixels. The proposed method has been compared with different state-of-the-art segmentation methods and evaluated with the accuracy measure (ACC) and the dice coefficient (D), using a test set of 350 CT images, obtaining an average dice coefficient $D_\mu = 0.6969$ and a mean accuracy $ACC_\mu = 0.9843$.

6.1 Introduction

Computed tomography, along with other non-invasive techniques, has been widely used in clinical practice for evaluation or diagnostic purposes. Bone structures appear on almost any CT slice, and therefore it can be used for the physicians as a reference point [1]. The non-homogenous structure, formed by the combination of the solid-textured cortical and the spongy-textured trabecular bone (Figure 6.1a), besides the fact that bone represents a small fraction of the image content with respect to the background (Figure 6.1b), along with the radiological similarities between the trabecular bone and the surrounding structures (Figure 6.1c), and the closeness of other bone structures (Figure 6.1d), are among the challenges faced in the automatic bone segmentation process.

Bone segmentation has become an essential part of computed assisted surgery [32, 24], prosthesis modeling and evaluation [28, 6], and pathology detection [20, 7]. However, automatic bone segmentation may present a poor performance because of variations in the anatomy of the object analyzed or restrictions in the image quality. This fact drives us to the creation of semi-automatic methods for cases when the segmentation result is used for planning or executing a medical process, in order to keep it safe [4, 12].

In the literature, this task has been carried out by several methods, such as: active contours [23, 27], region growing [17], adaptive thresholding [15], and border enhancement [40]. These methods rely on homogeneous bone structures, generating errors for non-homogeneous regions or including elements from the surrounding regions. Thus, region growing and active contour methods may end up stuck in local minima regions or may merge two or more separable bone structures [15, 13]. On the other hand, while thresholding methods do not require the use of spatial pixel connexity, they require closely

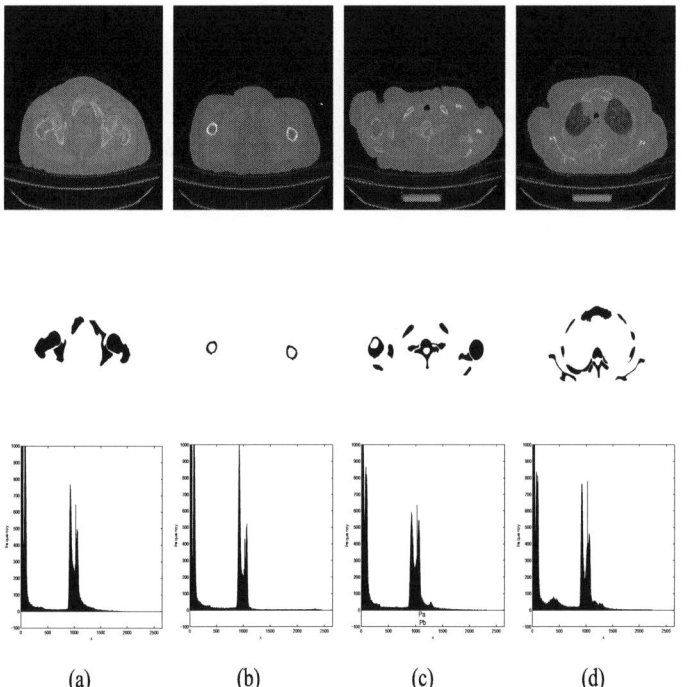

FIGURE 6.1: Challenges in bone segmentation over CT images (top), its gold standard (center), and gray level histogram (bottom) for: (a) non-homogenous structure, (b) reduced size, (c) radiological similarity, (d) closeness of other bones or organs

related intensity values for the bone pixels; nevertheless, soft tissue often possesses the same, or at least similar, intensity as some bone pixels, especially those in the trabecular bone, leading to misclassified pixels regardless of the algorithm [19]. Thereby, an algorithm that reduces the texture and intensity variations across the bone structures and its surroundings, while preserving its shape, is highly recommended before any bone segmentation algorithm.

Usually, smooth filtering with or without border preserving, such as bilateral or Gaussian filters, are applied to the image before the main algorithm is executed [11]. Although these filters improve texture and intensity separation, they work for each element within the image, preserving the same problem for non-isolated regions. Therefore, a method that enhances the bone structure while it reduces the surrounding elements is needed when these algorithms are used, in order to reduce the possibility of misclassified or unexpected results.

Another approach used for bone segmentation is based on the statistical properties of the image. These methods model the image content as Gaussian functions, and apply several classification techniques over them. Several examples of these processes can be found in the literature: Zhang et al. proposed the iterative generation of two Gaussian models by the Bayes theorem until the convergence is achieved [41], Zhao et al. applied a thresholding based on the intersection of the two models created from a gray-level histogram with the auto-thresholding algorithm and a 3D region-growing process, to obtain the bone structures [42], while Aguirre et al. defined a threshold from the assumption that 84% of one Gaussian model's data growing belong to bone class [9]. All these methods show that the use of statistics for bone classification is a simple yet promising area.

The remaining content of this work is presented as follows: Section 6.2 introduces the background for the proposed method components, defining the concept of spiral optimization and its applications, as well as an introduction to Gabor filtering. Section 6.3 describes the operation of each element within the methodology used and the considerations for its application. Section 6.4 presents the experimental results and assessment of the algorithm. Finally, in Section 6.5, the conclusions for this work are given.

6.2 Background

This section gives the user a general approach to the mathematics behind the method. The theory for spiral optimization and gabor filters are described, and the effects of the modification of its parameters in the solution of the sphere function are illustrated.

6.2.1 Spiral Optimization

SO is a metaheuristic method based on the spiral phenomena produced in nature; whirling currents, hurricanes, nautilus shells, some galaxies' shapes or plants' disposition, just to mention some of them, are examples of such phenomena (Figure 6.2). Originally proposed by Tamura and Yasuda in 2011 [34] as a two dimensional continuous optimization technique, so was later generalized to n-dimensions [35].

The general idea of SO is to search for the global minimum (or maximum) of an n-dimensional objective function $f_{Obj}^{(n)}(X^{(n)})$, considering each search point (current spiral position) as a possible solution and searching for the best value; however, since nature spirals end at their center or origin, at least two search points (two spirals) are necessary in order to get a value close to the actual solution. Thereby, one spiral will carry the best value, while the others will try to move toward that point, increasing the searching area and getting rid of the origin convergence problem, since the spiral "origin" can change along the process.

(a)　　　　　　　　(b)　　　　　　　　(c)

FIGURE 6.2: Spirals in nature: (a) spiral galaxy [21], (b) aloe polyphylla [29], and (c) hurricane [22]

The spiral model considered in this work is the n-dimensional approach proposed in [35], and shown in Equation 6.1, where $0 \leq \theta_{ij} \leq 2\pi$ and $0 < r < 1$ are a rotation angle (or angles) around the origin and a convergence rate of distance between a point and the origin at each iteration k, respectively, and $M^{(n)}(\theta_{ij})$ is an n-dimensional rotation matrix.

$$X^{(n)}(k+1) = rM^{(n)}(\theta_{ij})X^{(n)}(k) \quad := S_n(r,\theta)X^{(n)}(k) \qquad (6.1)$$

Any n-dimensional rotation matrix needs $n(n-1)/2$ parameters, one for each possible pair of axes specifying a plane of rotation, so that the concatenation in some order of all these rotations forms the rotation matrix. A rotation in the plane $x_i - x_j$ can be written as the matrix $R_{ij}(\theta_{ij})$ in Equation 6.2.

$$R_{ij}(\theta_{ij}) = \begin{bmatrix} 1 & \cdots & 0 & 0 & \cdots & 0 & 0 & \cdots & 0 \\ \vdots & \ddots & \vdots & \vdots & \ddots & \vdots & \vdots & \ddots & \vdots \\ 0 & \cdots & \cos\theta_{ij} & 0 & \cdots & 0 & -\sin\theta_{ij} & \cdots & 0 \\ 0 & \cdots & 0 & 1 & \cdots & 0 & 0 & \cdots & 0 \\ \vdots & \ddots & \vdots & \vdots & \ddots & \vdots & \vdots & \ddots & \vdots \\ 0 & \cdots & 0 & 0 & \cdots & 1 & 0 & \cdots & 0 \\ 0 & \cdots & \sin\theta_{ij} & 0 & \cdots & 0 & \cos\theta_{ij} & \cdots & 0 \\ \vdots & \ddots & \vdots & \vdots & \ddots & \vdots & \vdots & \ddots & \vdots \\ 0 & \cdots & 0 & 0 & \cdots & 0 & 0 & \cdots & 1 \end{bmatrix} \qquad (6.2)$$

A rotation in the plane $x_i - x_j$ means that only the elements (x_i, x_j) on the $x_i - x_j$ plane will be rotated by $\theta_{i,j}$ around the origin, letting the other elements be invariant. Then, the n-dimensional rotation matrix can be calculated as the product of the $n(n-1)/2$ different $R_{ij}(\theta_{ij})$ matrices, as defined in Equation 6.3.

$$M^{(n)}(\theta_{ij}) = \prod_{i<j} R_{ij}(\theta_{ij}) \qquad (6.3)$$

For example, the 2-dimensional rotation matrix $M^{(2)}(\theta_{12})$ shown in Equation 6.4 needs only one parameter, since $2(2-1)/2 = 1$; however, for $n = 3$, $3(3-1)/2 = 3$ parameters, one for each rotation in the planes $x_i - x_j$ for $i, j = 1, 2, 3$ (θ_{12}, θ_{13}, and θ_{23}), are necessary to form the rotation matrix $M^{(3)}(\theta_{12}, \theta_{13}, \theta_{23})$, from Equation 6.5.

$$M^{(2)}(\theta_{12}) = R_{12}(\theta_{12}) = \begin{bmatrix} \cos\theta_{12} & -\sin\theta_{12} \\ \sin\theta_{12} & \cos\theta_{12} \end{bmatrix} \quad (6.4)$$

$$M^{(3)}(\theta_{12}, \theta_{13}, \theta_{23}) = R_{12}^{(3)}(\theta_{12}) R_{13}^{(3)}(\theta_{13}) R_{23}^{(3)}(\theta_{23}),$$

$$R_{12}^{(3)}(\theta_{12}) = \begin{bmatrix} \cos\theta_{12} & -\sin\theta_{12} & 0 \\ \sin\theta_{12} & \cos\theta_{12} & 0 \\ 0 & 0 & 1 \end{bmatrix}$$

$$R_{13}^{(3)}(\theta_{13}) = \begin{bmatrix} \cos\theta_{13} & 0 & -\sin\theta_{13} \\ 0 & 1 & 0 \\ \sin\theta_{13} & 0 & \cos\theta_{13} \end{bmatrix} \quad (6.5)$$

$$R_{23}^{(3)}(\theta_{23}) = \begin{bmatrix} 1 & 0 & 0 \\ 0 & \cos\theta_{23} & -\sin\theta_{23} \\ 0 & \sin\theta_{23} & \cos\theta_{23} \end{bmatrix}$$

All the angles θ_{ij} can differ in value from the other ones depending on the nature of the problem, Figure 6.3 shows the position of the 3-dimensional point $X^{(3)} = (10, 0, 0)$ after several rotations by $M^{(3)}(0, \pi/18, 0)$ (Figure 6.3a), $M^{(3)}(\pi/18, \pi/18, 0)$ (Figure 6.3b), and $M^{(3)}(\pi/18, \pi/18, \pi/18)$ (Figure 6.3c). It can be observed that the rotation of the point by θ_{13} only affects the values for the planes x_1 and x_3, while keeping the position of the point on the plane x_2 at every moment, proving the property that any rotation by θ_{ij} affects only the points over a $x_i - x_j$ plane.

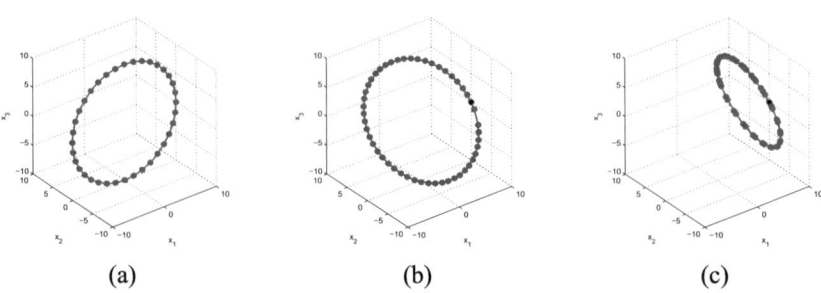

(a) (b) (c)

FIGURE 6.3: Trajectory of $X^{(3)} = (10, 0, 0)$ after multiple rotations by $M^{(3)}(\theta_{12}, \theta_{13}, \theta_{23})$ matrix for: (a) $M^{(3)}(0, \pi/18, 0)$, (b) $M^{(3)}(\pi/18, \pi/18, 0)$, and (c) $M^{(3)}(\pi/18, \pi/18, \pi/18)$

Because of its origin convergence property, the model from Equation 6.1 does not have enough flexibility for optimization problems; nevertheless, the origin can be translated to an arbitrary point X^* of dimension equal to $X^{(n)}$, so that the spiral can move toward the solution. The complete model that includes the center transformation is described in Equation 6.6, where I_n is the identity matrix.

$$X^{(n)}(k+1) = S_n(r,\theta)X^{(n)}(k) - (S_n(r,\theta) - I_n) X^* \qquad (6.6)$$

The change or update of the center point X^* will allow the SO to reach an area close to the actual solution and search intensively on it; in that sense, X^* can be considered the best solution at each iteration, and its closeness to the actual solution will depend on the maximum iteration number used. A new prospect for the new center, X^*_{k+1}, can be found using Equation 6.7, where i_g represent the search point index with the best fitness value, i.e., the minimum value of the objective function assessment, and s means the search points' (spiral) number. However, X^* will only change its value if the new prospect represents a better solution than the previous point, as is shown in Equation 6.8; this consideration was not mentioned in the original spiral optimization proposal but was implemented in this work.

$$X^*_{k+1} = X^{(n)}_{i_g}(k+1) \quad i_g = \arg\min_i f_{Obj}(X^{(n)}_i(k+1)), \quad i = 1, 2, \ldots, s \qquad (6.7)$$

$$X^* = \begin{cases} X^*_{k+1} & if \ f^{(2)}_{Obj}(X^*(k+1)) < f^{(2)}_{Obj}(X^*(k)) \\ X^*_k & otherwise \end{cases} \qquad (6.8)$$

6.2.2 Implementation Details

Using Equation 6.6, only two parameters need to be adjusted for the solution of a optimization problem for each spiral: $0 < r < 1$ and $0 \leq \theta \leq 2\pi$. Nonetheless, two additional values should be considered, the spiral number s and the maximum iteration number k_{max}. The improper selection of all these parameters can result in a deficient or incorrect solution. A review on the effects of the each parameter is presented in the following sections.

The convergence rate to the origin (r): This parameter modifies the distance of the search point (current spiral position) to its origin or center of the spiral; values of r close to 1 will keep a trajectory where each point slowly reduces its distance to the origin generating a search near previous searched points, while smaller values will produce a loose trajectory, focusing the search around the spiral origin. Common values for r are selected in the interval defined by $r \in [0.9, 0.99]$ [3]. Figure 6.4 shows the effect of different values of r for a spiral with rotation angle $\theta = \pi/4$ and an initial value $X^{(2)}(0) = (75, 75)$. For each value, the spiral shape is preserved.

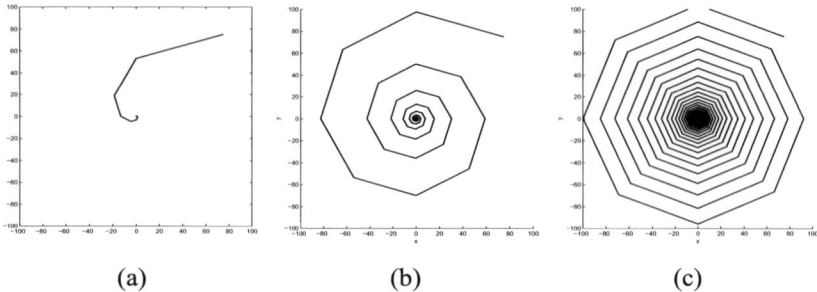

FIGURE 6.4: Trajectories for a spiral with $\theta = \pi/4$ and an initial point $X_1^{(2)}(0) = (75, 75)$ for: (a) $r = 0.5$, (b) $r = 0.92$, and (c) $r = 0.98$

The rotation angle (θ): Parameter θ modifies the way in which the search point moves, modifying the spiral shape. Values of θ close to zero will draw a smooth circular trajectory keeping the rotated point close to its previous value, while upper values will produce a less-circular shape with abrupt hops at each iteration; however, a search path with abrupt hops can free a point trapped on a local minimum. Common values used for the rotation angle include $\theta = \{\pi/2, \pi/4\}$ [36]. Figure 6.5 shows the trajectories for a search point with convergence rate to the origin $r = 0.95$ and three different rotation angles $\theta = \{\pi/6, 2\pi/3, 5\pi/6\}$, with the initial point at $X^{(2)}(0) = (75, 75)$. Contrary to the behavior of r, a different value of θ can change the final shape of the spiral.

Additional settings: the spiral number (s) and the maximum iteration number (k_{max}): As a general rule, SO processes with a small spiral number will need many more iterations to achieve results similar to those obtained with higher values of s in a fraction of the same iterations. In that sense, it is possible to obtain good results using a small number of spirals

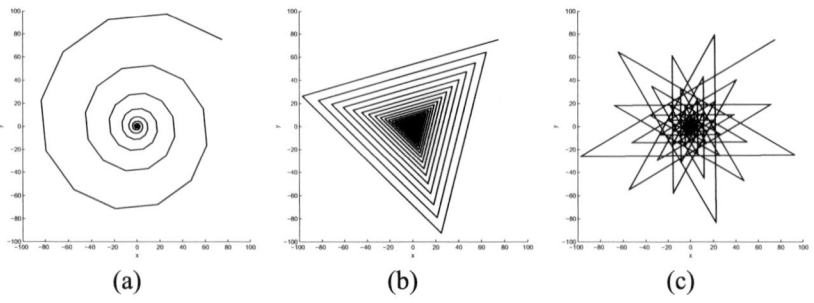

FIGURE 6.5: Trajectories for a search point with $\rho = 0.95$ from initial point $X^{(2)}(0) = (75, 75)$ for: (a) $\theta = \pi/6$, (b) $\theta = 2\pi/3$, and (c) $\theta = 5\pi/6$

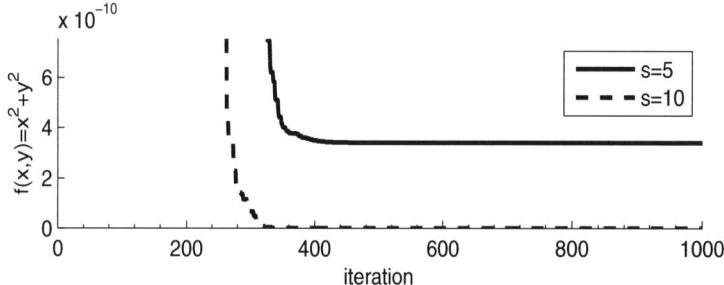

FIGURE 6.6: Sphere function assessment for $s = 5$ and $s = 10$ with $S(0.95, \pi/4)$ after 1000 iterations

if k_{max} is big enough, and the initial points are close to the actual solution; nonetheless, it is feasible that such results will still be far away from the actual solution, mostly due to the arbitrary position of the spirals at the beginning of the process and the lack of diversity in the searching process.

Figure 6.6 shows an example of this effect; in this figure, a minimum value $f_{Obj}^{(2)}(X^*) \approx 4 \times 10^{-10}$ for the evaluation of a certain objective function, with global minimum $f_{obj}^{(2)}(X^*) = 0$, is achieved after 470 iterations setting $s = 5$, while a similar result is obtained at around 200 iterations by setting $s = 10$ for a given $S(r, \theta)$. After a given point ($k \approx 700$) the search with $s = 5$ will not change its value considerably, so that the remaining iterations are not needed, showing that k_{max} plays an important role in a problem's solution: Too many iterations may not be necessary, but an insufficient number can break the process before the best solution can be found.

If the previous statement is true, it means that there exists a relation between s and k_{max}; however, such a relation is not as simple as it sounds, since an initial position close to the actual solution will need much fewer iterations to reach the best point than an initial value far away for it. Besides, k_{max} is usually used as a stop condition for the SO algorithm, and therefore it should be set as large as possible if an alternative to the algorithm end, like a stable value after several iterations, is not considered; suggested values for this parameter lie between $k_{max} \in [100, 1000]$.

Diversification and intensification: From Figure 6.4 and Figure 6.5, the fact that SO search satisfies the diversification and intensification properties is presented in a metaheuristic method. First, a search for a better solution is made by examining a wide region in the early phase, and then, a search around a good point is made at the late phase. Figure 6.7 shows the paths described for two search points and the optimal paths marked by the best point; such results were found after 100 iterations for four different runs, by setting $s = 2$ and $S(0.95, \pi/4)$. Although a limited number of positions are shown, the diversification and intensification properties can be viewed easily in each figure.

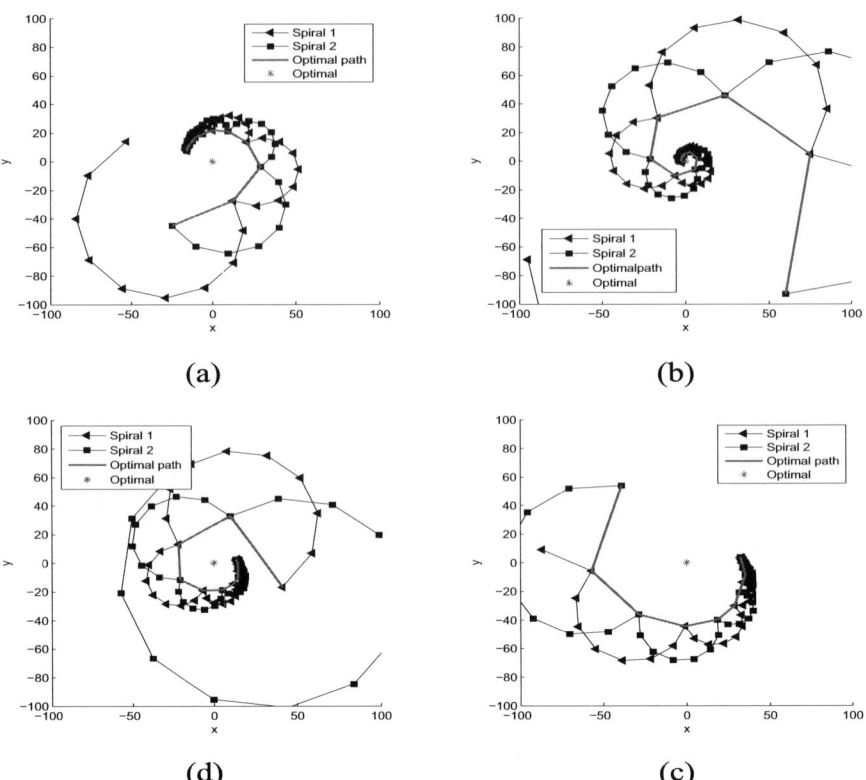

FIGURE 6.7: Different SO search paths using two spirals $S(0.95, \pi/6)$ after 100 iterations

The following algorithm shows the pseudocode for the SO search strategy. In the initial stage, the values of $S(r, \theta)$, the number of spirals (s), and the maximum iteration number (k_{max}) are set; then s n-dimensional arbitrary initial points are selected according to the problem's nature. After a preliminary evaluation, a center X^* is defined and the algorithm start's updating the values of $X_i^{(n)}$ and X^* until the maximum iteration number is reached.

6.2.3 Example: Optimization of Sphere Function

The sphere function (Equation 6.9) is an n-dimensional, continuous, unimodal, and convex function, commonly used to test metaheuristic algorithms; its global minimum is $f_{sphere}^{(n)}(X^*) = 0$ at $X^* = (0, \ldots, 0)$. Figure 6.8 shows its

Algorithm [Spiral Optimization]

Input: Objective function $(f_{Obj}^{(n)}(X))$, spiral number (s), convergence rate to the origin (ρ), rotation angle (θ), and maximum iteration number (k_{max})
Output: Best value (X^*)

1. Set s, k_{max}, ρ, and θ values
2. Set arbitrary initial points for each spiral $X_i^{(n)}(0) \in \mathbb{R}^{(n)}$, $\forall i = 1, 2, \cdots, s$
3. Set center X^* as $X^* = X_{i_g}^{(n)}(0)$, $i_g = \arg\min_i f_{Obj}^{(n)}(X_i^{(n)}(0))$, $\forall i = 1, 2, \cdots, s$
4. Set iterator $k = 0$
5. While $k < k_{max}$ do
   (a) Update $X_i^{(n)}$ as $X_i^{(n)}(k+1) = S_n(\rho, \theta)X_i^{(n)}(k) - (S_n(\rho, \theta) - I_n)X^*$, $\forall i = 1, 2, \ldots, s$
   (b) Get $X_{k+1}^* = X_{i_g}^{(n)}(k+1)$, $i_g = \arg\min_i f_{Obj}^{(n)}(X_i^{(n)}(k+1))$, $\forall i = 1, 2, \ldots, s$
   (c) If $f_{Obj}^{(n)}(X_{k+1}^*) < f_{Obj}^{(n)}(X^*)$, then
       i. Update $X^* = X_{k+1}^*$
   (d) else
       i. preserve X^*
   (e) increase k as $k = k + 1$

2-dimensional form over the range $X_i \in [-100, 100]$ for $i = 1, 2$.

$$f_{sphere}^{(n)}(x) = \sum_{i=1}^{n} x_i^2 \quad i = 1, \cdots, n \qquad (6.9)$$

To test the performance of the SO algorithm for solving the sphere function, the code in the SO search algorithm was implemented and tested over the function considering three different cases: I) s and r are fixed, while θ varies; II) s and θ are fixed, while r varies, and III) r and θ are fixed while s varies, any other combination can be inferred from all the experiments. The

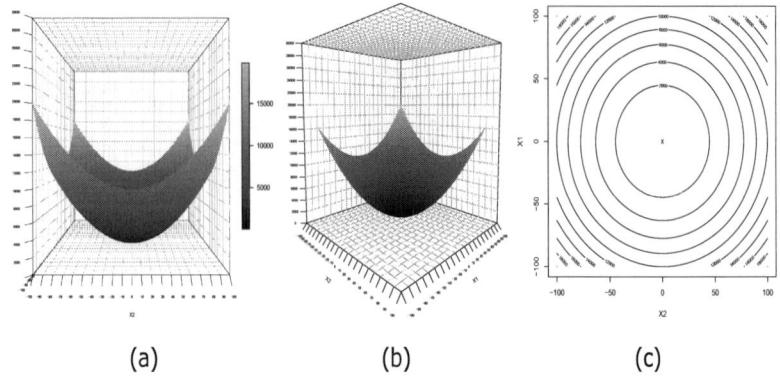

FIGURE 6.8: Sphere function $f^2_{Obj}(x) = x_1^2 + x_2^2$ for $x_i \in [-100, 100]$: (a) isometric view, (b) lateral view, and (c) level plot

process was then repeated 50 times and a statistical analysis was made over the obtained data; for all the runs, no constraint conditions were considered, and the initial position for the spirals were randomly set using a uniform distribution between $X_i^{(2)}(0) \in [-100, 100]$ and $\forall i = 1, \cdots, s$.

Case I (s and r are fixed, while θ varies): For the first case, the only parameter that changes is the rotation angle θ. Figure 6.9 shows the sphere function evaluation after 700 iterations, using three different values of θ: $\theta_1 = \pi/6$, $\theta_1 = \pi/4$, and $\theta_1 = \pi/2$, with $r = 0.95$. From this figure it can be appreciated that for θ_2 a $f_{sphere}^{(2)}(x) < 1 \times 10^{-25}$ is reached using a lower number of iterations, while the angle that needs more time to achieve the same result is θ_3. This tendency can be analyzed from the data in Table 6.1, where the best value at each k_{max} is observed by $\theta = \pi/4$. These results

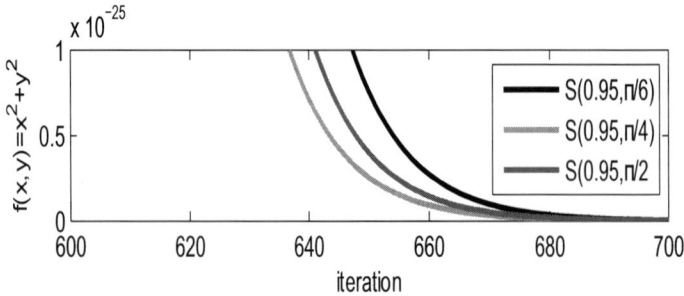

FIGURE 6.9: Sphere function evaluation for $s = 6$, $r = 0.95$, and $\theta = \{\pi/6, \pi/4, \pi/2\}$, after 700 iterations

TABLE 6.1: Statistics for the sphere evaluation after 50 runs, using $s = 6$, $r = 0.95$, for different rotation angles $\theta = \{\pi/6, \pi/4, \pi/2\}$ and maximum iteration number $k_{max} = \{50, 100, 400, 700, 1000\}$.

$2f(x,y)$	2θ	k_{max}				
		50	100	400	700	1000
max	$\pi/6$	1.436e+02	2.621e+01	1.170e+01	3.604e+01	1.879e+00
	$\pi/4$	1.757e+01	8.700e-02	3.984e-15	1.346e-28	6.201e-42
	$\pi/2$	1.550e+01	1.129e-01	3.610e-15	2.290e-28	7.730e-42
min	$\pi/6$	4.526e-01	6.742e-03	2.723e-16	1.537e-29	9.537e-43
	$\pi/4$	7.084e-02	1.121e-03	1.906e-17	2.135e-31	2.547e-44
	$\pi/2$	2.336e-01	2.047e-04	4.396e-17	5.158e-31	1.951e-43
μ	$\pi/6$	1.955e+01	6.661e-01	2.380e-01	7.213e-01	3.799e-02
	$\pi/4$	4.161e+00	2.084e-02	7.386e-16	2.259e-29	1.349e-42
	$\pi/2$	3.623e+00	2.202e-02	1.050e-15	4.644e-29	2.012e-42
median	$\pi/6$	1.395e+01	7.178e-02	3.266e-15	2.178e-28	1.928e-35
	$\pi/4$	2.434e+00	1.457e-02	4.666e-16	1.476e-29	1.045e-42
	$\pi/2$	2.714e+00	1.743e-02	8.246e-16	2.958e-29	1.741e-42
σ	$\pi/6$	2.297e+01	3.695e+00	1.654e+00	5.097e+00	2.657e-01
	$\pi/4$	3.634e+00	2.011e-02	7.975e-16	2.601e-29	1.366e-42
	$\pi/2$	3.331e+00	2.234e-02	8.625e-16	4.887e-29	1.608e-42

can conclude that for the definition of θ values, a nearly abrupt change or no change at all should be avoided.

Case II (s and θ are fixed, while r varies): In this case, r is the only parameter that is modified; these experiments showed that smaller values of r will lead to better results, since the intensification stage lasts longer for a given k_{max}. Figure 6.10 shows the evaluation of the sphere function after one

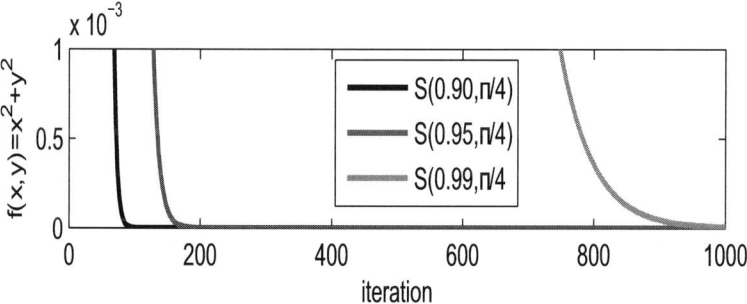

FIGURE 6.10: Sphere function assessment for $s = 6$, $\theta = \pi/4$, and $r = \{0.90, 0.95, 0.99\}$ after 1000 iterations

TABLE 6.2: Statistical analysis of the sphere function evaluation after 50 runs, using $s = 6$, $\theta = \pi/4$, for different convergence rates $r = \{0.20, 0.50, 0.90, 0.95, 0.99\}$ and maximum iteration numbers $k_{max} = \{50, 100, 400, 700, 1000\}$.

$2f(x,y)$	$2r$	k_{max}				
		50	100	400	700	1000
max	0.20	2.778e+03	4.011e+03	2.508e+03	3.871e+03	4.543e+03
	0.50	1.175e+03	1.026e+03	2.980e+03	3.843e+03	3.425e+03
	0.90	3.592e+00	4.108e-01	1.451e+00	1.842e+00	7.926e-12
	0.95	1.634e+01	8.735e-02	4.727e-15	1.847e-28	3.279e-42
	0.99	8.515e+02	1.842e+02	6.413e-01	1.082e-03	3.715e-06
min	0.20	8.580e+00	3.598e+00	2.041e-01	4.564e-02	1.980e-01
	0.50	3.447e-04	5.755e-08	7.691e-09	1.580e-11	4.373e-07
	0.90	2.734e-03	4.220e-08	0.000e+00	1.235e-62	4.165e-90
	0.95	1.144e-01	7.028e-04	6.312e-18	7.486e-31	2.609e-44
	0.99	2.000e+00	1.000e+00	3.683e-03	1.677e-05	3.563e-08
μ	0.20	6.017e+02	7.769e+02	7.375e+02	6.769e+02	7.121e+02
	0.50	9.625e+01	8.182e+01	2.552e+02	3.226e+02	1.967e+02
	0.90	1.718e-01	8.996e-03	3.103e-02	3.687e-02	1.596e-13
	0.95	4.083e+00	1.598e-02	6.834e-16	2.692e-29	7.461e-43
	0.99	1.989e+02	4.838e+01	1.251e-01	3.062e-04	9.124e-07
median	0.20	3.577e+02	5.459e+02	5.192e+02	2.877e+02	3.354e+02
	0.50	8.632e+00	6.948e+00	1.109e+01	4.122e+01	1.034e+01
	0.90	4.169e-02	1.188e-06	9.954e-34	2.356e-61	9.001e-89
	0.95	2.353e+00	9.454e-03	4.270e-16	1.459e-29	5.124e-43
	0.99	1.122e+02	3.686e+01	9.306e-02	2.777e-04	6.209e-07
σ	0.20	6.525e+02	8.070e+02	7.415e+02	9.298e+02	9.725e+02
	0.50	1.957e+02	1.837e+02	5.515e+02	7.070e+02	5.610e+02
	0.90	5.328e-01	5.810e-02	2.052e-01	2.605e-01	1.121e-12
	0.95	4.222e+00	1.982e-02	8.517e-16	3.683e-29	6.616e-43
	0.99	2.177e+02	4.198e+01	1.383e-01	2.273e-04	8.624e-07

run; it is clearly understood that a value of r close to 1 will need many more iterations in order to reach a value less than $f_{sphere}^{(n)}(x) < 1 \times 10^{-3}$, since higher values will lead to a slower approach to the origin reduction, and therefore it will take more iterations to reach the same value. On the other hand, $r < 0.90$ values will get results far away from the actual solution. Table 6.2 shows the results after 50 runs. From this table it can be observed that a value close to 1 will lead to a deficient value, and values close to 0 will keep the result far away from the actual solution; in this particular case, for a mean value of $f_{sphere}^{(2)}(x) = 7 \times 10^{-7}$, it takes the process around 1000 iterations for $r \approx 1$, while values of $r > 0.90$ have reached values less than $1 \times 10-40$ by the same amount of iterations. Table 6.2 also shows that values for r should be above 0.9.

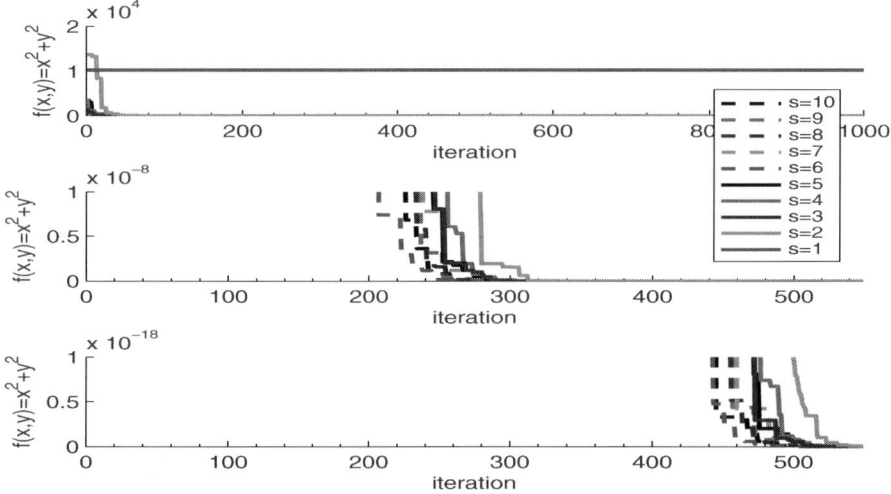

FIGURE 6.11: Sphere function evaluation for different spiral numbers $s = \{2, 3, 5, 6, 10\}$ with $S(0.95, \pi/6)$ after 1000 iterations

Case III (r and θ are fixed, while s varies): The final case observes the effect of the change in the search point number. Figure 6.11 shows the evaluation results for the sphere function, for $k_{max} = 1000$ after one run. For different values of s and a given $S_2(r, \theta)$, it can be observed that with $s = 6$, less than 500 iterations are needed in order to reach an assessed value of $f_{Obj}^{(2)}(X^*) < 1 \times 10^{-15}$; therefore setting s and k_{max} to bigger values may not be necessary. Table 6.3 shows the statistics for this case after 50 runs. From this data, it can be appreciated that for a minimum value $s = 3$, an average value $\mu(f_{sphere}^{(2)}(x)) < 1 \times 10^{-28}$ is reached after 700 iterations with $\sigma < 1 \times 10^{-28}$. Also, for values $k_{max} < 100$, the best solution found is still far away from the actual solution for any value of s. Besides, the table shows that results for values $s > 5$ do not present a considerable improvement.

SO modification (the φ model): We have considered the center modification in the spiral model as a function of the best value X^*, and the matrix $S_n(r, \theta)$ only; however, the addition of a random variable can lead to superb results in the evaluation of the 2-dimensional sphere function. This variable will displace the center to a point closer to the best solution at each iteration, accelerating the end of the diversification stage and letting the intensification stage last longer over a narrow area in the same amount of iterations. The modified spiral model, named the φ-model, is described in Equation 6.10, where φ represents an n-dimensional random variable with a certain distribution. For the case of $\varphi = 1$, the original spiral model is obtained; in that

TABLE 6.3: Statistics for the $f_{sphere}^{(2)}(x)$ assessment after 50 runs for a different combination of k_{max} and s, with $S(0.95, \theta = \pi/4)$.

$2f(x,y)$	$2s$	\multicolumn{7}{c}{k_{max}}						
		30	50	100	200	400	700	1000
max	2	1.045e+03	3.340e+03	6.490e+03	6.371e+03	2.284e+03	4.504e+03	1.406e+03
	3	5.240e+02	2.541e+02	8.603e-01	1.369e+00 3	1.431e-14	4.156e-28	7.483e+01
	5	1.322e+02	3.075e+01	1.378e-01	5.951e-06 5	4.610e-15	2.211e-28	6.892e-42
	6	3.341e+02	1.025e+01	1.398e-01	2.724e-06	4.529e-15	9.362e-29	1.108e-41
	10	8.167e+01	6.556e+00	4.888e-02	1.370e-06	1.239e-15	5.995e-29	1.678e-42
min	2	4.273e+00	1.196e+00	2.961e-02	2.828e-07	5.787e-16	1.625e-29	1.204e-42
	3	1.213e+01	3.193e-01	3.119e-03	5.182e-08	5.346e-17	2.763e-30	1.484e-43
	5	2.953e+00	5.332e-01	1.610e-03	3.861e-08	2.118e-17	2.124e-31	6.637e-44
	6	1.367e+00	1.339e-01	6.294e-04	5.238e-08	1.998e-17	1.113e-30	1.594e-43
	10	8.282e-01	6.656e-02	1.540e-04	4.511e-09	5.523e-18	9.413e-32	3.354e-44
μ	2	2.547e+02	1.278e+02	2.825e+02	1.749e+02	1.092e+02	2.520e+02	6.329e+01
	3	1.365e+02	1.828e+01	1.022e-01	2.737e-02	2.795e-15	9.746e-29	1.497e+00
	5	3.992e+01	6.447e+00	2.989e-02	8.823e-07	1.024e-15	3.897e-29	1.818e-42
	6	3.722e+01	2.803e+00	1.952e-02	6.155e-07	5.515e-16	2.266e-29	1.970e-42
	10	2.016e+01	1.407e+00	8.434e-03	2.286e-07	2.603e-16	1.368e-29	6.568e-43
median	2	1.418e+02	2.422e+01	1.453e-01	5.565e-06	6.259e-15	2.717e-28	1.008e-41
	3	1.013e+02	8.034e+00	5.338e-02	1.776e-06	1.788e-15	6.972e-29	3.136e-42
	5	2.293e+01	4.860e+00	1.759e-02	5.786e-07	6.913e-16	2.592e-29	1.004e-42
	6	2.011e+01	1.835e+00	1.317e-02	3.177e-07	3.509e-16	1.873e-29	1.292e-42
	10	1.311e+01	1.134e+00	5.737e-03	1.661e-07	1.931e-16	1.184e-29	6.032e-43
σ	2	2.560e+02	4.840e+02	1.233e+03	9.250e+02	4.479e+02	9.376e+02	2.571e+02
	3	1.200e+02	3.704e+01	1.504e-01	1.936e-01	3.283e-15	9.930e-29	1.058e+01
	5	3.680e+01	6.471e+00	3.256e-02	1.098e-06	1.044e-15	4.215e-29	1.751e-42
	6	5.325e+01	2.524e+00	2.309e-02	6.628e-07	7.462e-16	2.056e-29	2.236e-42
	10	1.928e+01	1.172e+00	8.941e-03	2.691e-07	2.580e-16	1.230e-29	4.612e-43

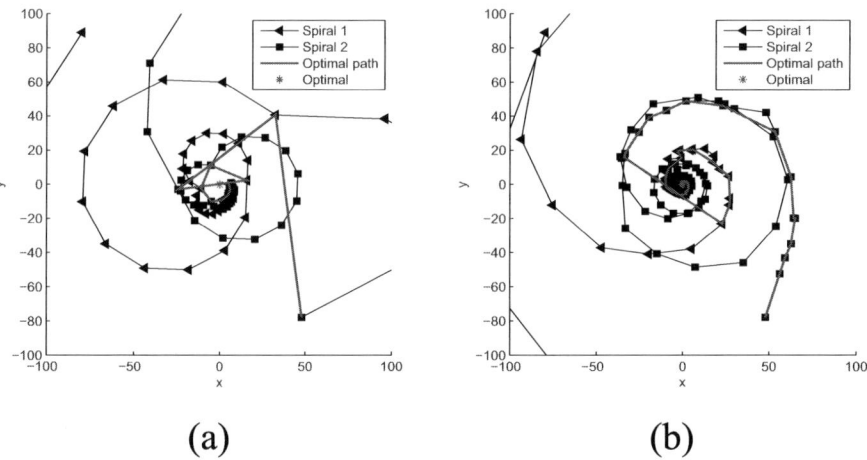

FIGURE 6.12: Spiral and optimal paths for sphere function evaluation with $s = 2$, $S_2(0.95, \pi/6)$, for two different random variables: (a) $\varphi_1 = U(0,1)$ and (b) $\varphi_2 = N(0.5, 0.25)$, after 100 iterations

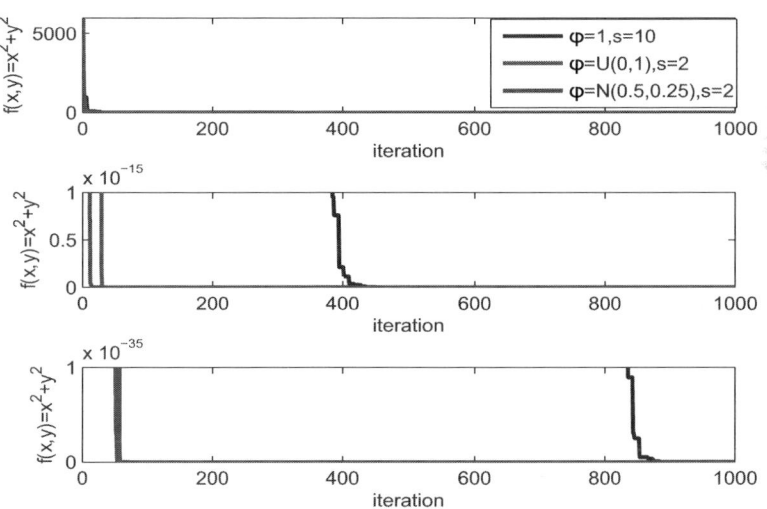

FIGURE 6.13: Sphere function evaluation after one run with an SO model with $S(0.95, \pi/6)$, for $\varphi = \{1, U(0,1), N(0.5, 0.2)\}$ with $s = \{10, 2, 2\}$, respectively, after 1000 iterations

sense, this model can be considered a general case of the spiral model.

$$X(k+1) = S_n(r,\theta)\varphi X - (S_n(r,\theta) - I_n)\varphi X^* \qquad (6.10)$$

For the evaluation of the sphere function, two distributions for φ were considered: a uniform distribution between $[0, 1]$ ($\varphi_1 = U(0, 1)$), and a normal distribution with mean 0.5 and standard deviation 0.25 ($\varphi_2 = N(0.5, 0.25)$); both cases keep the variation of X^* at about $(0, X^*)$. Figure 6.12 shows the trajectories described by two search points with $S_2(0.95, \pi/6)$ and its optimal path, using the modified model with both distributions φ_1 and φ_2. Contrary to the trajectories show in Figure 6.7, these trajectories shown a shorter diversification stage, reaching the intensification stage by the 100-th iteration.

Using the φ-model, by setting $s = 2$, comparable results can be found to those obtained by applying the traditional model with $s = 10$. Figure 6.13 shows the $f_{sphere}^{(2)}(X^*)$ evaluation for the sphere function at different iteration numbers using the φ-model with $S(0.95, \pi/4)$; it can be appreciated that the use of two search points (two spirals) generates competitive results from those obtained through the classical model with ten spirals, getting a $f_{sphere}^{(2)}(X^*) < 1 \times 10^{-15}$ value by its 50-th iteration, while the same result is achieved around 400 iterations later by the classical model using 10 search points.

Table 6.4 show the results for the SO search in the evaluation of the 2-dimensional sphere function, using the φ-model with $S(0.95, \pi/4)$ after 50 runs, for different values of k_{max} and s; data show that using one single spiral gives results on the order of 10^{-24} for the maximum value of $f_{sphere}^{(2)}(X^*)$, reached using $k_{max} = 50$ for both random variables φ_1 and φ_2, and showing the good performance of the φ-model in the evaluation of the sphere function.

6.2.4 Gabor Filters

Gabor filters have been widely used to characterize and extract texture features [8, 16]; they have also showed good performance for medical image processing applications, such as: microcalcification, edge, and tumor detection in mammographies [2, 5], and segmentation of coronary arteries [9], just to mention some of them. These filters are obtained through a sine function modulated by a Gaussian envelope; they represents the best arrangement between the spatial and frequency location as they are measured by the product between its spatial extension and frequency bandwidth [14]. An attractive property of these elements is that they fulfill the inferior limit of the uncertainty principle for its 1-dimensional and 2-dimensional approaches, given by Equation 6.11 and Equation 6.12, respectively, where Δt and Δf are the time and frequency resolutions for the 1-dimensional approach, and the pairs $(\Delta x, \Delta y)$ and $(\Delta u, \Delta v)$ represent the spatial and frequency resolutions for the 2-dimensional approach.

$$\Delta t \Delta f \geq \frac{1}{4\pi} \qquad (6.11)$$

TABLE 6.4: Sphere function evaluation for different k_{max} and s values using the ϕ-model with $S(0.95, \pi/4)$, for the two random variables $\varphi_1 = U(0,1)$ and $\varphi_2 = N(0.5, 0.25)$, after 50 runs.

$2*S\#$	$2*\phi$	$1*f(x,y)$	k_{max} 30	k_{max} 50	100	200	k_{max} 400	700
φ_1	1	max	1.774e-12	1.053e-27	1.214e-69	2.927e-149	8.154e-309	0.000e+00
		min	5.760e-38	1.001e-50	7.018e-100	1.930e-203	0.000e+00	0.000e+00
		μ	3.549e-14	2.149e-29	3.666e-71	9.504e-151	1.631e-310	0.000e+00
		median	3.002e-22	5.495e-40	2.470e-83	1.511e-164	0.000e+00	0.000e+00
		σ	2.509e-13	1.489e-28	1.870e-70	4.833e-150	0.000e+00	0.000e+00
φ_1	2	max	2.655e-15	5.008e-25	7.438e-68	3.757e-146	7.016e-322	0.000e+00
		min	2.097e-33	1.060e-59	4.773e-104	2.681e-209	0.000e+00	0.000e+00
		μ	5.828e-17	1.002e-26	2.098e-69	7.687e-148	1.482e-323	0.000e+00
		median	1.191e-23	1.004e-41	1.853e-84	5.864e-172	0.000e+00	0.000e+00
		σ	3.754e-16	7.083e-26	1.111e-68	5.311e-147	0.000e+00	0.000e+00
φ_1	3	max	4.526e-17	6.458e-31	1.924e-72	1.683e-148	0.000e+00	0.000e+00
		min	1.370e-31	2.464e-61	3.510e-105	1.689e-214	0.000e+00	0.000e+00
		μ	2.036e-18	1.471e-32	6.373e-74	3.387e-150	0.000e+00	0.000e+00
		median	4.699e-24	1.362e-41	4.147e-89	1.071e-178	0.000e+00	0.000e+00
		σ	8.445e-18	9.176e-32	3.224e-73	2.380e-149	0.000e+00	0.000e+00
φ_2	1	max	4.424e-11	7.754e-24	2.931e-56	3.817e-126	2.917e-250	0.000e+00
		min	5.148e-24	9.897e-42	4.183e-80	3.212e-155	4.364e-295	0.000e+00
		μ	1.009e-12	1.864e-25	1.145e-57	7.678e-128	5.844e-252	0.000e+00
		median	7.377e-17	1.794e-31	1.747e-66	4.381e-136	2.572e-276	0.000e+00
		σ	6.257e-12	1.114e-24	5.109e-57	5.398e-127	0.000e+00	0.000e+00
φ_2	2	max	1.711e-13	1.682e-24	7.246e-64	4.265e-130	1.607e-268	0.000e+00
		min	1.287e-24	8.786e-39	4.377e-86	9.293e-156	6.823e-311	0.000e+00
		μ	4.072e-15	3.367e-26	2.037e-65	8.531e-132	3.217e-270	0.000e+00
		median	1.070e-20	5.003e-33	4.926e-70	1.830e-145	1.009e-289	0.000e+00
		σ	2.417e-14	2.379e-25	1.045e-64	6.032e-131	0.000e+00	0.000e+00
φ_2	3	max	9.216e-14	2.833e-26	4.160e-60	6.620e-130	2.275e-274	0.000e+00
		min	9.648e-27	8.791e-42	8.941e-87	1.253e-157	2.866e-322	0.000e+00
		μ	1.952e-15	5.919e-28	8.352e-62	1.721e-131	4.550e-276	0.000e+00
		median	3.968e-19	1.320e-33	1.582e-71	5.033e-144	3.839e-293	0.000e+00
		σ	1.302e-14	4.075e-27	5.882e-61	9.586e-131	0.000e+00	0.000e+00

$$\Delta x \Delta y \Delta u \Delta v \geq \frac{1}{16\pi^2} \quad (6.12)$$

Due to its localized yet ondulatory behavior, they have been recognized as a suitable model for the receptive fields of simple cells in the visual cortex of certain mammals. Since certain parameters will cause a similar behavior in the spatial profile of the simple cells, where several interleaved regions of excitatory and inhibitory influences are modulated by an envelope, the image analysis by Gabor filtering is made to simulate human visual system perception [10, 30].

Equation 6.13 defines the complex kernel for Gabor filters oriented at $\theta = -\pi/2$, where σ_x and σ_y are the x-axis and y-axis standard deviations for the Gaussian envelope, respectively, u and v are the central frequencies of the filter in the frequency plane, and ϕ is the filter phase.

$$G^c(x,y) = \frac{1}{2\pi\sigma_x\sigma_x} \exp\left(-\frac{1}{2}\left(\frac{x^2}{\sigma_x^2} + \frac{y^2}{\sigma_y^2}\right)\right) \exp\left(j2\pi(ux+vy)+\phi\right) \quad (6.13)$$

The frequencies of the Gabor kernel in Cartesian coordinates (u,v) can be converted to polar coordinates (f,ω) through Equation 6.14, where f and ω represent the magnitude and direction.

$$\begin{aligned} u &= f\cos(\omega) \\ v &= f\sin(\omega) \end{aligned} \quad (6.14)$$

Since it is possible to fix one of these frequencies and preserve all the Gabor properties, this work considers only the first term of the second exponential, by setting $v = 0$, $\phi = 0$, and applying the polar coordinates conversion $u = f\cos(\omega = 0)$, reducing the kernel expression to Equation 6.15, where f is measured in cycles/pixel.

$$G^c(x,y) = \frac{1}{2\pi\sigma_x\sigma_y} \exp\left(-\frac{1}{2}\left(\frac{x^2}{\sigma_x^2} + \frac{y^2}{\sigma_y^2}\right)\right) \exp\left(j2\pi f x\right) \quad (6.15)$$

Using the Euler theorem, the expression in Equation 6.15 can be divided into its real and imaginary parts, generating Equation 6.16 and Equation 6.17.

$$G^r(x,y) = \frac{1}{2\pi\sigma_x\sigma_y} \exp\left(-\frac{1}{2}\left(\frac{x^2}{\sigma_x^2} + \frac{y^2}{\sigma_y^2}\right)\right) \cos(2\pi f x) \quad (6.16)$$

$$G^i(x,y) = \frac{1}{2\pi\sigma_x\sigma_y} \exp\left(-\frac{1}{2}\left(\frac{x^2}{\sigma_x^2} + \frac{y^2}{\sigma_y^2}\right)\right) \sin(2\pi f x) \quad (6.17)$$

With these kernel functions, the design parameters described by Rangayyan in [25] can be used in order to create the filter. Such parameters assure the correct operativity of the Gabor filter, since they include the constraints for the sinusoid plane and the Gaussian envelope; such constraints are described below:

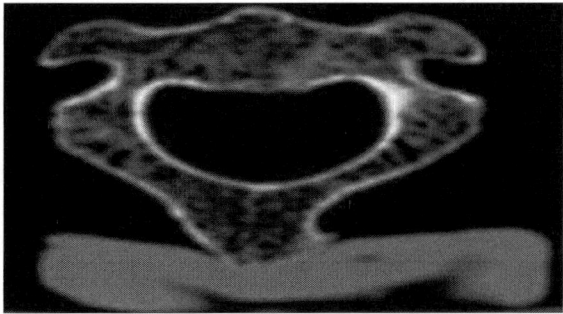

FIGURE 6.14: Details of a cervical vertebrae from a CT image [1]

- Let τ be the full-width at half-maximum of the Gaussian term in Equation 6.16 along the x-axis. Then, $\sigma_x = \tau/(2\sqrt{2ln2}) = \tau/2.35$.

- The cosine term is designed to have a period of τ; therefore $f = 1/\tau$.

- The value of σ_y is defined as $\sigma_y = \lambda \sigma_x$, where λ, named the aspect ratio, determines the elongation of the Gabor filter in the y-direction, as compared to the extension of the filter in the x-direction.

The previous equations work on single oriented features; however, a rotation of either the real or imaginary kernel coordinates (x, y) to $(\hat{x}_\alpha, \hat{y}_\alpha)$ by α radians can be calculated using the $M^{(2)}(\alpha)$ matrix described in Equation 6.4, with $\alpha \in [-\pi/2, \pi/2)$.

6.2.5 Optimal Parameter Selection Using Gabor Filter

In this section a small review on the effects of the different configuration parameters for the Gabor filter are described. All the gabor response images result from the filtering of the DICOM image in Figure 6.14 with a given filter kernel; the image represents a section of a cervical vertebrae from a CT image.

The parameter ρ: Parameter ρ defines the extension of the filter window by setting its coordinates $(x, y) \in [-\rho, \rho]$ with $(x, y, \rho) \in \mathbb{Z}$, so that, for any ρ value, a squared $2\rho + 1 \times 2\rho + 1$ filter window will be created. The ρ value should be big enough to cover all the effective coefficients but at the same time remain as small as possible, to avoid an overpopulation of near-to-zero coefficients. Figure 6.15 shows an example of the effect of different window sizes (given by $\rho_1 = 10$, $\rho_2 = 30$, and $\rho_3 = 60$) over the same Gabor filter for a certain combination of τ, λ, and α. It can be observed that for ρ_3 a close result to that with ρ_2 is obtained, but in a bigger amount of time due to the near-to-zero coefficients around the filter effective area. No recommended values for ρ are provided, since they are related to the τ and λ values; however, with $\rho > 1.2740\lambda\tau$, most of the effective coefficients are covered.

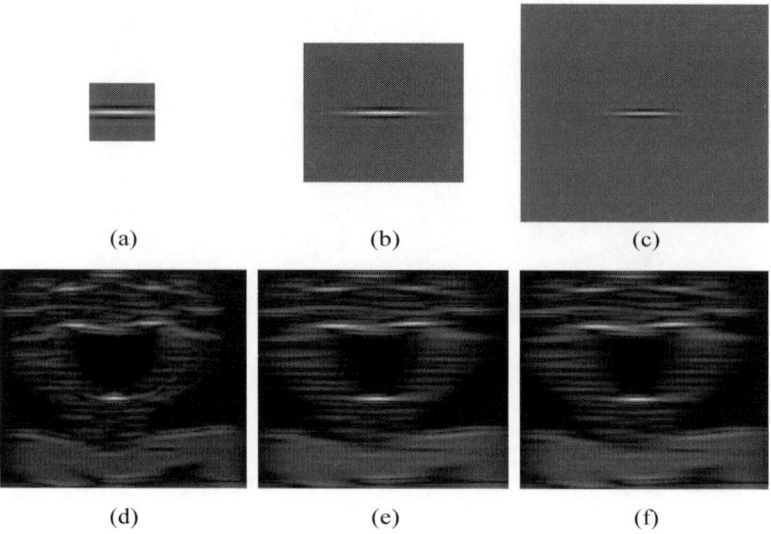

FIGURE 6.15: Real part of a Gabor filter with $\tau = 4$, $\lambda = 6$, and $\alpha = 0$, for different window size: (a) 21×21, 61×61, and (c) 121×121, and their respective effects over a medical image (d–f)

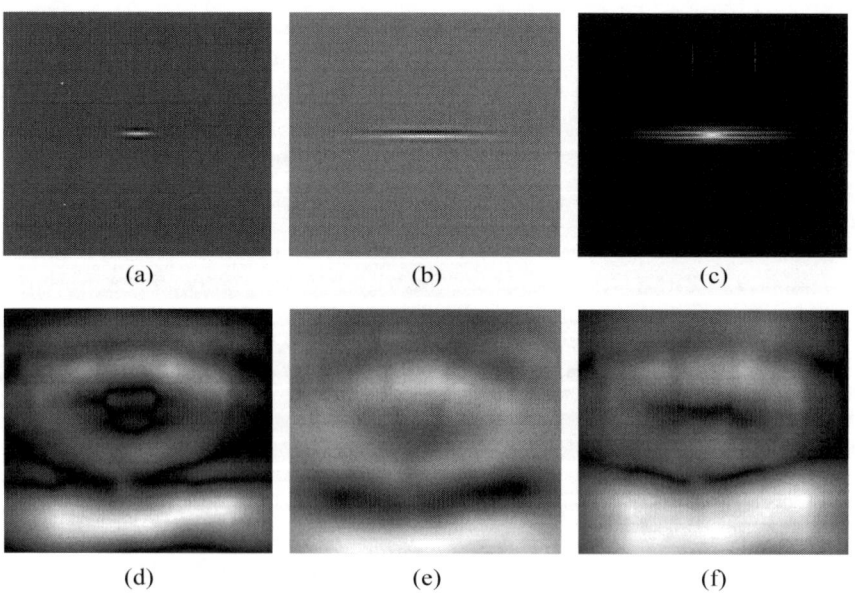

FIGURE 6.16: Magnitude of the (a) real, (b) imaginary, and (c) complex kernel for a Gabor filter, with $\tau = 50$, $\lambda = 1.5$, $\alpha = 0$, and $\rho = 20$, and their respective effects over a medical image (d–f)

The real and complex kernel: Using either the complete or the real part only from the Gabor kernel is a common practice for texture discrimination; however, for medical image processing, the real kernel is the most common option since its response possesses sharper edges than the results of the complex kernel for a given configuration; besides, due to the lack of imaginary parts, the number of operations required is reduced. Figure 6.16 shows the 2-dimensional representation of the magnitude for the real, imaginary, and complex kernel for a Gabor filter with $\tau = 50$, $\lambda = 1.5$, $\alpha = 0$, and $\rho = 20$, and the result of the filtering process over the image from Figure 6.14; it can be observed that the use of the real kernel generates a sharper structure for the vertebrae body than the use of the imaginary or the complex kernel.

The aspect ratio λ: The aspect ratio, or λ, defines the ellipticity of the Gaussian factor; from the setting of this variable, more or less points along the α direction will be considered during the filtering process. When $\lambda = 1$, a circular filter is defined, i.e., the filter extension at any axis will be the same. Figure 6.17 shows three filters for a given τ, α, and ρ combination, and aspect ratios $\lambda_1 = 1$, $\lambda_2 = 4$, and $\lambda_3 = 6$; it can be observed that the extension along the x-axis is preserved for any λ_i value, while the elongation at the y-axis will become larger for bigger values of λ. The effect on each image is appreciated as the detection of larger structures oriented at α for big values of λ. Typical values of the receptive fields of simple cells lie between 0.2 and 1; however, a greater value is commonly used.

The sinusoid period τ: This parameter specifies the wavelength of the sinusoid factor of the Gabor function, and it is given in cycles/pixels. This parameter is related to the σ_x and σ_y values of the Gaussian factor; and therefore, its value will modify the effective size of the filter, i.e., the non-near-to-zero coefficients from both the x and the y-axis, so, a small value will create a filter with more near-to-zero coefficients, while an upper value generates a larger effective area. Figure 6.18 shows three different filters created through the combination of given λ, α, and ρ values, and three different sinusoid periods, $\tau_1 = 2$, $\tau_2 = 4$, and $\tau_3 = 8$; while the shape of each filter is preserved, a smaller effective size is obtained with a small value of τ; similar to λ behavior, a greater value will detect larger non-local features. Commonly used values for τ lie between $[2, 256]$.

The rotation angle α: The parameter α, given in radians or degrees, defines the orientation of the filter's main axis, i.e., the one with the longer extension, with respect to the x-axis; in this way, a filter oriented at a certain angle α_i will enhance all the features aligned at that direction. Regardless of λ's value, the shape of the filter along the x and y-axes is preserved. Any filter can be rotated between $[-\pi/2, \pi/2)$, for the approach used. Figure 6.19 shows a filter with $\tau = 3$, $\lambda = 8$, and $\rho = 30$, oriented at directions $\alpha_1 = -\pi/4$, $\alpha_2 = 0$, and $\alpha_3 = \pi/4$ radians, and the effect of its application over the image from Figure 6.14; the bright structures detected correspond to the features oriented at each α_i direction.

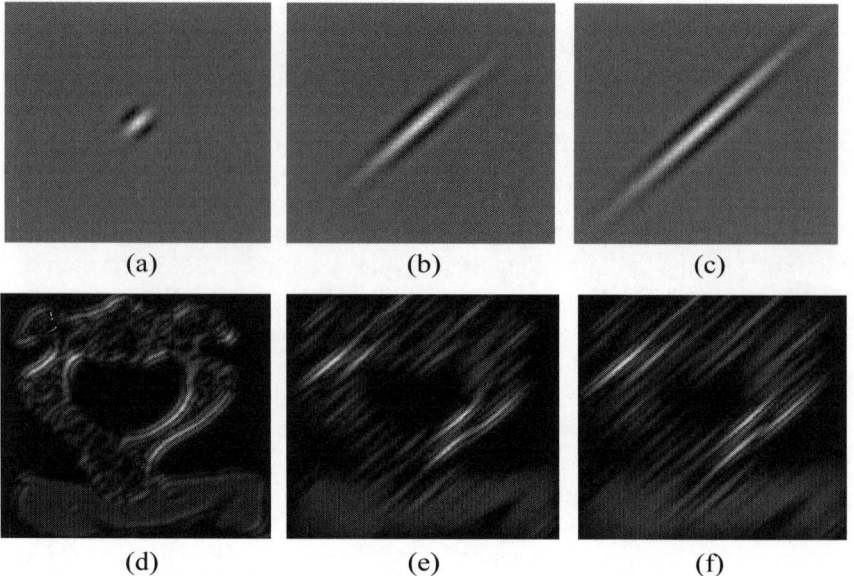

FIGURE 6.17: Real part of a Gabor filter with $\tau = 50$, $\alpha = \pi/4$, and $\rho = 25$ for three different aspect ratios: (a) $\lambda_1 = 1$, (b) $\lambda_2 = 4$, and (c) $\lambda_3 = 6$, and their respective effects over a medical image (d–e)

6.2.6 Filter Bank Definition

A Gabor filter bank (GFB) is formed by the set of N Gabor filters oriented at different angles α_m for $m = 1, \ldots, N$. Given a certain combination of τ, λ, and α, any element from the Gabor filter bank $G_m(x, y|\tau, \lambda, \alpha_m, \rho)$ is calculated from Equation 6.18, where $G^{type}(x, y|\tau, \lambda, \alpha_m, \rho)$ is either the rotated version of the real or complex Gabor kernel for the given τ, λ, α_m, and ρ values. For any filter window, the central value corresponds to the $(0, 0)$ coordinates.

$$G_m(x, y|\tau, \lambda, \alpha_m, \rho) = G^{type}(x, y|\tau, \lambda, \alpha_m, \rho)$$

$$\forall x, y = -\rho, \ldots, 0, \cdots, \rho; \forall m = 1, \cdots, N$$

(6.18)

Each element of the filter bank is then applied over the input image and its magnitude is calculated to form the m-Gabor response image $\|G_m(x, y|\tau, \lambda, \alpha_m, \rho)\|$; each of these responses are then added and a Gaussian filtering with mean μ_g and standard deviation σ_g is applied over the result to produce the final Gabor filter response image defined by Equation 6.19, where $*$ means a convolution, and $N_f(\mu_g, \sigma_g)$ is the Gaussian post-filter kernel. A study on the effect of the Gaussian post-filter can be found

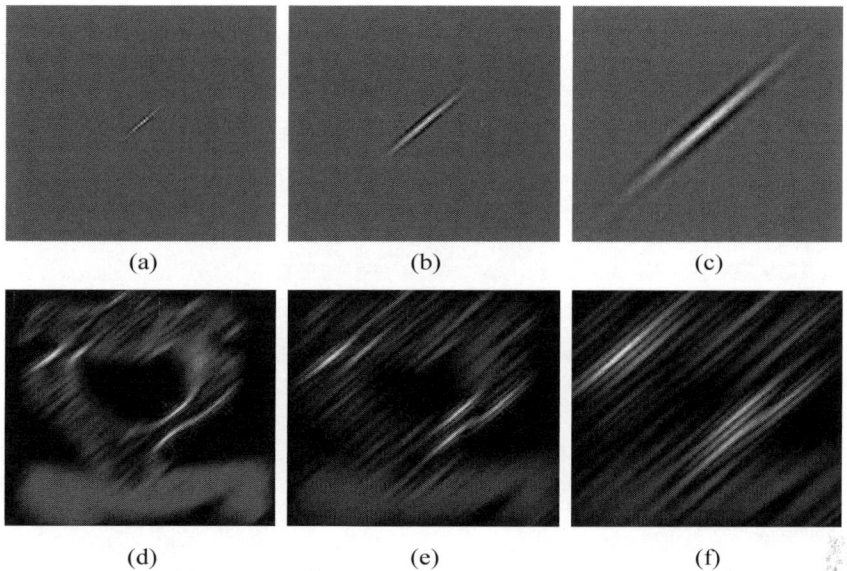

FIGURE 6.18: Real part of a Gabor filter with $\lambda = 6$, $\alpha = \pi/4$, and $\rho = 50$, for three different sine periods: (a) $\tau_1 = 2$, (b) $\tau_2 = 4$, and (c) $\tau_3 = 8$, and their respective effects over a medical image (d–f)

in [37].

$$F(x,y) = N_f(\mu_g, \sigma_g) * \sum_{m=1}^{N} ||G_m(x,y|\tau, \lambda, \alpha_m, \rho)|| \qquad (6.19)$$

6.3 The SO Initialized Gaussian Threshold Algorithm

The flow chart of the proposed method, SO initialized Gaussian threshold (SOIGT), is depicted in Figure 6.20. First the input CT image is pre-processed in order to delete unnecessary data; then, a 30-element Gabor filter bank is applied over the corrected image, and the Gabor response image $F(x,y)$ is obtained. Afterwards, four points are automatically selected by spiral optimization over the Gabor response image, and a seed-based region growing (SBRG) process, with these four points as input, is carried out over the Gabor response image. Next, two classes are built based on the SBRG result: class p for the bone pixels and class n for non-bone pixels. Then, from these classes, the mean and standard deviation pairs, (μ_p, σ_p) and (μ_n, σ_n), are iteratively calculated and used to generate the Gaussian models $N_p(\mu_p, \sigma_p)$ and

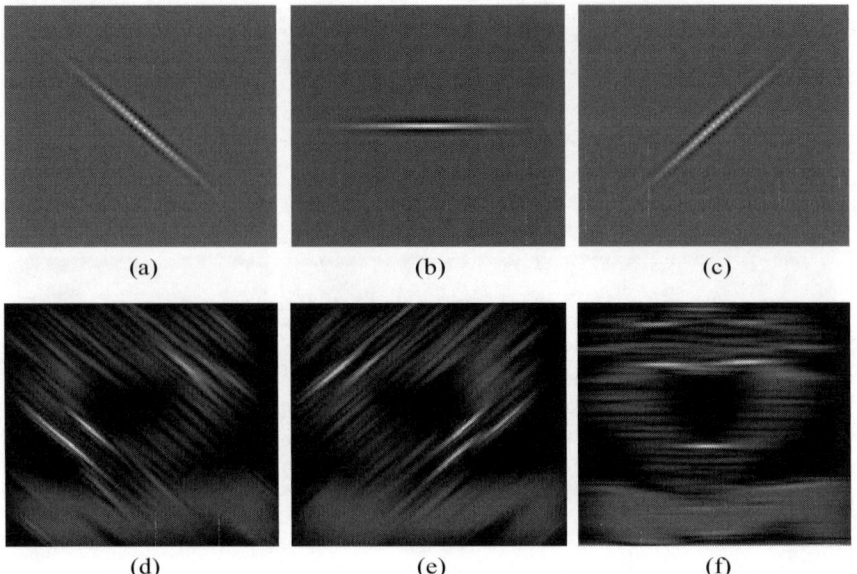

FIGURE 6.19: Real part of a Gabor filter with $\tau = 3$, $\lambda = 8$, and $\rho = 30$, for three different rotation angles: (a) $\alpha_1 = -\pi/4$, (b) $\alpha_2 = 0$, and (c) $\alpha_3 = \pi/4$, and its effects over a medical image (e-f)

$N_n(\mu_n, \sigma_n)$, and a threshold U, until μ_p stabilizes; finally, a post-processing, based on image erosion, *in-filling*, and border correction are executed prior to the method's end.

Although the bone classification is based on two Gaussian models, the SOIGT differs from those defined in [41, 42] in at least one of the following characteristics: the selection of the initial region, the considerations for the choice of the regions that generate the model statistics, or the lack of any probability rule to define the pixel classification, i.e., Bayes' theorem or probability of membership. In the following sections, a more detailed explanation for each part of the process is given.

6.3.1 Image Preprocessing

Prior to any process, the input image may be masked by DR and PM matrices as shown in Figure 6.21; the first one avoids the use of the blank space regions on the image corners, since it is not part of the actual CT image data, while the second one can be used to remove a fixed element present over the entire image set, e.g., the patient table, and therefore can change its value depending

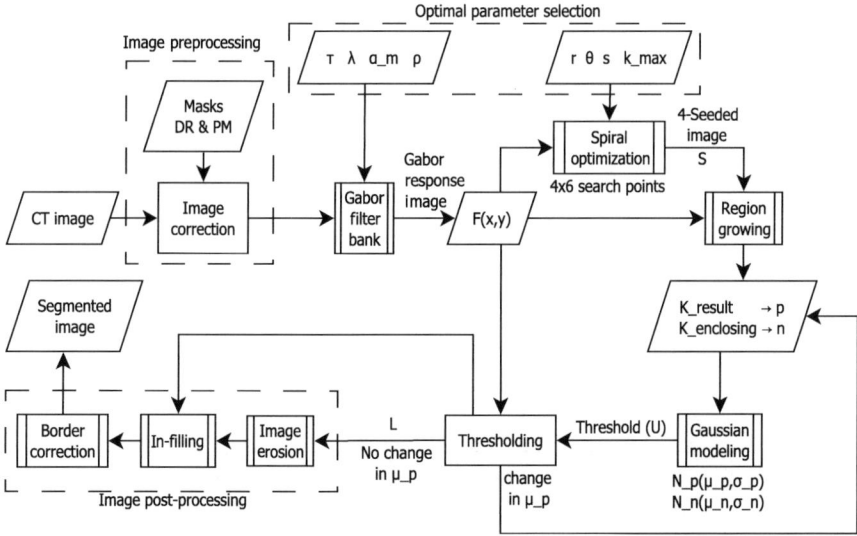

FIGURE 6.20: The proposed method: A threshold U is iteratively set, based on the intersection of two Gaussian models generated from an SO-seeded grown region over the Gabor filter response of an input image, until the models converge

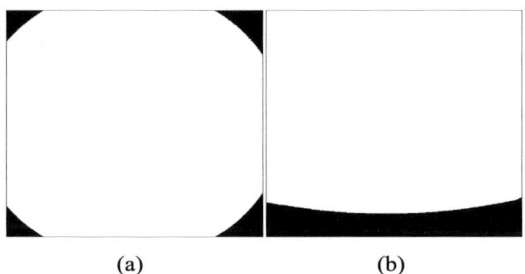

FIGURE 6.21: Example of the masks used on image preprocessing: (a) DR mask, and (b) PM mask

on the object that should be removed. If a non-fixed element is present, the PM-masking can be omitted.

6.3.2 The Gabor Filter Bank

This stage enhances and removes certain oriented features within the image, so that the changes between one texture and another, e.g., the skip from one bone region to soft tissue regions, becomes computationally more noticeable. The kernel function used for the definition of each GFB element is the real part

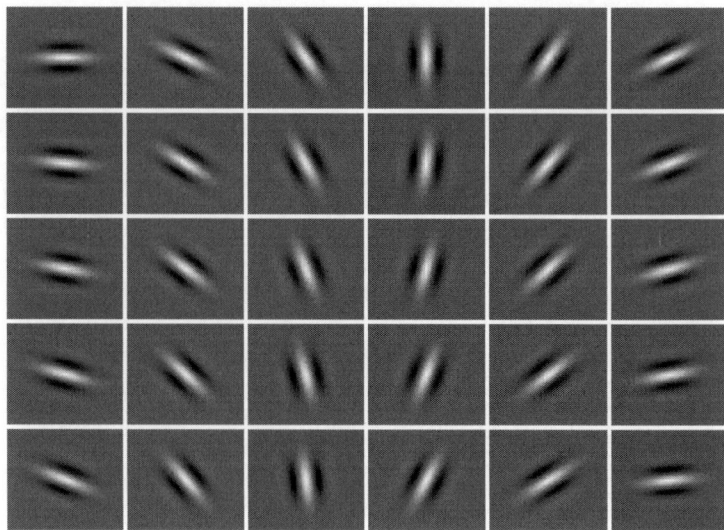

FIGURE 6.22: Real Gabor filter kernels used, in the GFB defined with $\tau = 8$ and $\lambda = 1.5$, and oriented at different angles, $\alpha_m \in (-\pi/2, \pi/2)$, with steps of $\alpha = \pi/30$

of the Gabor kernel shown in Equation 6.16; all the settings for the GFB were experimentally determined. For the Gaussian post-filtering, a mean $\mu_g = 0$ and a standard deviation $\sigma_g = \sigma_x$ were used.

For this work a 30-element GFB was used. Each member is composed of a 33×33 Gabor filter with $\tau = 8$ and $\lambda = 1.5$ at orientation α_m for $\alpha_m = \{-\pi/2, \ldots, -\pi/30, 0, \pi/30, \ldots, 7\pi/15\}$. Figure 6.22 shows the Gabor filter bank kernels used in this work.

6.3.3 SO Seeds

The SO technique is used to select diverse points from image $F(x, y)$ that will be used later on in the process. The image exploration is performed by the running of multiple spiral sets over the image $F(x, y)$; nonetheless, the behavior of any set is independent from any other. A spiral set is formed by the union of several spirals (search points), where one carries the optimal value while the others try to move toward it.

The output from the SO algorithm for any spiral set is the coordinate pair (x, y); this result is obtained by the solution of Equation 6.20, where nx represent the maximum value of $F(x, y)$. For this work, four coordinate pairs (four spiral sets) are used; this number was selected in order to increase the diversity of the system solutions.

$$\bar{S}_i = \arg\min_{x,y} |nx - F(x,y)|; \quad i = 1, \cdots, 4$$
$$\text{sucht that} \quad (x,y) \in \mathbb{N} \quad (6.20)$$
$$(x,y) \mapsto F$$

The initial position for each spiral, regardless of the set number, was arbitrarily selected using a uniform distribution, while its movement depends on the spiral model from Equation 6.6 and the SO algorithm. However, since images exist on the discrete domain, the spiral behavior is subject to two constraints: the integer condition $(x, y) \in \mathbb{N}$, and the border restriction $(x, y) \mapsto F$. These constraints guarantee that for any pair (x, y) a $F(x, y)$ value exists.

The settings for the spiral configuration were experimentally determined based on the observations from the sphere function solution: A matrix $S(0.95, \pi/4)$ was used and the number of spirals was stated to $s = 6$ for each set. Meanwhile, for the value of k_{max}, experimental results showed that using the constraint conditions for the objective function, a value slightly above the lower limit for the recommended maximum iteration number ($k_{max} = 110$) was enough to cease the optimal value update in most cases.

In addition to the k_{max} stop condition, this work incorporates a *stable value* break action, so that if the optimal position has not changed over a certain amount of iterations ($k = 12$), the entire set concludes its operation, and the conclusion of one set does not affect the behavior of the remaining search processes.

6.3.4 Seed-Based Region Growing

SBRG stage generated a new region based on the similarity of the points around a given initial area; this area can be a single point or a set of points, and the process is defined as it follows.

Let Z be an input image, and K_k a binary image, with ones at each seed point. The SBRG algorithm will attach all the pixels that satisfy the rule given in Equation 6.21 to binary image K_{k+1} until there exist no more points to add. At any iteration, $\vee$ means the logical OR operation, C_k is a binary image with ones at the outer contour of the current K_k image, $E[Z(K_k)]$ is the average value of all the correspondent points in Z where $K_k = 1$, and T represents a closeness threshold value that decides whether or not one point should be attached to the current grown region. In this work, a value of $T = 0.04 \max(Z)$ is used, e.g., for an 8-bit image $T = 0.04 * 255 \approx 10$.

$$K_{k+1} = K_k \vee (|Z \times C_k - E[Z(K_k)]| < T) \quad (6.21)$$

The SBRG process implemented uses the coordinate pairs S_i, from the previous stage as the initial value to generate two outputs: the actual grown region, named the *k_result*, and the enclosure of the grown region, named the *k_enclosing*. An example of these outputs, for a given seed point, is shown

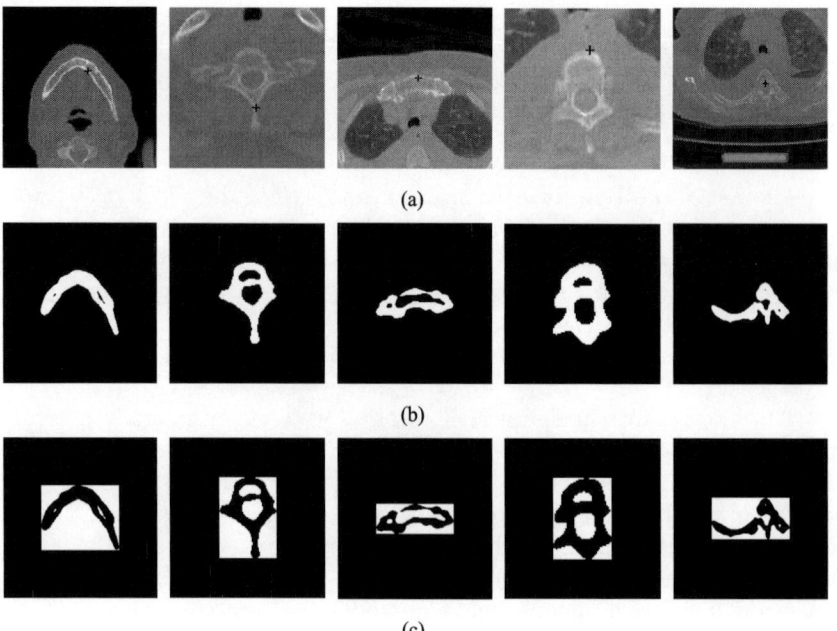

FIGURE 6.23: (a) Single-point seed for different images in row (b) and their respective SBRG outputs: *k_result* (c) and *k_enclosing*

in Figure 6.23, where the first output (Figure 6.23b), contains "1" at every position of the grown region and "0" at any other point, while the second output (Figure 6.23c), contains "1" at every position of the enclosure of the grown region, but the actual grown region and "0" are outside. The results for each S_i are then combined to be used in the following phase.

6.3.5 Gaussian Modeling

From the previous stage outputs, two classes are defined: class p, formed by all the "1" pixels in the SBRG result, and class n, constituted by all the "1"-valued pixels from the enclosing output. These two classes are then used to create two Gaussian models, $N_p(\mu_p, \sigma_p)$ and $N_n(\mu_n, \sigma_n)$, respectively, by using Equation 6.22, where μ_p and σ_p are the mean and standard deviation of class p, and μ_n and σ_n are the mean and standard deviation of class n. Class p is assumed to have a higher mean than class n, and therefore it contain the bone pixels, and will be considered the bone class for now on.

$$N_{cl}(\mu_{cl}, \sigma_{cl}) = \frac{1}{\sigma_{cl}\sqrt{2\pi}} \exp\left(-\frac{1}{2}\left(\frac{x - \mu_{cl}}{\sigma_{cl}}\right)^2\right) \quad \forall x \in \mathbb{N} \qquad (6.22)$$

Then, a threshold U is computed as the intersection point found between

Bone Tissue Segmentation Using SO and Gaussian Thresholding

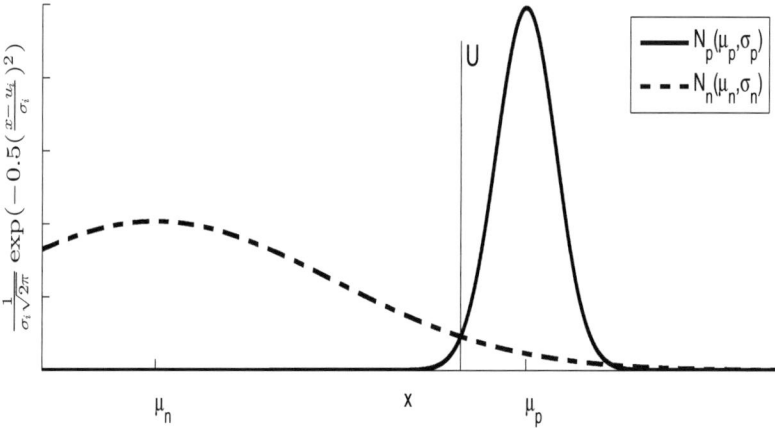

FIGURE 6.24: Selection of threshold U as the intersection of the Gaussian models $N_p(\mu_p, \sigma_p)$ and $N_n(\mu_n, \sigma_n)$

the means of the two models, as described in Figure 6.24; this point is estimated as the point x, which gives the minimum difference between the two models for $x \in [\mu_n, \mu_p]$. With threshold U, a threshold image L can be determined as in Equation 6.24.

$$U = \arg\min_x |N_p(\mu_p, \sigma_p) - N_n(\mu_n, \sigma_n)| \quad x \in [\mu_n, \mu_p] \tag{6.23}$$

$$L = F(x, y) \geq U \tag{6.24}$$

6.3.6 Image Post-Processing

After threshold application, it is possible that isolated non-bone structures are still present on the threshold image L. Besides this, due to the effect of the filtering process on the input image, it is probable that the borders of image L are distorted with respect to the untreated image; in order to reduce these effects, a post-processing stage is performed.

The post-processing consists of three operations: image erosion, *in-filling*, and border correction. On the erode process, an input image, like any in Figure 6.25b, is eroded using a disk-shaped structural element of radius 5. Examples of the erode process with the given structural element are shown in Figure 6.25c.

The second operation, *in-filling*, removes all the isolated structures while preserving the original shape from the input image. In this process, one binary image, named *source*, starts to grows until it reaches the bounds of another binary image, named *target*, that contains it; that is to say, only the sets from *source* and *target* that have at least one common point are preserved. Several examples of the result of an *in-filling* process are shown in Figure 6.25d, where

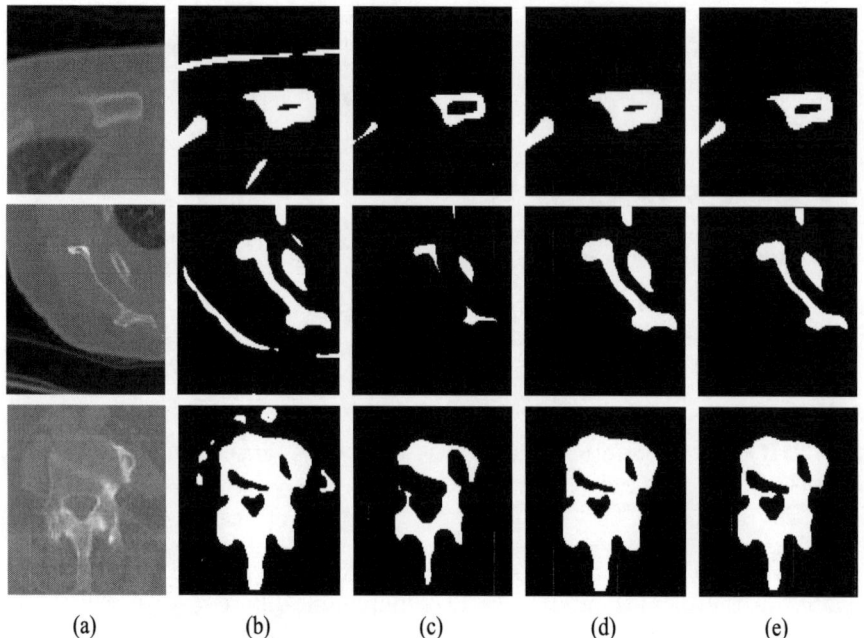

FIGURE 6.25: (a) Different stages of post-processing for images in column (b) result after threshold, (c) erode, (d) *in-filling*, and (e) border correction.

erode results from Figure 6.25c act as *source*, and the images from Figure 6.25b serve as *target*; contrary to a morphological opening, the *in-filling* preserves the shape of the original image, but does not separate connected structures.

The third part of the post-processing, border correction, eliminates the inner contour of the *in-filling* binary result, by a simple subtraction. The inner contour is obtained by a Laplacian filter and subtracted from the input image, creating an image like those shown in Figure 6.25e.

6.4 Computational Experiments

In this section, the results of the application of the proposed method in the segmentation of bone structures is shown. The effects produced by the lack of either the SO or the GFB components are described, and the importance of both components is stated.

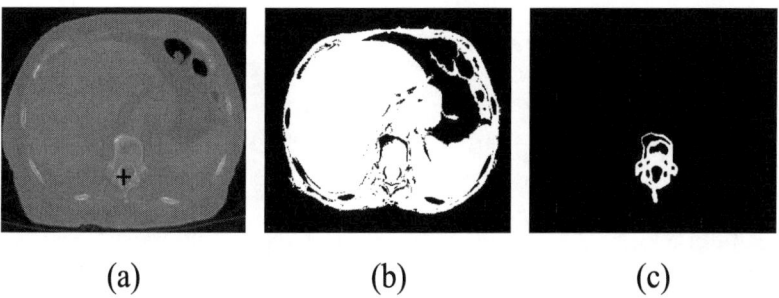

(a) (b) (c)

FIGURE 6.26: (a) Seed-based region growing results for the seed shown, (b) over an image without Gabor filtering, (c) and an image with Gabor filtering

6.4.1 Enhancement of CT Images

Gabor filters have previously been used to separate textures and enhance oriented features. The application of the GBF over the CT images increases the distance from one textured region to another, e.g., the change from the bone to soft tissue, a separation that may not be present on the original image. Figure 6.26 shows the effect of the GFB over a SBRG process, with the initial seed positioned at the vertebrae lamina, the lower part of the vertebrae, as shown in Figure 6.26a.

Since any grown region exhibits similar features for all the points attached to it, it can be used to show that a separation of textured regions is achieved through the application of the Gabor filter bank. Figure 6.26b shows that the use of the intensity only (untreated image) will lead to a improper selection of the vertebrae; however, after the application of the filter bank, the grown region is confined to the vertebrae structures, as shown in Figure 6.26c. So that, the GFB can be used to enhance the bone and non-bone structures by increasing the contrast between its boundaries.

These kinds of results will depend on the correct position of the seeds. In this work, the use of SO for the initial position solution shows a good performance, compared to the use of other initial conditions. Figure 6.27 depicts an example of the method using different alternatives for the initial position selection for the region growing. These alternatives include: random seeds (Figure 6.27b), all the points over 50% of the maximum value of the GFB (Figure 6.27c), all the points over 75% of the maximum value of the GFB (Figure 6.27d), the points with the maximum value of the GFB (Figure 6.27e), and finally, the points obtained using SO (Figure 6.27e); it seems that a initial region given by a threshold value, as well as the use of random seeds, do not lead to favorable results; on the other hand, the lack of diversity in the use of the points with the highest values as seed retains the region growth over a small area close to those points. It also can be observed that SO tends

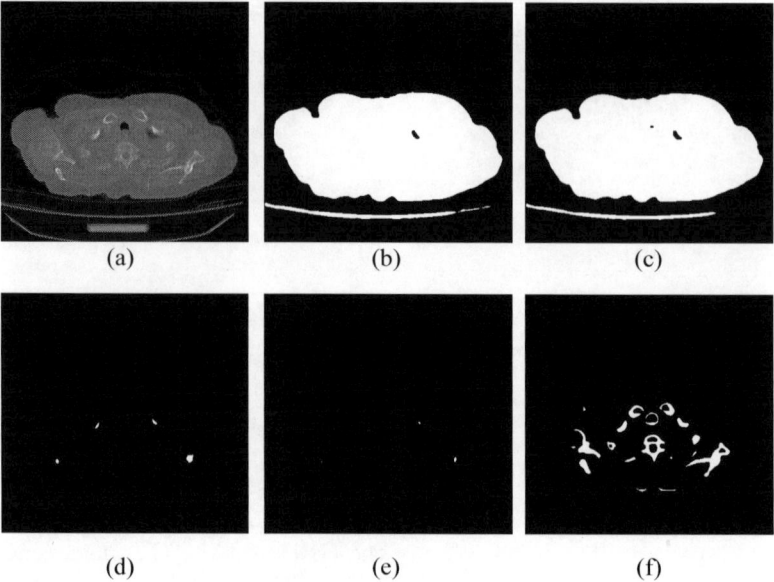

FIGURE 6.27: SOIGT results for image (a) considering: (b) random seeds, (c) seeds $= (x,y) > 0.5 \max(F(x,y))$, (d) seeds $= (x,y) > 0.75 \max(F(x,y))$, (e) seeds $= \max(F(x,y))$, and (f) seeds obtained by SO

to select an adequate initial region, thanks to the variety that gives the image exploration from the spiral process. In that sense, SO plays a crucial stage in the appropriate performance of the method.

6.4.2 Segmentation of Bone Tissue

The SOIGT algorithm is assessed through the Dice coefficient (D) and the accuracy (ACC) metrics; they represent a measure of the similarity between two sets A and B, whose values lie between $[0,1]$, 1 being the best matching. Equation 6.25 defines the Dice coefficient, while Equation 6.26 describes the accuracy. In these equations, $|\cdot|$ represents the variable cardinality, and $\bar{A}$ and $\bar{B}$ are the negated sets of A and B, respectively.

$$D(A,B) = \frac{2|A \cap B|}{|A| + |B|} \tag{6.25}$$

$$ACC(A,B) = \frac{|A \cap B| + |\bar{A} \cap \bar{B}|}{|A| + |\bar{A}|} \tag{6.26}$$

SOIGT is also compared with state-of-the-art threshold and clustering techniques, including the Kittler and Illingworth's minimum error thresholding method [18] ($K_{classes}$), Ridler and Calvard thresholding using an iterative selection [26] ($R_{classes}$), Otsu [33] ($O_{classes}$), K-means [38] ($KM_{classes}$), and fuzzy c-mean [31] ($FCM_{classes}$), for $classes = 2$ for Kittler, and Ridler and Calvard, and $classes \in [2, 7]$ for Otsu, K-means, and fuzzy c-mean.

Using an argument similar to that which forms the Gaussian models, it is assumed that the class with the highest mean, in multi-class approaches, includes the bone pixels, and therefore, such a class will be considered as class p for evaluation purposes. All the methods are evaluated using the Gabor response image $F(x, y)$; however, for the Kittler and Otsu methods, this image is converted to its 256-levels version, $F(x, y) \in [0, 255]$ previous to the method application. Table 6.5 contains the statistic analysis for the method evaluation over the image database for both the Dice and the accuracy metrics; table data include the maximum (max), minimum (min), average (μ), median (median), and standard deviation (σ) values for the proposed and the comparative methods after their application over the test database.

It can be observed that similar results for R_2, K_2, and O_2 are obtained; such behavior is due to the fact that there exists a relationship between these methods as is shown in [39], and therefore similar results will be achieved under certain circumstances. It is also appreciable that regardless of the comparative method, at least three classes are necessary in order to obtain competitive results, due mostly to the effect of the background pixels over each method performance; however, a larger number of classes may not be necessary, leading to an overclassification, and the reduction of the method effectiveness, e.g., $ACC_\mu(O_{>5}) < ACC_\mu(O_5)$.

By analyzing the Dice coefficient results in Table 6.5, it can be observed that the SOIGT algorithm outperforms the comparative methods in almost every category, only being surpassed at the standard deviation by K_2, R_2, O_2, $KM_{2,3}$, and $FCM_{2,3}$. However, by analyzing the correspondent average values for all these cases, it can be appreciated that its evaluation results are lower than the assessment obtained by the method used, and those with average values close to the proposed method imply higher standard deviations. In that sense, in order to be taken as a representative variable for any method performance, the standard deviation should also consider the method average value. Thereby, if the relation μ-σ is considered, the SOIGT observes a better performance than the comparative methods for all categories.

On the other hand, by using the accuracy metric, the method obtains better results for all the categories, outperforming all the comparison methods; however, O_5 shows a competitive performance, being slightly outperformed by the SOIGT. Despite this condition, the framework used still observes a better performance than O_5, as shown in Figure 6.28, where several examples for the bone pixel assigned for each method with the best performance, given by its D_μ and ACC_μ values, are shown; thereby, only results for K_2, R_2, O_5, KM_7, and FCM_7 are displayed. As expected, results for K_2 and R_2

TABLE 6.5: Comparative analysis of the proposed and the comparative methods, using statistical measures.

	D				
	max	min	μ	median	σ
K_2	0.1676	0.0479	0.1159	0.1133	**0.0322**
R_2	0.1963	0.0651	0.1422	0.1459	0.0338
O_2	0.1964	0.0651	0.1422	0.1457	0.0338
O_3	0.7967	0.0698	0.2720	0.1774	0.2342
O_4	0.8009	0.0012	0.5425	0.6657	0.2348
O_5	0.7992	0.0009	0.5331	0.6278	0.2746
O_6	0.7573	0.0005	0.2677	0.3135	0.2182
O_7	0.7215	0.0007	0.1046	0.0209	0.1618
KM_2	0.1963	0.0651	0.1422	0.1459	0.0338
KM_3	0.2574	0.0367	0.1538	0.1586	0.0427
KM_4	0.7758	0.0711	0.3048	0.1844	0.2285
KM_5	0.8028	0.0712	0.4265	0.3446	0.2667
KM_6	0.8026	0.0726	0.4835	0.6129	0.2482
KM_7	0.8025	0.0770	0.5492	0.6615	0.2109
FCM_2	0.1962	0.0651	0.1421	0.1455	0.0338
FCM_3	0.3227	0.0695	0.1661	0.1665	0.0513
FCM_4	0.8009	0.1370	0.3883	0.3478	0.1940
FCM_5	0.8025	0.1433	0.4596	0.3822	0.2032
FCM_6	0.8034	0.1654	0.5183	0.5938	0.1983
FCM_7	0.8038	0.1657	0.5723	0.6400	0.1903
SOIGT	**0.8229**	**0.4892**	**0.6969**	**0.7023**	0.0792

exhibit similar behavior for a two-class classifier; nonetheless, with the use of a higher number of classes and the highest mean condition, soft tissue pixels are removed, affecting the segmentation accuracy results.

Otsu results seems competitive, but for some images, only a small portion of the bone structures are classified as such; these results show that other considerations should be taken in account in order to use these methods for bone classification. Although its higher D_μ and ACC_μ evaluation, the results for KM_7 and FCM_7 seem to be less accurate that Otsu's outcome, grouping some soft tissue pixels with the bone pixels, and showing a class-dependency due to the large number of textures typically present on medical images.

6.5 Conclusions

This work presents a novel method for bone tissue segmentation in CT images based on the iterative search for a threshold value; such a process seeks the optimal separation of two classes modeled as Gaussian distributions given an ini-

TABLE 6.6: Comparative analysis of the proposed and the comparative methods, using statistical measures.

			ACC		
	max	min	μ	median	σ
K_2	0.6818	0.5417	0.6080	0.6100	0.0293
R_2	0.7825	0.6465	0.6919	0.6819	0.0358
O_2	0.7820	0.6464	0.6919	0.6819	0.0357
O_3	0.9919	0.6891	0.7609	0.7068	0.1154
O_4	0.9923	0.8315	0.9528	0.9769	0.0516
O_5	0.9921	0.9641	0.9825	0.9831	0.0066
O_6	0.9925	0.9641	0.9778	0.9751	0.0078
O_7	0.9911	0.9625	0.9757	0.9738	0.0077
KM_2	0.7825	0.6465	0.6919	0.6819	0.0358
KM_3	0.8210	0.0187	0.7089	0.7070	0.0975
KM_4	0.9907	0.6950	0.8074	0.7179	0.1159
KM_5	0.9925	0.6960	0.8557	0.9029	0.1319
KM_6	0.9925	0.7053	0.8952	0.9772	0.1182
KM_7	0.9925	0.7023	0.9403	0.9769	0.0791
FCM_2	0.7819	0.6479	0.6918	0.6818	0.0355
FCM_3	0.8602	0.6879	0.7370	0.7085	0.0540
FCM_4	0.9923	0.8465	0.8974	0.8812	0.0499
FCM_5	0.9924	0.8520	0.9206	0.8947	0.0532
FCM_6	0.9927	0.8583	0.9394	0.9773	0.0524
FCM_7	0.9928	0.8535	0.9558	0.9802	0.0486
SOIGT	**0.9941**	**0.9736**	**0.9859**	**0.9861**	**0.0046**

tial classification. The selection of this class is made through the combination of spiral optimization and a seed-based region growing process. The method was tested over a set of 350 images, and compared with several state-of-the-art thresholding methods, obtaining an average Dice coefficient $D = 0.6969$ and a mean accuracy $ACC_\mu(PM) = 0.9843$, outperforming the assessment obtained by the comparative methods.

The effectiveness of the method depends mostly on two aspects: the Gabor filter bank, and the initial region given by SO. The first one achieves the image enhancement and the texture class separation, while the second one establishes a suitable star point for the optimal threshold value searching, taking advantage of the separation achieved by the GFB. Although other alternatives for the initial class were considered (threshold, maximum values, and random points), SO+SBRG seems to be the best solution for the establishment of the initial classification. Other alternatives like threshold may be subject to the image's nature. The use of maximum values will trap the region over a small area around them, while random points may or may not lead to good results; meanwhile, SO will explore the image previous to the region's establishment,

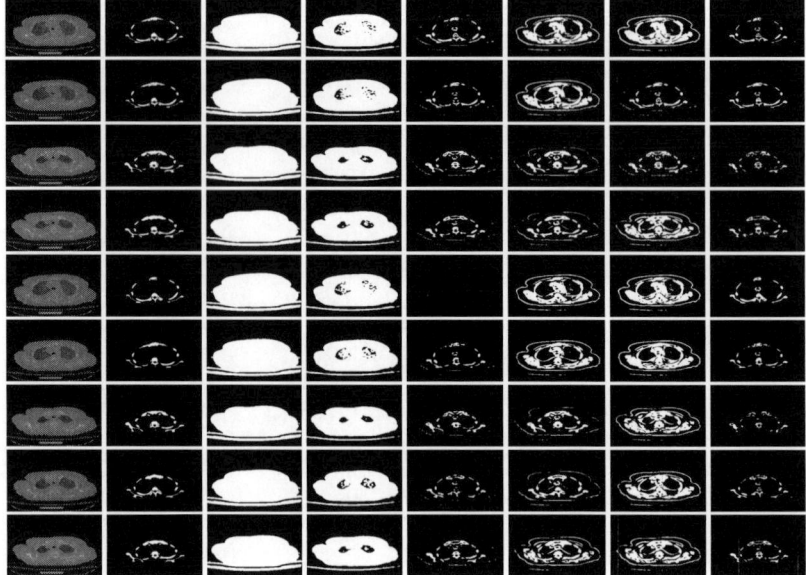

FIGURE 6.28: Bone pixels from image in column one. From column two to seven: gold standard, K_2, R_2, O_5 (best for O), KM_7 (best for KM), FCM_7 (best for FCM), and the $SOIGT$

increasing the feature diversity and keeping the region away from a narrow area of convergence.

Results for bone segmentation on CT images show that more than three classes are necessary in order to start separating the bone tissue pixels from the surrounding soft tissue elements, demonstrating that there exists a class dependency between the bone tissue segmentation in CT images and the state-of-the-art comparative methods, due mostly to the large number of textures usually found on CT images, which are a product of the soft tissues, organs, and other internal structures. The proposed method, nonetheless, takes advantage of the fact that the bone pixels usually possess the highest values, and starts the class separation on the right side of the histogram, avoiding the necessity of several classes by grouping all the pixels into only two classes, *Bone* (higher values) and *Non-Bone* elements (lower values).

The use of SO in image processing is a new approach for the SO algorithm until now, this technique has been used on power system planning, economic problem solution, design of digital filters, and other optimization problems not related to the image processing. Furthermore, its use in optimal initialization has not yet been explored, and its application for other problems, such as tumor segmentation, may be an interesting topic, since several current image processes are still user-interactive for initial conditions selection; in that sense,

SO can be exploited for this problem's solution. Meanwhile, the modification made to the spiral model by a random variable has shown an improvement on the solution for the sphere function, increasing the method's effectiveness and reducing the number of search points and iterations needed, compared to those required on the original model. Further investigation, however, should be done in order to settle the efficacy of this modification for other optimization problems.

Acknowledgments

This work has been supported by the National Council of Science and Technology of Mexico (CONACYT) under the Grant no. 473661/273068, and Project Catedras-CONACYT No. 3150-3097. The authors would like to thank Jorge Mario Cruz Duarte, Ph.D.(c), for his valuable help in the development of this work. They also would like to thank the Engineering Division (DICIS) of the University of Guanajuato (UG) and the Center for Research in Mathematics (CIMAT) for the facilities given for the realization of this research.

Bibliography

[1] Bone and joint ct-scan data. Laboratory of Human Anatomy and Embryology, University of Brussels (ULB), Belgium, http://isbweb.org/data/vsj/.

[2] Fábio J. Ayres, Rangaraj M. Rangayyan, and J.E. Leo Desautels. Analysis of oriented texture with applications to the detection of architectural distortion in mammograms. *Synthesis Lectures on Biomedical Engineering*, 5(1):1–162, 2010.

[3] Lahouaria Benasla, Abderrahim Belmadani, and Mostefa Rahli. Spiral optimization algorithm for solving combined economic and emission dispatch. *International Journal of Electrical Power & Energy Systems*, 62:163–174, 2014.

[4] Helen R. Buie, Graeme M. Campbell, R. Joshua Klinck, Joshua A. MacNeil, and Steven K. Boyd. Automatic segmentation of cortical and trabecular compartments based on a dual threshold technique for in vivo micro-ct bone analysis. *Bone*, 41(4):505–515, 2007.

[5] J.W. Byng, N.F. Boyd, E. Fishell, R.A. Jong, and M.J. Yaffe. Automated analysis of mammographic densities. *Physics in Medicine and Biology*, 41(5):909, 1996.

[6] Bernard Chalmond. Individual hip prosthesis design from ct images. *Pattern Recognition Letters*, 8(3):203–208, 1988.

[7] F. Chevalier, A.M. Laval-Jeantet, M. Laval-Jeantet, and C. Bergot. Ct image analysis of the vertebral trabecular network in vivo. *Calcified Tissue International*, 51(1):8–13, 1992.

[8] David A. Clausi and M. Ed Jernigan. Designing gabor filters for optimal texture separability. *Pattern Recognition*, 33(11):1835–1849, 2000.

[9] Ivan Cruz-Aceves, Faraz Oloumi, Rangaraj M. Rangayyan, Juan G. Aviña-Cervantes, and Arturo Hernandez-Aguirre. Automatic segmentation of coronary arteries using Gabor filters and thresholding based on multiobjective optimization. *Biomedical Signal Processing and Control*, 25:76–85, 2016.

[10] John G. Daugman. Uncertainty relation for resolution in space, spatial frequency, and orientation optimized by two-dimensional visual cortical filters. *JOSA A*, 2(7):1160–1169, 1985.

[11] D.H. Davies and D.R. Dance. Automatic computer detection of clustered calcifications in digital mammograms. *Physics in Medicine and Biology*, 35(8):1111, 1990.

[12] R. Eliashar, J.Y. Sichel, M. Gross, E. Hocwald, I. Dano, A. Biron, A. Ben-Yaacov, A. Goldfarb, and J. Elidan. Image guided navigation system - a new technology for complex endoscopic endonasal surgery. *Postgraduate Medical Journal*, 79(938):686–690, 2003.

[13] P. Furnstahl, T. Fuchs, Andreas Schweizer, Ladislav Nagy, Gábor Székely, and Matthias Harders. Automatic and robust forearm segmentation using graph cuts. In *Biomedical Imaging: From Nano to Macro, 2008. ISBI 2008. 5th IEEE International Symposium on*, pages 77–80. IEEE, 2008.

[14] Dennis Gabor. Theory of communication. Part 1: The analysis of information. *Journal of the Institution of Electrical Engineers-Part III: Radio and Communication Engineering*, 93(26):429–441, 1946.

[15] Wil G.M. Geraets, Paul F. van der Stelt, Coen J. Netelenbos, and Petra J.M. Elders. A new method for automatic recognition of the radiographic trabecular pattern. *Journal of Bone and Mineral Research*, 5(3):227–233, 1990.

[16] Anil K. Jain and Farshid Farrokhnia. Unsupervised texture segmentation using Gabor filters. *Pattern Recognition*, 24(12):1167–1186, 1991.

[17] Yan Kang, Klaus Engelke, and Willi A. Kalender. A new accurate and precise 3-d segmentation method for skeletal structures in volumetric ct data. *IEEE Transactions on Medical Imaging*, 22(5):586–598, 2003.

[18] Josef Kittler and John Illingworth. Minimum error thresholding. *Pattern Recognition*, 19(1):41–47, 1986.

[19] Mihails Kovalovs and Aleksandrs Glazs. Trabecular bone segmentation by using an adaptive contour. *Rigas Tehniskas Universitates Zinatniskie Raksti*, 14:6, 2013.

[20] A. Mundinger, B. Wiesmeier, E. Dinkel, A. Helwig, A. Beck, and J. Schulte Moenting. Quantitative image analysis of vertebral body architecture: Improved diagnosis in osteoporosis based on high-resolution computed tomography. *The British Journal of Radiology*, 66(783):209–213, 1993.

[21] NASA, and ESA, and the Hubble Heritage (STScI/AURA)-ESA/Hubble Collaboration. Spiral galaxy, 2015. http://www.nasa.gov/sites/default/files/images/ 611265main_hubble_holidaywreath_full.jpg.

[22] NOAA. Satellite image for larger view of hurricane Isabel northeast of Puerto Rico taken on Sept. 13, 2003, at 11:45 a.m. edt, 2003. http://www.noaanews.noaa.gov/stories/images/ isabel091303-1545zb2.jpg.

[23] Xose Manuel Pardo, María J. Carreira, A. Mosquera, and Diego Cabello. A snake for ct image segmentation integrating region and edge information. *Image and Vision Computing*, 19(7):461–475, 2001.

[24] W.C.G. Peh. Ct-guided percutaneous biopsy of spinal lesions. *Biomedical Imaging and Intervention Journal*, 2(3):e25, 2006.

[25] Rangaraj M. Rangayyan and Fábio J. Ayres. Gabor filters and phase portraits for the detection of architectural distortion in mammograms. *Medical and Biological Engineering and Computing*, 44(10):883–894, 2006.

[26] T.W. Ridler and S. Calvard. Picture thresholding using an iterative selection method. *IEEE Transactions on Systems, Man and Cybernetics*, 8(8):630–632, 1978.

[27] Hilmi Rifai, Isabelle Bloch, Seth Hutchinson, Joe Wiart, and Line Garnero. Segmentation of the skull in mri volumes using deformable model and taking the partial volume effect into account. *Medical Image Analysis*, 4(3):219–233, 2000.

[28] Douglas D. Robertson, Peter S. Walker, John W. Granholm, Philip C. Nelson, Peter J. Weiss, Elliot K. Fishman, and Donna Magid. Design of custom hip stem prostheses using three-dimensional ct modeling. *Journal of Computer Assisted Tomography*, 11(5):804–809, 1987.

[29] S/A. Aloe polyphylla, 2014. https://pixabay.com/en/aloe-succulent-aloe-polyphylla-510113/.

[30] Peter H. Schiller, Barbara L. Finlay, and Susan F. Volman. Quantitative studies of single-cell properties in monkey striate cortex. i. Spatiotemporal organization of receptive fields. *Journal of Neurophysiology*, 39(6):1288–1319, 1976.

[31] Shan Shen, William Sandham, Malcolm Granat, and Annette Sterr. MRI fuzzy segmentation of brain tissue using neighborhood attraction with neural-network optimization. *IEEE Transactions on Information Technology in Biomedicine*, 9(3):459–467, 2005.

[32] Wolfgang Sörgel, Sabine Girod, Martin Szummer, and Bernd Girod. Computer aided diagnosis of bone lesions in the facial skeleton. In *Bildverarbeitung für die Medizin 1998*, pages 179–183. Springer, 1998.

[33] Jung-Min Sung, Dae-Chul Kim, Bong-Yeol Choi, and Yeong-Ho Ha. Image thresholding using standard deviation. In *IS&T/SPIE Electronic Imaging*, pages 90240R–90240R. International Society for Optics and Photonics, 2014.

[34] Kenichi Tamura and Keiichiro Yasuda. Primary study of spiral dynamics inspired optimization. *IEEJ Transactions on Electrical and Electronic Engineering*, 6(S1):S98–S100, 2011.

[35] Kenichi Tamura and Keiichiro Yasuda. Spiral optimization-a new multipoint search method. In *IEEE International Conference on Systems, Man, and Cybernetics (SMC), 2011*, pages 1759–1764. IEEE, 2011.

[36] Chun-Wei Tsai, Bo-Chi Huang, and Ming-Chao Chiang. A novel spiral optimization for clustering. In *Mobile, Ubiquitous, and Intelligent Computing*, pages 621–628. Springer, 2014.

[37] Thomas P. Weldon, William E. Higgins, and Dennis F. Dunn. Efficient Gabor filter design for texture segmentation. *Pattern Recognition*, 29(12):2005–2015, 1996.

[38] Jinhua Xu and Hong Liu. Web user clustering analysis based on kmeans algorithm. In *2010 International Conference on Information, Networking and Automation (ICINA)*, 2010.

[39] Jing-Hao Xue and Yu-Jin Zhang. Ridler and Calvards, Kittler and Illingworths and Otsus methods for image thresholding. *Pattern Recognition Letters*, 33(6):793–797, 2012.

[40] Weiguang Yao, Purang Abolmaesumi, M. Greenspan, and Randy E. Ellis. An estimation/correction algorithm for detecting bone edges in ct images. *IEEE Transactions on Medical Imaging*, 24(8):997–1010, 2005.

[41] Jing Zhang, C.-H. Yan, C.-K. Chui, and S.-H. Ong. Fast segmentation of bone in ct images using 3d adaptive thresholding. *Computers in Biology and Medicine*, 40(2):231–236, 2010.

[42] Kai Zhao, Bin Kang, Yan Kang, and Hong Zhao. Auto-threshold bone segmentation based on ct image and its application on cta bone-subtraction. In *2010 Symposiums Photonics and Optoelectronics on (SOPO)*, pages 1–5. IEEE, 2010.

Chapter 7

Digital Image Segmentation Using Computational Intelligence Approaches

S. Vijayakumar
VIT University, Vellore, India

V. Santhi
VIT University, Vellore, India

7.1	Introduction	206
7.2	Background of Swarm Intelligence	206
	7.2.1 Ant Colony Optimization Algorithm	207
	7.2.2 Particle Swarm Optimization Algorithm	209
7.3	Swarm Intelligence with Fuzzy Clustering Algorithms	210
	7.3.1 Ant Colony Optimization with Fuzzy c-mean Algorithms	211
	7.3.2 Particle Swarm Optimization with Fuzzy c-mean Algorithms	214
7.4	Swarm Intelligence and Fuzzy Morphological-Based Fusion Algorithms	217
	7.4.1 Ant Colony Optimization with Fuzzy Morphological-Based Fusion Algorithm	217
	7.4.2 Particle Swarm Optimization with Fuzzy Morphological-Based Fusion Algorithm	218
7.5	Results and Discussion	218
7.6	Conclusion	221
	Bibliography	224

Image segmentation is considered to be one of the important tasks in image processing. It is applied to get regions of interest from images. The image segmentation process subdivides an image into its constituent objects or regions. In general, the image segmentation process is carried out after the application of image preprocessing techniques. The segmented regions are used to extract attributes or features for further analysis of images. Swarm intelligence (SI) is one of the modern and incipient digital image segmentation approaches inspired by nature. In this chapter, an overview of various ways to carry out image segmentation using swarm intelligence (SI) approaches are presented with their performance evaulations.

7.1 Introduction

In general, image processing is considered to be of the important tasks performed in the modern media processing era by digital computers. Digital image processing is carried out to improve the pictorial representation of image data for human perception, to process it for automation in machine applications, and for efficient storage and transmission of it. The process that divides an image into multiple segments is called image segmentation, and it is considered to be one of the essential and complex tasks of an image processing system. The output of the image segmentation process is smaller regions or segments that could be combined to form regions of interest. In real-world applications, image segmentation is considered an important tasks to extract features for classification of objects.

In order to carry out the image segmentaion process, many intelligence techniques exists. In this chapter, two predominantly used tehniques are considered for segmentation of regions of interest in images. They are the ant colony optimization and particle swarm optimization techniques. These tecniques are combined with the Fuzzy C-Mean (FCM) clustering algroithm to enhance the performance of the proposed approach. The FCM algorithm is widely used due to its simplicity and ability to retain more information in images. In general, the FCM algorithm works on the pixel intensity; thus, segmentation is decided based on pixel intensities. The concept of the FCM algorithm was developed based on the Euclidean metric distance in [2]. The fuzzy morphological-based fusion (FMF) algorithm is proposed in [22] and it could also be used for image segmentation. It uses morphological operations such as dilation, erosion, opening, and closing for enhancing image features. In this chapter a new hybrid algorithm that combines both swarms intelligence and the fuzzy C-mean algorithm is proposed. In the subsequent sections, the background of ant colony optimization and particle swarm optimization algorithms are presented in detail.

7.2 Background of Swarm Intelligence

In order to obtain optimized solutions, many intelligence-based techniques inspired by the behavior of natural systems need to be developed. In general, evo-

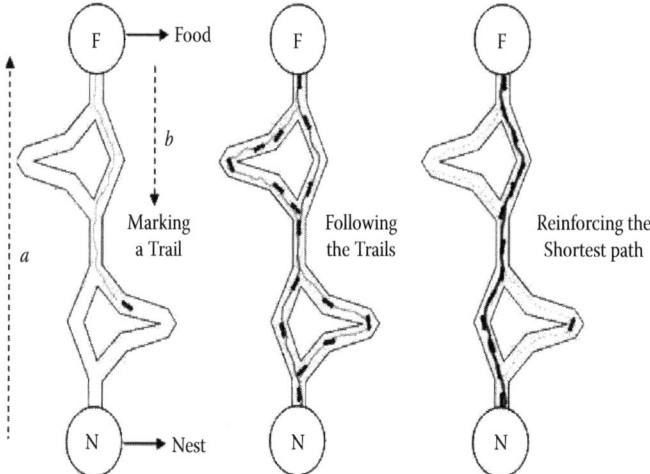

FIGURE 7.1: Shortest path identification by an ant to reach its destination

lutionary computational algorithms have increasingly become popular in the last few years. Evolutionary computing algorithms include the ant colony optimization algorithm, particle swarm optimization algorithm, and bee colony optimization algorithms. These algorithms could be used to carry out clustering and segmentation processes when combined with the FCM algorithm.

7.2.1 Ant Colony Optimization Algorithm

In 1992, the ant colony optimization (ACO) was proposed by Marco Dorigo et al. It is a population-based metaheuristic approach used to find the appropriate solution for complex optimization problems. The ant colony optimization algorithm mimics ants' foraging behavior, with simulated ants walking around the graph searching for the optimal solution. In this algorithm, each route followed by the simulated ants is considered as a membership solution of a given problem. During the travel, ants deposit phenomena called pheromone, and its thickness identifies the optimal solution. In general, ants searching for food select a route with a greater capacity of pheromone than the route with less [7, 8]. The optimization technique is applied successfully on classical optimized problems and on stochastic discrete optimization problems with dynamic mechanism.

The swarm intelligence algorithm is largely used in computer networks—in particular, the traveling salesman problem—to find out the shortest path. Similarly, ACO is used to solve many real-world problems. The ACO is used to solve continuous and mixed-variable optimization problems, and is one of the examples of artificial swarm intelligence systems for real-world problems. The behavior of ants is shown in Figure 7.1.

As shown in Figure 7.1, if ants have two or more fixed paths, it is observed that they select the shortest path to reach the destination, and return back quickly using the same path. During the course of time, the selected path will be reinforced by pheromone and it becomes the selected path for all the other ants [16]. In an ant colony optimization-based problem, the best solution is obtained based on numbers of ants using the selected path. In this way, the ant colony optimization algorithm is used to obtain the shortest path. The ant colony optimization algorithm was initially applied to the traveling salesman problem to obtain the shortest path. In this application, for any pair of nodes V_i and V_j on the graph H, the edge (V_i, V_j) is provided with weight. The trail membership function and heuristic are stored on the selected edges from the graph. Artificially, the ant colony algorithm uses n ants that are normally used for getting the best result. The probability of selected ants is computed using Equation 7.1.

$$p_{ij}^k = \begin{cases} \frac{\tau_{ij}^\alpha \eta_{ij}^\beta}{\sum_{j \in U} \tau_{ij}^\alpha \eta_{ij}^\beta} & j \in U \\ 0 & otherwise \end{cases} \quad (7.1)$$

Here, p_{ij}^k is the probability of selected path (V_i, V_j) for ant k of graph H. The parameters τ_{ij}^α and η_{ij}^β are the mammal and investigative statistics assigned from the edge (V_i, V_j), respectively, α and β are constants that determine the relative influence of the mammal and heuristic information, and U denotes the set of possible paths and specified problem limitations. The ant k moves and deposits a mammal on the trial test, as defined in Equation 7.2.

$$\Delta \tau_{ij}^k = \begin{cases} \frac{R}{D_k} & \text{if } ant(k) \text{ uses connection } (V_i, V_j) \text{ in its solution} \\ 0 & otherwise \end{cases} \quad (7.2)$$

Here, R denotes number of positive connections and D_k denotes the weight of the path used by ant k. The Euclidean distance (2-D vector) of each edge from V_i to V_j is d_{ij} for all values of i and j, and it is given by

$$d_{ij} = \sqrt{(V_{i1} - V_{j1})^2 + (V_{i2} - V_{j2})^2} \quad (7.3)$$

$$\eta_{ij} = \frac{1}{d_{ij}} \quad (7.4)$$

The mammal on each vertex is updated according to the following equation:

$$\tau_{ij}(t+1) = (1-\rho)\tau_{ij} + \rho \Delta \tau_{ij} \quad (7.5)$$

Here, ρ is the dehydration factor ($0 \leq \rho \leq 1$), which can be the initial mammal level over the iterations. Hence, the corresponding mammal depends on the next solution of the earlier mammal, where the initial corresponding undesired solution can be found. From Equation 7.5, ρ is the mammal vanishing co-efficient, where $\Delta \tau_{ij}$ denotes the amount of information that the k_{th} ant leaves in the path ij.

$$\Delta \tau_{ij} = \sum_{k=1}^{m} \Delta \tau_{ij}^{k} \tag{7.6}$$

7.2.2 Particle Swarm Optimization Algorithm

The particle swarm optimization (PSO) algorithm is a population-based stochastic optimization technique for the solution of continuous optimization problems. It is based on the behavior of social insects such as ants/ant colonies. In PSO, a set of pixels are called as particles, and their movements are used for finding good solution to a given continuous problem. In order to find a solution for a given problem, each particle uses the experience of neighboring particles. In the initial phase, each particle is given a random initial position and an initial velocity. The individual particle's rate is modified based on its distance from its specified location. In each iteration, every particle updates itself by tracking two optimum values: The first is the optimal solution found by the particle itself and it is called personal best, the other is the best current solution sought out by the whole population called global best. When finding these two optimal values, the particles update their own speeds and arrive at new locations. The final position of the particle represents the optimum solution for the given problem. The initial position of the particle is represented as A, its velocity is represented as V, position B represents the major solution of the problem, and it is the best position (local best) of the particle visiting from the initial point, and $\bar{B}$ represents the entire best position (global best) of the swarm. At a particular position, the specified function is evaluated for arriving at the best position using Equation 7.7. The entire best position is obtained by using Equation 7.8, whereas in each iteration of the PSO algorithm, the value of V and A are updated using Equations 7.9 and 7.10.

$$B(t+1) = \begin{cases} B(t) & if f(A(t+1)) \geq f(B(t)) \\ A(t+1) & if f(A(t+1)) < f(B(t)) \end{cases} \tag{7.7}$$

$$\bar{B} \in B_0, B_1, \ldots, B_n = min\{f(B_0(t)), f(B_1(t)), \ldots, f(B_n(t))\} \tag{7.8}$$

$$V(t+1) = W * V(t) + \left((B(t) - A) + (\bar{B}(t) - A(t))\right) \sum_{l=1}^{n} c_l * r_l(t) \tag{7.9}$$

$$A(t+1) = A(t) + V(t+1) \tag{7.10}$$

Here, W denotes the weight, which is selected from the previous velocity. The intellectual element of $(B(t)-x)$ represents the particle expectation of the best solution. The social element of $(\bar{B}(t) - x)$ represents the entire swarm's best solution, C_l denotes the acceleration constant, and $r_l(t) \approx LI(0,1)$ denotes the random number between 0 and 1. The PSO algorithm is repeated until the velocity becomes zero. The ability of membership function is used to evaluate the optimal solution. The average of maximum Euclidean distance of particles to be associated with clusters is calculated by using Equation 7.11, and the minimum Euclidean distance between any pair of clusters is calculated using Equation 7.12.

$$D_{MAX}(Z, A) = MAX\{\sum_{\eta \in C} D(\eta, m)/|C|\} \qquad (7.11)$$

$$D_{MIN}(A_j) = MIN\{d(m_j 1, m_j 2)\}, \& \forall j_1, j_2 \neq j_2 \qquad (7.12)$$

The particle swarm optimization algorithm is a population-based search algorithm with random initialization, and there are interactions among population members. In particle swarm optimization, each particle flies through the solution space and has the ability to remember its previous best position, surviving from generation to generation. The steps used in the PSO algorithm are briefly discussed below [1].

1. Cluster centers of each particle are randomly initialized.
2. Assign each particle a pixel of clusters that can minimize the distance between the cluster heads.
3. Calculate the membership function of each particle and then find the best solution.
4. Update the cluster centers using Equations 7.9 and 7.10.
5. Repeat the process until the final criterion is reached.

7.3 Swarm Intelligence with Fuzzy Clustering Algorithms

The fuzzy c-mean algorithm (FCM) was developed by J. C. Dunn in 1973 and implemented by J. C. Bezdek in 1981. The FCM algorithm is an unsupervised clustering algorithm that divides a data set into a certain number of clusters. The FCM algorithm, minimizes the distance between each data pixel in a cluster with its cluster center, and maximizes the distance between cluster centers, as shown in Figure 14.2.

The separation from unlabeled data to discrete and fine sets generates a method called clustering analysis, to work towards unsupervised learning with

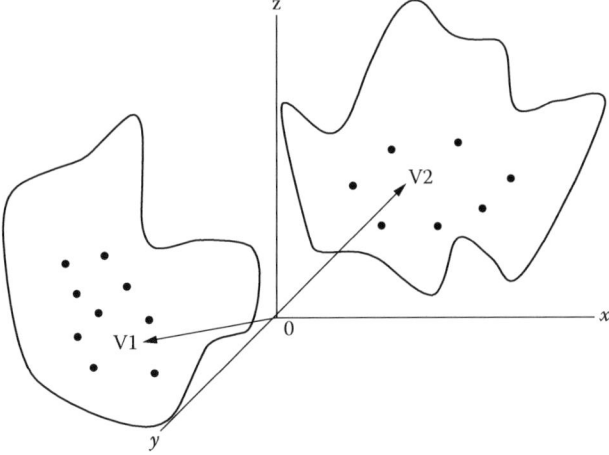

FIGURE 7.2: Representation of clusters using fuzzy c-mean algorithm

a better pattern recognition technique. In the existing k-mean conventional clustering method, all data belong to one cluster only, which does not appear to be realistic in few applications. It is considered to be a limitation of the k mean clustering method, and it is eliminated by the fuzzy version of the k mean algorithm. This fuzzy c-mean algorithm (FCM) includes data in every cluster with a separate degree of membership [5]. The FCM algorithm works as in [16, 18] and as follows:

1. Initialize C to initial cluster centers randomly.

2. Compute each data pixel distance from the cluster centers and then assign data pixel value to a cluster that has the minimum distance to its center.

3. Compute the average of data pixel values in each cluster and create a new cluster center.

4. Go to Step 2, then repeat the process until the cluster center converges to existing cluster centers.

7.3.1 Ant Colony Optimization with Fuzzy c-mean Algorithms

Ant colony optimization is combined with the fuzzy c-mean algorithm to enhance the performance of clustering. In this work, the algorithm chooses a number of clusters and a random initial cluster center for each cluster. The role of ant colony optimization is to assign each pixel to a cluster. This process depends upon the probability of pixel's similarity to the cluster centers and a variable. The variable represents a social behavior level and it is defined to be proportional to the minimum distance between each pair of cluster centers, and inversely proportional to the distances between each pixel and its cluster center [9]. The FCM algorithm combined with ant clustering by minimizing the membership function is defined in Equation 7.13. This is the weighted sum of squared errors with each cluster, where $0 \leq \psi_{mn} \leq 1$ or $\psi_{mn} = 1$, or $\psi_{mn} > 0$.

$$Min(U,C) = \psi_{11}^k X d_{11}^2 + \psi_{11} k X d_{12}^2 + \psi_{11}^k d_{13}^2 + \cdots + \psi_{mn}^k X d_{mn}^2 \quad (7.13)$$

for all $m = 1, 2, 3, \cdots, M$ and $n = 1, 2, 3, \cdots, N$

Here, ψ_{mn} denotes the membership degree of the pixel intensity n to the intensity of the cluster center m, while $U = [\psi_{mn}]_{RXC}$ is the fuzzy partition matrix, $d = ||C_m - \infty_n||$ represents the intensity comparisons between the cluster head m and n pixel, and $C_m = c_1, c_2, c_3, \cdots, c_m$ is the set of intensities of cluster centers [16]. In order to carry this out, some of the conditions are necessary to minimize updates using Equation 7.14.

$$\psi_{mn} = \frac{1}{\sum_{k=1}^{K} \frac{d_{mn}}{d_{kn}}} \quad (7.14)$$

$$C_m = \frac{\sum_{n=1}^{N} \mu_{mn}^k * \infty_n}{\sum_{n=1}^{N} \mu_{mn}^k} \quad (7.15)$$

From Equation 7.14, the FCM frequently optimizes the membership function $Min(U,C)$ by updating ψ_{mn} and C with the expectation of $||U(t+1) - U(t)|| \leq \varepsilon, \forall -1 < \varepsilon < 1$. Hence, the FCM method could be used for image segmentation. The relationship of each data point does not always reflect its actual relationship to the clusters very well, and may perhaps be inaccurate in a noisy environment. For better clustering, the ant colony optimization could be combined with the FCM algorithm.

In general, in the initial stage of cluster center selection, the ant colony optimization can attractively introduce the optimization method with clustering problems due to its parallel searching capabilities. Subsequently as to each image pixel (i.e., ant) should be predictable, the distances and the amount of social behavior-connected paths. The running cost of ant colony optimization can be quite high in image segmentation. In order to carry out image segmentation, the appropriate number of clusters and their center head need to be identified using the FCM algorithm. The manual process determines the number of clusters by investigating the histogram of image gray scale. In order to reduce time complexity, ant colony optimization would be more appropriate to obtain the correct number of clusters.

Suppose we make the M partitions with 256 gray levels and assign image pixels to each division, while the set of divisions $U = u_1, u_2, u_3, \cdots, u_M$ contains assigned image pixels. Let $|U_k|$ be large than the threshold and $V = v_1, v_2, v_3, \cdots, v_M$, and calculate v_t by the following equation.

$$v_\tau = \frac{1}{|U_k|} \sum_{n \in d_k} \infty_n, \& n = 1, 2, 3, \cdots, N \qquad (7.16)$$

Here, M denotes the number of partitions, v_τ denotes the cluster center, and N denotes the number of pixels in an image. The time complexity of ant colony optimization is almost $O(N^2)$ with the above preprocessing step, and to obtain the cluster partitions of M number. The time complexity of ant colony optimization is reduced to almost $O(N)$ [11, 21]. Then the initial cluster centers should be more compact and optimal through the proposed ant colony optimization with FCM. The ant colony-FCM algorithm has a resilient essential capability for local searching. However, obtaining local optima is to be expected when the unsuitable initial numbers of the clusters and centroids are used. It substantially affects the overall segmentation accuracy and cluster compactness, and also decides the parameters of PCM, which affects the final segmentation results.

In this work, we adopt ACO-based clustering to provide the appropriate number of clusters and centroids automatically. Thus, the mitigating problem is getting trapped in local optima of FCM, from side to side. The ACO-based clustering has tentative initial cluster centers that could be more compact and optimal, as illustrated in Figure 14.3.

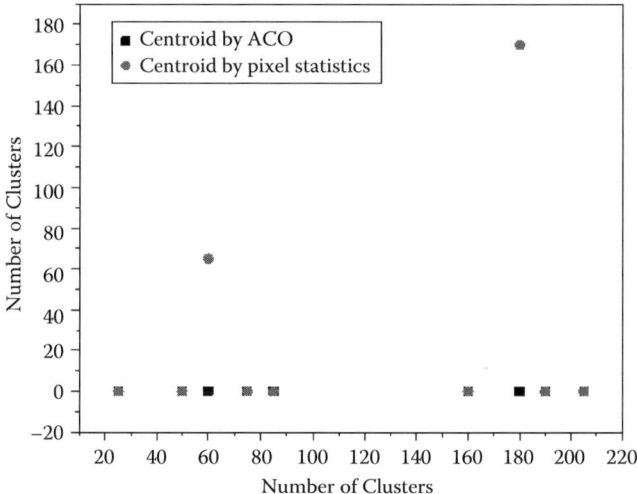

FIGURE 7.3: Representation of clusters using FCM algorithm

However, this algorithm provides an appropriate initialization of parameters; it does not solve the coincident clustering problem of FCM. To overcome this problem, preprocessing of pixels (ants) is derived from the ant colony-based clustering to FCM. The preprocessed pixels are composed of the categorized or uncategorized ants. The categorized ants are clustered by the ant colony optimization-based clustering, using its strong capability to converge to the optimum global results. All categorized ants with centroid information belong to individual cluster sets. The remaining members are defined as uncategorized ants. During the application of FCM, the categorized ants are assigned to play a key role as base pixels in preventing the coincident clustering problems. Also, the uncategorized ants will be positioned into any discovered cluster by means of FCM. The categorized ants are shown as their centroid value and the uncategorized ants are shown as white pixels. The proposed ant colony optimization method combined with the fuzzy c-mean algorithm is given below.

7.3.2 Particle Swarm Optimization with Fuzzy c-mean Algorithms

In general, the PSO is a population-based optimization tool that was introduced by Eberhart and Kennedy [10]. It can be employed and applied easily to solve various function optimization problems. Runkler and Katz [15] introduced two new methods for minimizing the reformulated objective functions of the FCM clustering model by PSO-V and PSO-U, to overcome shortcomings in fuzzy clustering algorithms (FCM), as discussed in [13] It uses the global searching capability of PSO to overcome the shortcomings of FCM. Thus, the appropriate cluster center searching mechanisms using a generalized FCM, optimized using the PSO algorithm, is proposed in [12].

The proposed PSO-FCM algorithm needs to be an input image of pre-estimated cluster number C. It is obtained from the desirable cluster partitions in a given image. The common cluster center is selected from the given set manually, and this can vary from an independent to a some what random process. Then, the selection of appropriate C from the various numerical approaches proposed in literature follows. In [3], Bezdek et al. proposed the rule of thumb $C \leq \chi^{1/2}$ where the upper bound must be determined based on knowledge or applications about the image. The other approach is used to validate the index as a measure of criterion about the data partition, as presented in the Davies–Bouldin [6], Xie–Beni [20], and Dunn [4] catalogs.

Therefore, the proposed model finds the optimized cluster center C in the entire image, to acquire the cluster partitions by considering compactness and inter-cluster separation, and then reduces the cluster center sensitivity value. As discussed in Section 7.2.2, particle swarm optimization (PSO)-based clustering is presented in [14]. Cluster centers are assigned to random initialization of particles. However, each pixel (i.e.,

Algorithm [ACOFCM Algorithm]

1. Initialization of the cluster center v_τ by Equation 7.16. V is the set of cluster centers.
2. Q_t is the pixel (ant) set that contains the members of v_τ.
3. Initialize the cluster iteration $t = 0$.
4. Compute each ant n in entire image I and its distance d_{mn} to every v_t.
5. If $l_{mn} = 0$, then the fuzziness $F_{mn} = 1$. Otherwise, if $l_{mn} \leq r$, calculate F_{mn} by using the following equation.

$$F_{mn} = \frac{T^\alpha_{mn}(t) \cdot \eta^\beta_{mn}(t)}{\sum_{u \in U} T^\alpha_{mn}(t) \cdot \eta^\beta_{mn}(t)} \& m, n \in U \tag{7.17}$$

where

$$T_{mn} = \begin{cases} 1 & l_{mn} \leq r \\ 0 & otherwise \end{cases} \tag{7.18}$$

6. Assign the pixel and update the social behavior activity by using the following Equation 7.19

$$T_{mn}(t) = \rho * T_{mn}(t) + \sum_{k=1}^{N} \delta T^k_{mn} \tag{7.19}$$

Otherwise, the pixel is left out into the uncategorized pixels set and would be given for the next iteration.

7. Update the cluster center using Equation 7.20.

$$v_\tau = \frac{1}{|Q_\tau|} \sum_{n \in Q_\tau} x_n, \& n = 1, 2, 3, \cdots, N \tag{7.20}$$

8. Calculate the distance between cluster centers. If the distance is less than the given threshold value T, merge the clusters and update the cluster centers by Equation 7.20.
9. Increase the cluster iteration value, i.e., $t = t + 1$.
10. Then update, repeating until $||V(t+1) - V(t)|| < \phi$.
11. Assign the value of v_τ to its categorized pixels and remaining uncategorized pixels.
12. Set the cluster number $M = |V(t+1)|$ and the cluster center $C = V(t+1)$.
13. After that, to initialize the fuzzy partition matrix, $U = [\psi_{mn}]_{RXC}$, and it is estimated η_i by the following equation:

$$\eta_i = K \frac{\sum_{n=1}^{N} \mu^k_{mn} * d^2_{mn}}{\sum_{n=1}^{N} \mu^k_{mn}} \tag{7.21}$$

14. Initialize $Z = 0$ and repeat $Z = Z + 1$.
15. Compute cluster centers $C(Z + 1)$ is using Equation 7.15.
16. Update fuzzy membership of pixels $U(Z + 1)$ by using Equation 7.14 till $||U(Z + 1) - U(Z)|| < \varepsilon$. 17. Finally, segment the image by matrix U, which depends on the cluster center majority of membership functions.

particle) is distributed based on the cluster center with minimum Euclidean distance. Then, the PSO is used to enhance the cluster centers using a fuzzification clustering technique. In this work, PSO with FCM algorithm is applied to all particles, and solutions are evaluated in a similar manner to evaluate the proposed ant colony optimization with FCM algorithm. The proposed PSO-FCM algorithm is accessed as follows.

Algorithm [PSO-FCM Algorithm]
1. Initialize number of clusters as C_{max} and C_{min} number of particles to P.
2. Initialize the P size of the cluster center C randomly assigned to P particles and iteration counter $t = 0$.
3. Create the partition matrix for all particles $U(k)$ using Equation 7.14.
4. Calculate the cluster center for each particle using Equation 7.15.
5. Calculate the optimized value of each particle using $F(P) = \frac{1}{Min(U,C)}$.
6. Calculate the pbest for each particle and gbest for each swarm.
7. Update the velocity and position of each particle by using the following equation

$$V(t+1) = wV(t) + c_1 r_1 (pbest(t) - P(t)) + c_2 r_2 (gbest(t) - P(t))$$
$$P(t+1) = P(t) + V(t+1).$$

8. Increment the iteration counter $t = t + 1$.
9. Repeat until PSO gets terminated.
10. Calculate the optimal threshold value α_n for each column of partition matrix $U(k)$ using Equation 7.22.

$$\alpha_n = arg_{\alpha_n} min\{|\sum_{\psi_{mn} \leq \alpha_n}(\psi_{mn}) + \sum_{\psi_{mn} \geq \psi_{n(max)}}(\psi_{mn} - \alpha_n) - card(\psi_{mn}|\alpha_n < \psi_m n < \psi_{n(max)} - \alpha_n)|\} \tag{7.22}$$

11. Update the $\psi_{mn}(1 \leq \eta \geq N)$ of nth cluster center according to α_n.
12. Calculate the cardinality M_i for each cluster on the basis of the cluster centers whose membership value is equal to $1 \leq i \geq C$.
13. Remove the cluster whose M_i ¡ ε and M_i is among $[RXC]$ lowest cardinality.
14. Update the cluster center value C.
15. Calculate the validity index $S_{XP}(k) = \frac{MIN(U,C)}{N*min||\beta_i - \beta_j||}$.
16. Update the image segmentation matrix $U(k)$.

7.4 Swarm Intelligence and Fuzzy Morphological-Based Fusion Algorithms

7.4.1 Ant Colony Optimization with Fuzzy Morphological-Based Fusion Algorithm

In this section, the ant colony optimization with the fuzzy morphological-based fusion (FMF) algorithm and its performance are presented. The clustering operation considers similarity of image pixels between the cluster center and image pixel intensity values. The difference of the clusters and compactness of each cluster is inversely propositional to the similarity of image pixels. At the end of the iteration process, the best optimized solution is selected the same way as ant colony optimization with FCM algorithm.

The ant colony optimization with FMF algorithm is given below. Let T be the threshold value of the input image for the Nth degree gray-level value. For example, a 2-D (M × N) output layer is defined for the algorithm, where M is chosen; then the expected number of the classes is less than or equal to N^2.

Algorithm [ACO-FMF Algorithm]
1. Initialize the number of clusters v_τ and the number of ants to m. Initialize the social behavior level assigned to each pixel to 1.
2. Initialize the m sets of v_τ different random cluster centers to be used by m ants.
3. Each ant assigns each pixel Q_t to one of the clusters v_τ randomly processed with the possible differentiating cluster centers F_{mn} and investigative information T_{mn} given in Equation 7.19.
4. For each input pixel the cluster center of the centroid is considered by using the fuzzy morphology. To calculate new cluster centers, use the following equation, where $\Delta(t)$ is decreasing membership function of the cluster center.

$$C_m(t+1) = C_m(t) + \Delta(t)(x - C_m(t)), m = 1, 2, 3, \cdots, M \quad (7.23)$$

5. Then calculate and save the optimized solution among the v_τ solutions found.
6. Update the social behavior level for each pixel according to the best solution. The social behavior value is updated using Equation 7.18.
7. Then assign the cluster center value of the best clustering solution to the cluster centers of all ants.
8. Terminate the process after getting optimized results. Otherwise go to step 3 and repeat the process.
9. Finally, the result of the output is the optimal solution.

7.4.2 Particle Swarm Optimization with Fuzzy Morphological-Based Fusion Algorithm

The particle swarm optimization with the fuzzy morphological-based fusion (PSO-FMF) algorithm is discussed below. Let L be the dimension of the input image, and N is selected as number of the classes, which is less than or equal to N^2. Here, the cost of cluster center values containing from the image pixels.

Algorithm [PSO-FCM Algorithm]
1. Initialize the number of clusters and the number of particles belonging to each cluster.
2. Initialize sets of different random cluster centers to be used by particles.
3. For each particle, let each pixel belong to a cluster in which it has the lowest Euclidean distance to the cluster centroid.
4. For each input pixel, calculate new cluster centers using Equation 7.10.
5. Save and update the best solution found for each particle, and also calculate the personal best solution.
6. Save the best solution for the entire personal best solution found, and also calculate the global best solution.
7. Update cluster centers of each particle according to the cluster center values and solution obtained using Equation 7.10.

7.5 Results and Discussion

The experimental results of FCM and FMF algorithms are discussed in this section in detail. The FMF is dependent on the morphological process with fusion, that is, $\Delta(t)$ in Equation 7.23; the performance depends on the experimental value selected for $\Delta(t)$. The $\Delta(t)$ is considered for decreasing the membership function of t and its value between 0 and 1; it is suggested as:

$$\Delta(t) = \frac{0.3}{t+1} \qquad (7.24)$$

where t and r denote number of iteration and a rate respectively. The valuae of rate is a constant value that is obtained by experiments. The experiments have been carrieed out using 30 trial runs on several different images, for r varies from 10 to 50 incrementing by 10. The experiments showed better results for $r=10$. Therefore, the experiment has been repeated with r values varying from 1 to 10 by incrementing it by 1. The experimental setup showed better results for value of r varying between 1 and 5, thus the value of r chosen to carry out the experiment is 2.

The ACO-FCM and ACO-FMF algorithms are dependent on parameterization as well. Parameters used in these algorithms other than r include k, Q, ρ, α, and β. Parameters α, β and k are used to keep values of τ and η in the same order. The parameter Q controls the added amount of social behavior and ρ eliminates the influence of the earlier added social behavior. The value of r should vary between 1 and 5; as per the previous experiment the value of r is selected as 2. The evaporation factor has been set to be $\rho = 0.8$. As per the performance of experiments, parameters k and Q are slightly influencing results, while α and β are more influential parameters. The parameter values tested are given as follows: $k = 1000$ and 10000, $Q = 10$ and 100, $\alpha = 0.1$ to 50 incrementing by 10, and $b = 0.1$ to 50 incrementing by 10. Each experiment has been performed for 20 trial runs on each image. If $b = 0.1$, the results are unacceptable, but value of $\alpha = 0.1$ gives better results. There are some sets of parameters that still did well for some images but not for all images. It is understood that α should be small while b should not be small. It is observed that the chosen parameters are as follows: $r = 2$, $\rho = 0.8$, $\alpha = 2$, $\beta = 5$, $k = 1000$, and $Q = 10$. The number of ants was chosen to be $m = 5$.

The PSO-FCM and PSO-FMF algorithms also include a set of parameters to be determined empirically. Those parameters have been chosen as suggested by [19], which resulted in good convergence. Parameters set are as follows: $c_1, c_2 = 1.49$ and $\omega = 0.72$. The number of ants selected is $m = 10$.

The proposed algorithm has been examined and compared with results of FCM and FMF, and is presented in Figure 7.4 to Figure 7.7. Test images used include MR brain images. The number of clusters found in all images is 3. The most predominant results of the algorithms over 30 different run trials are presented. The enhancement of the ACO and PSO on the FCM algorithm is obvious in all of the images tested.

The normal digital images have shown that the FCM algorithm produces unstable results and in some cases it misses some clusters, while the ACO-FCM and PSO-FCM algorithms are more stable and clearly recognize the clusters. In the river image, the results show that the FCM algorithm can generate stable results and the ACO-FCM algorithm seems to be less stable. However, it is apparent that even for this image, the ACO-FCM algorithm can enhance the classification results. This is also the case with the ACO-FMF and PSO-FMF algorithms as opposed to FMF in the ocean image. The results of the ocean image clearly show that the ACO and PSO algorithm can enhance the FMF in cases where the FMF algorithm is trapped local optima.

Thus, each algorithm is executed with a run of 30 times on the river image, shown in Figures 7.4(a) to 7.7(a). Then, comparing the classification results with the ground truth data, shown in Figures 7.4(g) to 7.7(g), the error matrix for each classified image is calculated. The results for best, worst, and average cases are shown in Figure 7.8. The immovability of the FMF algorithm over

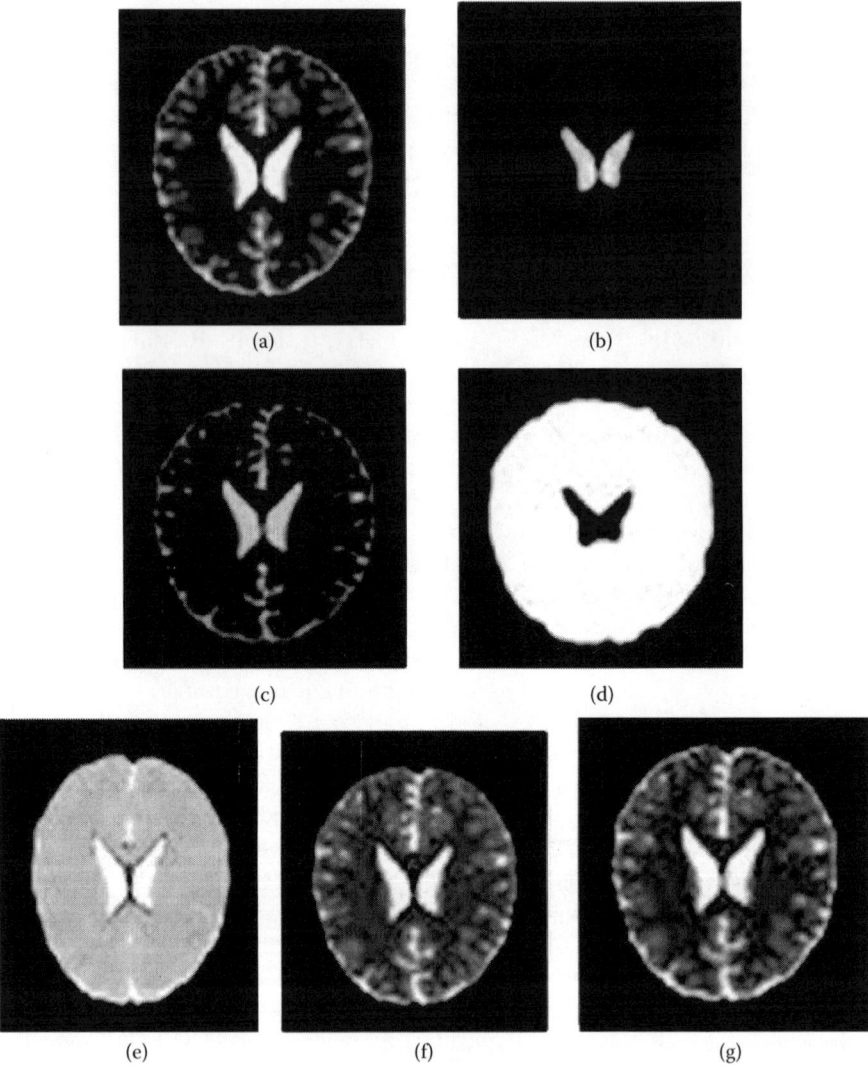

FIGURE 7.4: The most predominantly classified results among 30 trial runs: (a) original image, (b) FCM, (c) ACO-FCM, (d) PSO-FCM, (e) FMF, (f) ACO-FMF, (g) PSO-FMF

the ACO-FMF and the PSO-FMF algorithms can be inferred from Figure 7.8. The FCM algorithm is more stable than the ACO-FCM algorithm in the case of the river image. Results of the ACO-FCM algorithm include higher classification accuracy than those of the FCM algorithm.

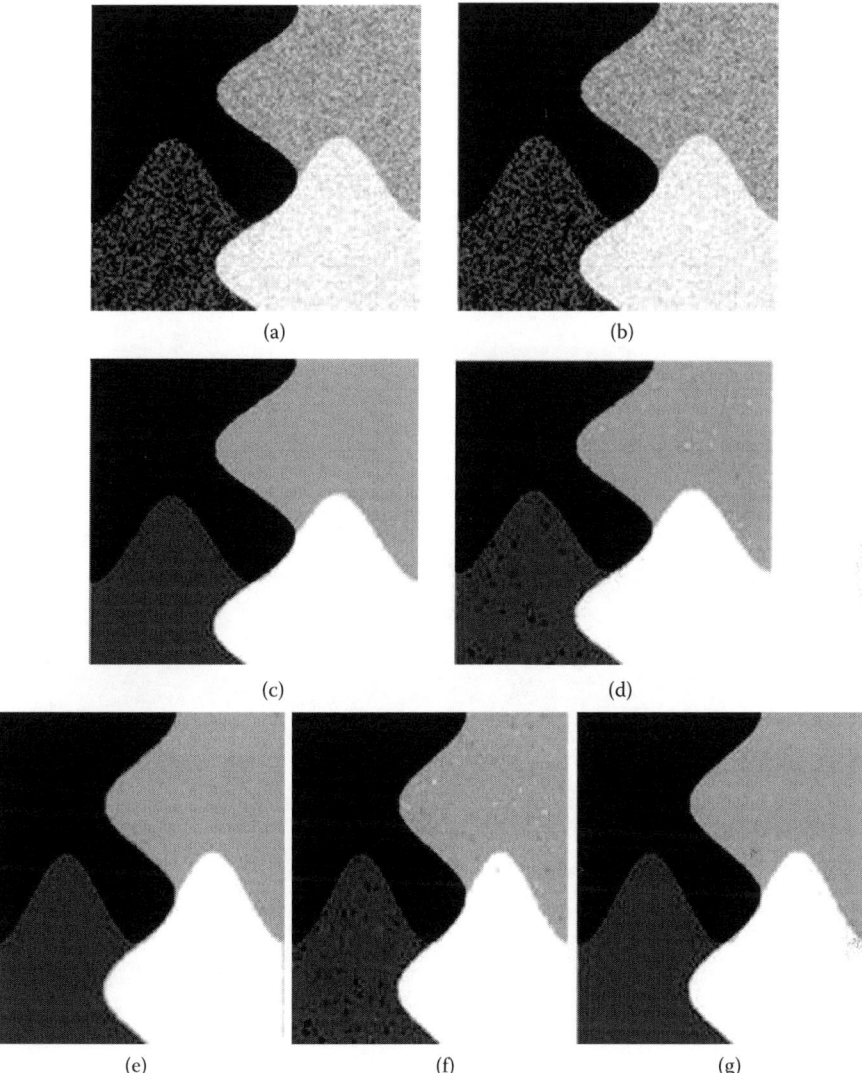

FIGURE 7.5: The most predominantly classified results among 30 trail runs: (a) original image, (b) FCM, (c) ACO-FCM, (d) PSO-FCM, (e) FMF, (f) ACO-FMF, (g) PSO-FMF

7.6 Conclusion

The experimental results show that swarm intelligence (SI) techniques can be combined to increase the performance of FCM and FMF algorithms in rec-

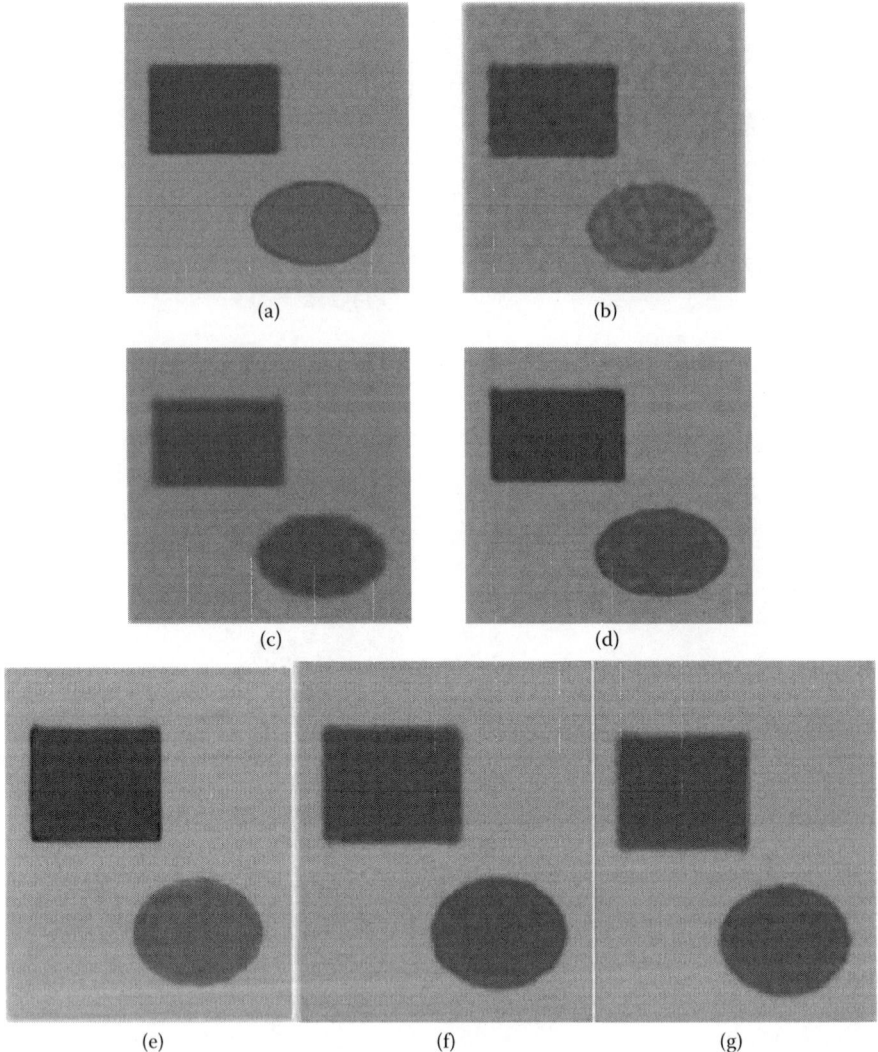

FIGURE 7.6: The most predominantly classified results among 30 trial runs: (a) original image, (b) FCM, (c) ACO-FCM, (d) PSO-FCM, (e) FMF, (f) ACO-FMF, (g) PSO-FMF

ognizing clusters. The FCM algorithm often fails to realize the clusters, since it depends on the cluster centers. The ACO-FCM and PSO-FCM algorithms provide a huge space search for comparing to the FCM algorithm. Due to these clustering algorithms, each initialized cluster center will be diminished over a total number of iterations.Therefore, these algorithms are slowly dependent on a random selection process to initialize image pixels that are clusters and

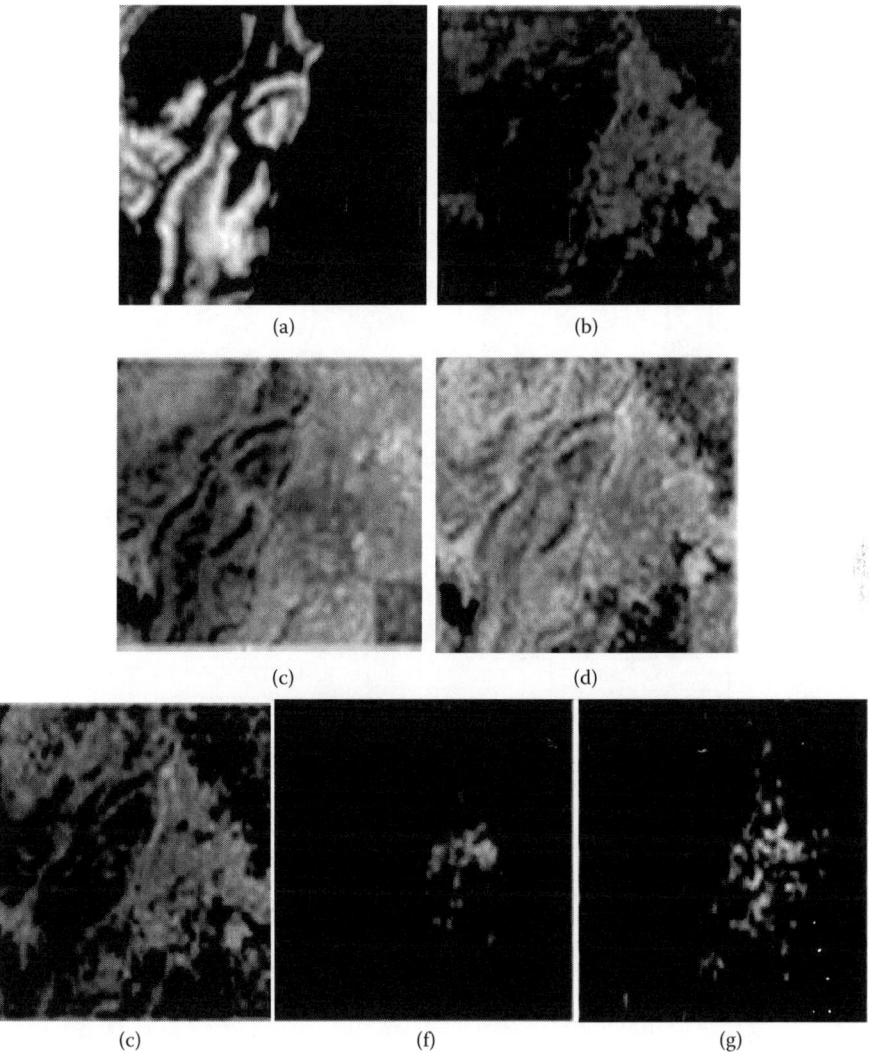

FIGURE 7.7: The most predominantly classified results among 30 trial runs: (a) original image, (b) FCM, (c) ACO-FCM, (d) PSO-FCM, (e) FMF, (f) ACO-FMF, (g) PSO-FMF

are likely to find the global best optimization solution. Also, it is shown that SI can be beneficial to the fuzzy morphological algorithm. Swarm Intelligence can help the fuzzy morphology technique find the global best optimization solution using the same set of parameters, and recognize the clusters where the FMF fails to do so in some cases.

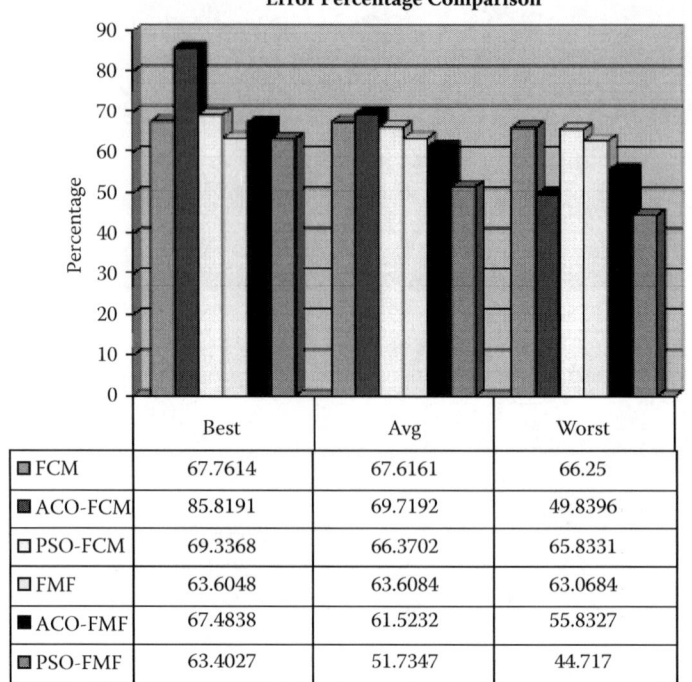

FIGURE 7.8: Error percentage comparison of various techniques over best, worst and average cases

---

Bibliography

[1] Debi Prasanna Acharjya and Ahmed P. Kauser. Swarm intelligence in solving bio-inspired computing problems: Reviews, perspectives, and challenges. *Handbook of Research on Swarm Intelligence in Engineering*, pages 74–98, 2015.

[2] A.N. Benaichouche, Hamouche Oulhadj, and Patrick Siarry. Improved spatial fuzzy c-means clustering for image segmentation using pso initialization, mahalanobis distance and post-segmentation correction. *Digital Signal Processing*, 23(5):1390–1400, 2013.

[3] James C. Bezdek, James Keller, Raghu Krisnapuram, and Nikhil Pal. *Fuzzy models and algorithms for pattern recognition and image processing*, volume 4. Springer Science & Business Media, 2006.

[4] James C. Bezdek and Nikhil R. Pal. Some new indexes of cluster validity. *IEEE Transactions on Systems, Man, and Cybernetics, Part B (Cybernetics)*, 28(3):301–315, 1998.

[5] Chiranji Lal Chowdhary and D.P. Acharjya. Clustering algorithm in possibilistic exponential fuzzy c-mean segmenting medical images. In *Journal of Biomimetics, Biomaterials and Biomedical Engineering*, volume 30, pages 12–23. Trans Tech Publications, 2017.

[6] David L. Davies and Donald W. Bouldin. A cluster separation measure. *IEEE transactions on pattern analysis and machine intelligence*, (2):224–227, 1979.

[7] Marco Dorigo, Mauro Birattari, and Thomas Stutzle. Ant colony optimization. *IEEE Computational Intelligence Magazine*, 1(4):28–39, 2006.

[8] Marco Dorigo and Thomas Stützle. The ant colony optimization metaheuristic: Algorithms, applications, and advances. In *Handbook of Metaheuristics*, pages 250–285. Springer, 2003.

[9] Joseph C. Dunn. A fuzzy relative of the isodata process and its use in detecting compact well-separated clusters, pages 32–57. Taylor & Francis, 1973.

[10] Russell Eberhart and James Kennedy. A new optimizer using particle swarm theory. In *Proceedings of the Sixth International Symposium on Micro Machine and Human Science, 1995. MHS'95.*, pages 39–43. IEEE, 1995.

[11] Yanfang Han and Pengfei Shi. An improved ant colony algorithm for fuzzy clustering in image segmentation. *Neurocomputing*, 70(4):665–671, 2007.

[12] Hesam Izakian and Ajith Abraham. Fuzzy c-means and fuzzy swarm for fuzzy clustering problem. *Expert Systems with Applications*, 38(3):1835–1838, 2011.

[13] Chia-Feng Juang, Che-Meng Hsiao, and Chia-Hung Hsu. Hierarchical cluster-based multispecies particle-swarm optimization for fuzzy-system optimization. *IEEE Transactions on Fuzzy Systems*, 18(1):14–26, 2010.

[14] Jiayin Kang and Wenjuan Zhang. Combination of fuzzy c-means and particle swarm optimization for text document clustering. In *Advances in Electrical Engineering and Automation*, pages 247–252. Springer, 2012.

[15] Thomas A. Runkler and Christina Katz. Fuzzy clustering by particle swarm optimization. In *2006 IEEE International Conference on Fuzzy Systems*, pages 601–608. IEEE, 2006.

[16] Sara Saatchi and Chih-Cheng Hung. *Swarm intelligence and image segmentation*. INTECH Open Access Publisher, 2007.

[17] Anita Tandan, Rohit Raja, and Yamini Chouhan. Image segmentation based on particle swarm optimisation technique. *International Journal of Science, Engineering and Technology Research*, 3(2):257–260, 2014.

[18] Julius T. Tou and Rafael C. Gonzalez. Pattern recognition principles. 1974.

[19] D.W. Van der Merwe and Andries Petrus Engelbrecht. Data clustering using particle swarm optimization. In *The 2003 Congress on Evolutionary Computation, 2003. CEC'03.*, volume 1, pages 215–220. IEEE, 2003.

[20] Xuanli Lisa Xie and Gerardo Beni. A validity measure for fuzzy clustering. *IEEE Transactions on Pattern Analysis and Machine Intelligence*, 13(8):841–847, 1991.

[21] Zhiding Yu, Oscar C Au, Ruobing Zou, Weiyu Yu, and Jing Tian. An adaptive unsupervised approach toward pixel clustering and color image segmentation. *Pattern Recognition*, 43(5):1889–1906, 2010.

[22] Liu Yucheng and Liu Yubin. An algorithm of image segmentation based on fuzzy mathematical morphology. In *International Forum on Information Technology and Applications, 2009. IFITA'09.*, volume 2, pages 517–520. IEEE, 2009.

Chapter 8

Digital Color Image Watermarking Using DWT SVD Cuckoo Search Optimization

S. Ganesh Babu
Robert Bosch Engineering and Business Solutions Limited, Coimbatore, India

B. Sarojini Ilango
Avinashilingam University, Coimbatore, India

8.1	Introduction		228
8.2	Fundamentals of Watermarking		228
	8.2.1	Importance of Digital Watermarking	230
	8.2.2	Classification of Digital Watermark	231
	8.2.3	Formulating the Watermarking Problem	231
8.3	Cuckoo Search: An Overview		232
	8.3.1	Cuckoo Breeding Behavior	232
	8.3.2	Building the Cuckoo's Primary Habitat	234
8.4	Cuckoo's Method of Laying the Eggs		234
	8.4.1	Cuckoo's Immigration	235
	8.4.2	Eliminating the Cuckoos in Inappropriate Areas	236
	8.4.3	Convergence of the Algorithm	236
8.5	Proposed Watermarking Method with Cuckoo Search Optimization		236
8.6	Experimental Results		239
8.7	Conclusion		242
	Bibliography		242

In every field there is enormous use of digital contents, and copying of digital content is not so difficult. Therefore, there is a great need for prohibiting such illegal copyright of digital media. The solution to this end is digital watermarking. In digital watermarking, a watermark is embedded into a cover image in such a way that the resulting watermarked signal is robust to certain distortion caused by either standard data processing in a friendly environment, or malicious attacks in an unfriendly environment. In this chapter, a new metaheuristic algorithm, called cuckoo search for based digital watermarking is presented.

8.1 Introduction

We are living in the era of information, where billions of bits of data is created in every fraction of a second, and with the advent of the Internet, creation and delivery of digital data (images, video and audio files, digital repositories and libraries, Web publishing) has grown many fold. Since copying digital data is very easy and fast, too issues like protection of rights of the content and proving ownership arise. Digital watermarking came as a technique and a tool to overcome shortcomings of current copyright laws for digital data. The specialty of a watermark is that it remains intact on the cover work even if it is copied. So, to prove ownership or copyrights of data, the watermark is extracted and tested. It is very difficult for counterfeiters to remove or alter watermarks. As such, the real owner can always have his data safe and secure. A number of improved algorithms have appeared since the literature [2, 5, 7, 4, 1, 8] put forward the watermarking algorithm based on SVD [6]. An SVD-based image-watermarking method put forward by Zude Zhou and others has achieved very good results, but the inadequacy is that it has small capacity for the watermark embedded: A 512×512 size image can only have a 64×64 size watermark image embedded, at most. A novel watermarking algorithm for digital images based on DWT, SVD, and cuckoo search optimization is proposed in this chapter, based on previous studies. A large number of experiments show that the algorithm is relatively robust and has enhanced the capacity and psycho-visual redundancy of the embedded watermark information.

8.2 Fundamentals of Watermarking

Although the art of papermaking had been invented in China over one thousand years earlier, paper watermarks did not appear until about 1282, in Italy. The marks were made by adding thin wire patterns to the paper molds. The paper would be slightly thinner where the wire was and hence more transparent. The meaning and purpose of the earliest watermarks are uncertain. They may have been used for practical functions such as identifying the molds on which sheets of paper were made, or as trademarks to identify the paper maker. On the other hand, they may have represented mystical signs, or might simply have served as decoration. By the eighteenth century, watermarks on paper made in Europe and America had become more clearly utilitarian. They were used as trademarks, to record the date the paper was manufactured, and to indicate the sizes of original sheets. It was also about this time that watermarks began to be used as anti-counterfeiting measures on money and other documents. The term watermark seems to have been coined near the end of the eighteenth century and may have been derived from the German term sermarke (although it could also be that the German word is derived from the English). The term is actually a misnomer, in that water is not especially important in the creation of the mark. It was probably given because the marks resemble the effects of water on paper. About the time the term watermark was coined, counterfeiters began developing methods of forging watermarks used to protect paper money. Counterfeiting prompted advances in watermarking technology. William Congreve, an Englishman, invented a technique for making color watermarks by inserting dyed material into the middle of the paper during papermaking. The resulting marks must have been extremely difficult to forge, because the Bank of England itself declined to use them on the grounds that they were too difficult to make. A more practical technology was invented by another Englishman, William Henry Smith. This replaced the fine wire patterns used to make earlier marks with a sort of shallow relief sculpture, pressed into the paper mold. The resulting variation on the surface of the mold produced beautiful watermarks with varying shades of grey. This is the basic technique used today for the face of President Jackson on the $20 bill. One hundred years later, in 1954, Emil Hembrooke of the Muzak Corporation filed a patent for "watermarking" musical works. An identification code was inserted in music by intermittently applying a narrow notch filter cantered at 1 kHz. The absence of energy at this frequency indicated that the notch filter had been applied, and the duration of the absence was used to code either a dot or a dash. The identification signal used Morse code. It is difficult to determine when digital watermarking was first discussed. In 1979, Szepanski described a machine-detectable pattern that could be placed on documents for anti-counterfeiting purposes. Nine years later, Holt described a method for embedding an identification code in an audio signal. However, it was Komatsu and Minaga, in 1988, who appear to have first used the term digital watermark. Still, it was probably not until the early 1990s that the term digital watermarking really came into vogue. About 1995, interest in digital watermarking began to mushroom. In addition, about this time, several organizations began considering watermarking technology for inclusion in various standards. The copy protection technical working group (CPTWG) tested watermarking systems for protection of video on DVD disks. The secure digital music initiative (SDMI) made watermarking a central component of their system for protecting music. Two projects sponsored by the European Union, VIVA and Talisman, tested watermarking for broadcast monitoring. The international organization for standardization (ISO) took an interest in the technology in the context of designing advanced MPEG standards. In the late 1990s, several companies were established to market watermarking products. More recently, a number of companies have used watermarking technologies for a variety of applications.

8.2.1 Importance of Digital Watermarking

The sudden increase in watermarking interest is most likely due to the increase in concern over copyright protection of content. The Internet had become user friendly with the introduction of Marc Andreessen's Mosaic Web browser in November 1993, and it quickly became clear that people wanted to download pictures, music, and videos. The Internet is an excellent distribution system for digital media because it is inexpensive, eliminates warehousing and stock, and delivery is almost instantaneous. However, content owners (especially large Hollywood studios and music labels) also see a high risk of piracy [9].

This risk of piracy is exacerbated by the proliferation of high- capacity digital recording devices. When the only way the average customer could record a song or a movie was on analog tape, pirated copies were usually of a lower quality than the originals, and the quality of second-generation pirated copies (i.e., copies of a copy) was generally very poor. However, with digital recording devices, songs and movies can be recorded with little, if any, degradation in quality. Using these recording devices and using the Internet for distribution, would-be pirates can easily record and distribute copyright protected material without appropriate compensation being paid to the actual copyright owners. Thus, content owners are eagerly seeking technologies that promise to protect their rights. The first technology content owners turn to is cryptography. Cryptography is probably the most common method of protecting digital content. It is certainly one of the best developed as a science. The content is encrypted prior to delivery, and a decryption key is provided only to those who have purchased legitimate copies of the content. The encrypted file can then be made available via the Internet, but would be useless to a pirate without an appropriate key. Unfortunately, encryption cannot help the seller monitor how a legitimate customer handles the content after decryption. A pirate can actually purchase the product, use the decryption key to obtain an unprotected copy of the content, and then proceed to distribute illegal copies. In other words, cryptography can protect content in transit, but once decrypted, the content has no further protection. Thus, there is a strong need for an alternative or complement to cryptography: a technology that can protect content even after it is decrypted. Watermarking has the potential to fulfill this need because it places information within the content where it is never removed during normal usage. Decryption, re-encryption, compression, digital-to-analog conversion, and file format changes—a watermark can be designed to survive all of these processes. Watermarking has been considered for many copy prevention and copyright protection applications. In copy prevention, the watermark may be used to inform software or hardware devices that copying should be restricted. In copyright protection applications, the watermark may be used to identify the copyright holder and ensure proper payment of royalties. Although copy prevention and copyright protection have been major driving forces behind research in the watermarking field, there are a number of other applications for which watermarking has been used or suggested. These include broadcast monitoring, transaction tracking, authentication, copy control, and device control.

8.2.2 Classification of Digital Watermark

Some of the important types of watermarking based on different watermarks are discussed below:

1. Visible watermarks: Visible watermarks are an extension of the concept of logos. Such watermarks are applicable to images only. These logos are inlaid into the image but they are transparent. Such watermarks cannot be removed by cropping the center part of the image. Further, such watermarks are protected from statistical analysis.

   The drawbacks of visible watermarks are degrading the quality of image and detection by visual means only. Thus, it is not possible to detect them by dedicated programs or devices. Such watermarks have applications in maps, graphics, and software user interface.

2. Invisible watermark: An invisible watermark is hidden in the content. It can be detected by an authorized agency only. Such watermarks are used for content and or author authentication and for detecting an unauthorized copier.

8.2.3 Formulating the Watermarking Problem

In digital watermarking, the watermark image is embedded into the cover image. In this process, the watermark image intensity will be reduced by selecting an alpha value. This will decide the reduction factor of the watermark intensity.

$$WE = B + \alpha W \tag{8.1}$$

In the above watermarking equation, W is the watermark image and B is the cover image. The watermark image will be embedded into the cover image with that multiplication factor α, and the result is the embedded watermarked image WE. This alpha value will decide how much watermark intensity we can add to the base image.

In watermarking algorithms, this value will be a constant value. For improving the watermarking embedding process and for best visual quality, the basic equation is modified little bit. For getting the best watermark embedding, we can adjust the α value based on the intensity value of the cover image, so that we can improve the psycho-visual redundancy of the watermarked image. For choosing the α value based on the cover image we have used the cuckoo search algorithm. A detailed description of the cuckoo search algorithms is discussed in Section 8.3, whereas the watermark extraction can be done by using Equation 8.2.

$$W = \frac{(WE - B)}{\alpha} \qquad (8.2)$$

8.3 Cuckoo Search: An Overview

In order to describe the cuckoo search more clearly, let us briefly review the interesting breeding behavior of certain cuckoo species. Then, we will outline the basic ideas and steps of the proposed algorithm [10, 3].

8.3.1 Cuckoo Breeding Behavior

Cuckoos are fascinating birds, not only because of the beautiful sounds they can make, but also because of their aggressive reproduction strategy. Cuckoos start laying the eggs in the nests within their own egg laying radius, as shown in Figure 8.1. This process continues until they reach the best place for laying eggs. This optimal location is where the greatest numbers of cuckoos are gathered. A cuckoo optimization algorithm flow chart is shown in Figure 8.2. Like other evolutionary algorithms, the cuckoo optimization algorithm begins with as initial population that consists of cuckoos. This population of cuckoos has eggs which they lay into the host birds' nests. Some of these eggs,

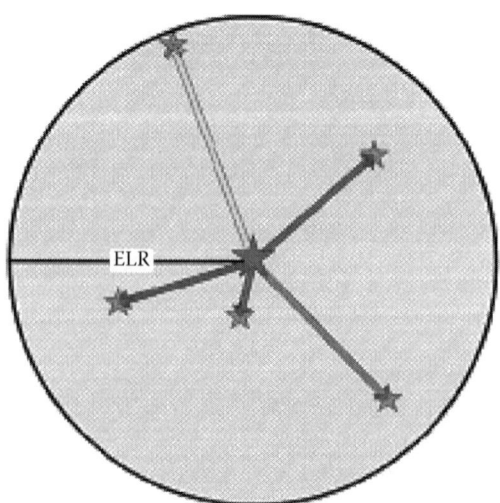

FIGURE 8.1: The egg-laying radius (ELR)

Digital Image Watermarking Using DWT SVD Cuckoo Search

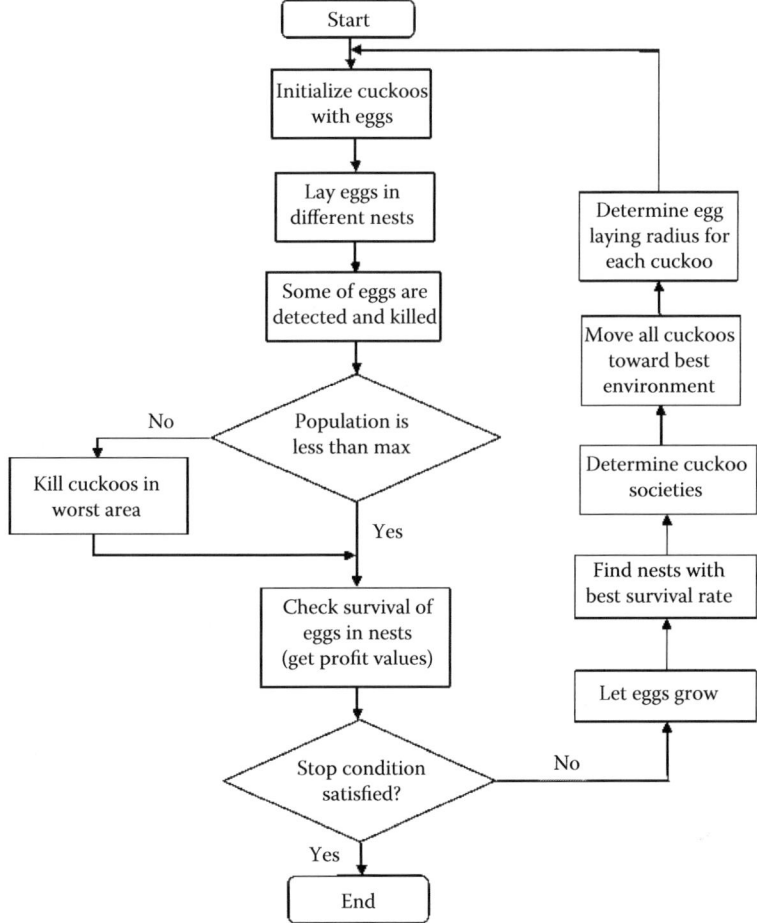

FIGURE 8.2: Cuckoo optimization algorithm flowchart

which are more similar to the host bird's eggs, have more of a chance to grow and become young cuckoos. Other eggs are recognized and disposed of by the host bird. The rate of growing nest eggs shows the suitability of the nests in that area. The more eggs in the environment that are able to survive and be saved, the more profit is assigned to that area. Therefore, a situation in which the greatest number of eggs are saved is the parameter that the cuckoo's optimization algorithm intends to optimize.

Cuckoos search for the best place for maximizing the survival of their own eggs. The communities and groups are created after the cuckoos' chicks get out of the eggs and change into the adult cuckoos. Each group has its own habitat. The best habitat for all groups will be the cuckoo's next destination in other groups. All groups migrate to the best existing region. Each group resides in an area close to the current position. Several egg-laying radii will be calculated and created by considering the number of eggs that each cuckoo will lay, in addition to the distance of cuckoos from the current optimal area for habitat.

8.3.2 Building the Cuckoo's Primary Habitat

For solving an optimal issue, it is necessary to form the variables at issue in the array form. These arrays are determined with the names "chromosome" in the case of the genetic algorithm (GA) and "particle position" in the case of particle swarm optimization (PSO). But this array is called the "habitat" in the cuckoo optimization algorithm.

In an optimization problem, the next N_{var} of a habitat will be an $1 \times N_{var}$ array that shows the current position of cuckoos. This array is defined as follows:

$$Habitat = [X1, X2, \cdots, X_{Nvar}] \qquad (8.3)$$

The suitable amount (or benefit rate) in current habitat is calculated by evaluating the profit function (f_p) in habitat as defined in Equation 8.4.

$$Profit = f_p(habitat) = f_p(X1, X2, \cdots, X_{Nvar}) \qquad (8.4)$$

As can be seen, the cuckoo optimization algorithm is an algorithm that maximizes the profit function. For applying the cuckoo optimization algorithm in minimization problem, just multiply cost function by -1.

A habitat matrix with the size $N_{pop} * N_{var}$ is built in order to start an optimization algorithm. A random numbers of eggs are assigned to each of these habitats. In the nature, each cuckoo lays 5 to 20 eggs. These numbers are used as higher and lower limits for assignment of eggs to each cuckoo in different iterations. Another habit is that they each cuckoo eggs in a specific domain. The maximum eggs its laying domain is called the egg-laying radius (ELR).

In an optimization problem, the higher limit of the variable is shown by var_{hi} and the lowest limit is shown by var_{low}. Each cuckoo has ELR, which is proportional to the total number of eggs, the number of current eggs as well as the higher and lower limits of issue variables. Therefore, ELR is defined as below, where α is the variable by which the maximum value of ELR is adjusted.

$$ELR = \alpha \times \frac{Number\ of\ current\ cuckoo's\ eggs}{Total\ number\ of\ eggs} \times (Var_{hi} - Var_{low}) \qquad (8.5)$$

8.4 Cuckoo's Method of Laying the Eggs

Each cuckoo randomly lays the eggs in the host birds' nests that are in the ELR. When all cuckoos lay their eggs, some of the eggs, which are less similar to the host bird's eggs, are identified and thrown out of the nest. Thus, after each egg-laying, p% of all eggs (usually 10%), which have a lower profit function are eliminated. The remaining chicks feed and grow in the host nests.

Another interesting point about the cuckoo chicks is that only one egg can grow in each nest, because when the cuckoo chicks get out of the eggs, they throw out the host bird's eggs from the nests. If the host chicks get out of the eggs sooner, the cuckoo chick eats the largest amount of food that the host bird brings (the chick pushes the other chick over with its 3-times, larger body). After a few days, the host bird's chicks will die of starvation and only the cuckoo's survives.

8.4.1 Cuckoo's Immigration

When the cuckoos' chicks are grown up and become mature, they live in their own groups and environments, but when the time of laying the eggs becomes close, they will migrate to better habitats where the chance of eggs' survival is higher. After forming the cuckoo groups in different geographical areas (searching for space is an issue), the group with the best position is selected for other cuckoos as the target location for migration. This is depicted in Figure 8.3.

When the mature cuckoos live in all parts of the environment, it is difficult to determine which cuckoo belongs to which group. For solving this problem, the cuckoos' grouping is performed by the classification method K-means (a k from 3 to 5 is usually sufficient).

When the cuckoos' groups are formed, the average profit of the group is calculated in order to obtain the relative optimality of group habitat. Then, the group with the highest average profit (optimality) is selected as the target group and other groups migrate to it. During the migration to the target

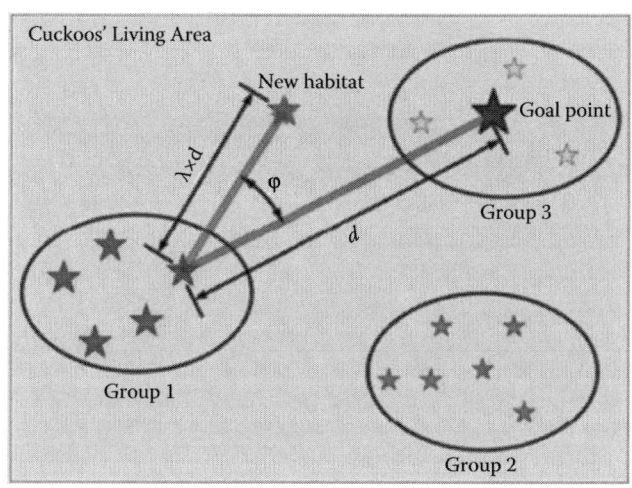

FIGURE 8.3: Cuckoo's immigration

point, the cuckoos do not pass all the way to the target point. They just pass along part of the path in which they also have the deviation. Such movement is clearly shown in Figure 8.3.

As shown in Figure 8.3, each cuckoo only passes $\lambda\%$ of the whole path towards the current ideal goal, and also has a deviation of φ radian. These two parameters help the cuckoo to seek a further environment, where λ is a random number from 0 to 1, and φ is a number from $-\pi/6$ to $\pi/6$. When all cuckoos have migrated to the target location and the new residential areas of each one is determined, each cuckoo owns a large number of eggs. Given the number of each cuckoo's eggs, an ELR is determined for the bird and then the egg-laying starts.

8.4.2 Eliminating the Cuckoos in Inappropriate Areas

Given the fact that there is always a balance for the population of birds in nature, a number like $N_{\max}$ controls and limits the maximum number of cuckoos who can live in an environment. This balance is established due to the bird's dietary restrictions, hunting by predators, and the inability to find a suitable nests for eggs.

8.4.3 Convergence of the Algorithm

After a few iterations, the whole cuckoo population reaches an optimal point with the maximal similarity of eggs to the host birds' eggs, as well as the location of the largest amount of food sources. This place will have the highest overall profit and the lowest number of eggs will be destroyed there. The convergence of more than 95% for all cuckoos towards a point causes the cuckoo optimization algorithm to be ended.

8.5 Proposed Watermarking Method with Cuckoo Search Optimization

In the proposed invisible digital watermarking, a secret image is embedded into the cover image with the limitations of imperceptibility and robustness. To achieve this we are limited to minimal modifications in the cover image pixels. If the modifications are increased from a certain level, the visual quality of the cover image will be degraded and the secret information will become visible to the human eye. The major focus of this study is to select optimal intensity value for the watermark in the cover image. With this good visual quality, a high payload is also achieved. Once the optimal intensity value is selected, then based on the application's requirements, one can choose any method for watermark embedding, so watermark intensity modification based on cover image is the most critical aspect in the whole process. We have used

Digital Image Watermarking Using DWT SVD Cuckoo Search

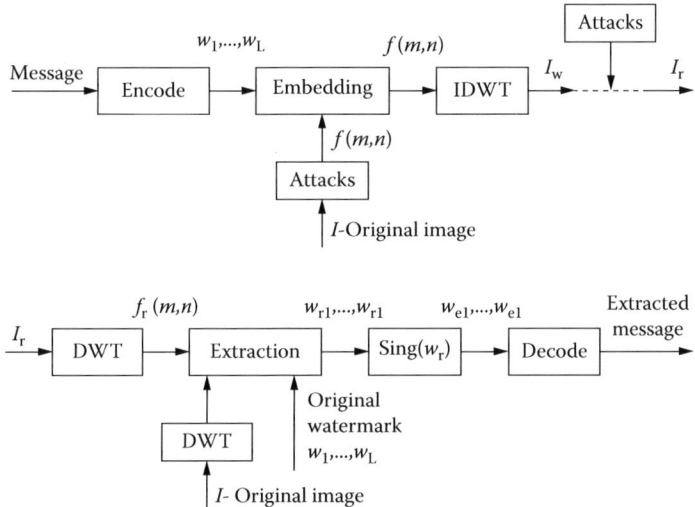

FIGURE 8.4: Watermarking system in DWT

the cuckoo search algorithm for selecting the optimal intensity value of the watermark image.

In the proposed watermarking method, the host image's color channels are separated. On each channel's host image, one level of wavelet transformation is applied. After transformation, the host image will separate the image into HH, HL, LH, and LL components. The LL component will be taken for watermark embedding. This will be given to the input for the SVD operation using Equation 8.6.

$$A = USV^T \tag{8.6}$$

In Equation 8.6, U and V are orthogonal matrix $U \in R^{M \times M}$ and respectively, while $V \in R^{M \times M}$, R is the image matrix. S is a matrix $S \in R^{M \times M}$ whose elements are 0 in a non-digital format and the elements in its diagonal satisfy: $\sigma_1 \geq \sigma_2 \geq \cdots \geq \sigma_r \geq \sigma_{r+1} = \cdots = \sigma_m = 0$, in which r is the rank of A, and is equal to the number of non-negative singular value. $\sigma_i; i = 1, 2, \cdots m$ is then called the singular value of matrix A, and is the square root of the characteristic value of AA^T. The watermarking system in DWT is depicted in Figure 8.4. After applying the SVD operation, the S component will be taken for embedding the watermark image. The DWT decomposition with two levels is depicted in Figure 14.5.

The SVD operation is performed at both the host image and the watermark image before the embedding process. The cuckoo searching method is used to select the α_k value in Equation 8.7.

$$LL_{wi,j} = LL_{i,j} + \alpha_k W_{ij}; i, j = 1, 2, \cdots, n \tag{8.7}$$

In above equation, $W_{i,j}$ is the watermark pixel value; α_k is the watermark intensity embedding factor, but this may be vary based on host image intensity

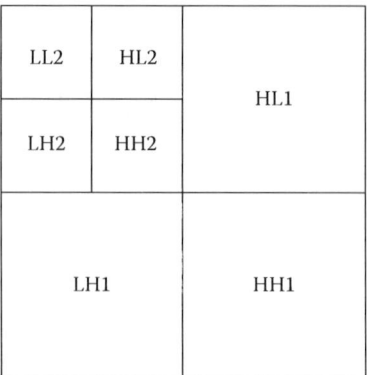

FIGURE 8.5: DWT decomposition with two levels

value $H_{i,j}$ if the host image is H. $LL_{i,j}$ is the host image's low-level component after the DWT decomposition. $LL_{wi,j}$ is the watermark's embedded low-level component. After the embedding process, the watermarked image is obtained by using Equation 8.8.

$$H_w = IDWT(LL_{wi,j}, LH, HL, HH) \tag{8.8}$$

In above equation, $IDWT$ is the inverse discrete wavelet transformation. $LL_{wi,j}$ is the modified watermark's embedded low-level band. LH, HL, and HH are vertical, horizontal, and diagonal high level bands. H_w is the watermarked host image.

The watermarked host image is transmitted to the receiving side, as shown in Figure 8.4. At the receiving side is the watermarked host image H_w. In the watermark extraction process, the watermarked image will be decomposed using wavelet transformation.

$$[LL_{wi,j}, LH, HL, HH] = DWT(H_w) \tag{8.9}$$

In Equation 8.9, H_w is the watermarked host image. DWT is the discrete wavelet transformation, while $LL_{wi,j}$ is the watermark's embedded low-level band. LH, HL, and HH are the vertical, horizontal, and diagonal high-level components. The low-level component $LL_{wi,j}$ then performs the SVD operation using Equation 8.5 at the watermark extraction side. The watermark image is extracted by using the reverse operation of watermark embedding with Equation 8.10.

$$W_{i,j} = (LL_{wi,j} - LL_{i,j})/\alpha_k \tag{8.10}$$

In Equation 8.10, $W_{i,j}$ is the extracted watermark's image. $LL_{i,j}$ and $LL_{wi,j}$ are the original and watermark embedded host images low band, and α_k is the watermark intensity embedding factor. This factor will be selected using the cuckoo search optimization algorithm, which will improve the psychovisual redundancy of the cover image after watermarking. The cuckoo search for the optimum intensity amplification factor for the watermark image is described below. The objective function is given as $f(x), x = (x_1, x_2, \cdots, x_d)$. The following pseudocode generates an initial population of n host nests.

Pseudocode

1. While $t <$ Max Generation or stop criterion.
2.   Get a cuckoo randomly i (say), and replace its solution by performing intensity values.
3.   Evaluate its quality / fitness F_i.
4.   Choose a nest among n (say j) randomly.
5.   If $(F_i > F_j)$
6.   Replace j by the new solution
7.   End if
8.   Abandon a fraction (p_a) of worse nests and build new ones in given intensity range.
9.   Keep the best solutions (or nests with quality solutions).
10.   Rank the solutions and find the current best.
11.   Pass the current best solutions to the next generation.
12. End While

8.6 Experimental Results

In order to verify the method proposed in this chapter, a large number of simulation experiments in the proposed watermarking algorithm are carried out. In the proposed algorithm, a host image of size 512 and a watermark image of size 256 × 256 are considered, as shown in Figure 8.6. Thus, the watermark image is half the size of the host image, and the watermarking capacity is high in this method. Figure 8.7 shows the watermarked image and the extracted watermark image. There is not much information lost from the original and extracted watermark image. This will also show the psycho-visual redundancy

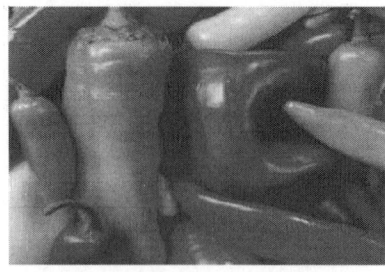 

(a) Host image (b) Watermark image

FIGURE 8.6: Host image and watermark image before embedding

strength of the algorithm. In addition to the human eye's subjective feelings, this chapter also takes peak signal-to-the noise ratio (PSNR) as the objective evaluation criterion of the watermarking image and extracted watermark image. The formula of PSNR is as follows where MSE is defined in Equation 8.13.

$$PSNR = 10\log^{(\frac{255^2}{MSE})} \qquad (8.11)$$

$$MSE = \frac{1}{M \times N} \sum_{i=1}^{M} \sum_{j=1}^{N} (W(i,j) - W'(i,j))^2 \qquad (8.12)$$

The formula of normalized correlation elected is:

$$NC(W,W') = \frac{\sum_{i=1}^{M} \sum_{j=1}^{N} W(i,j)W'(i,j)}{\sqrt{\sum_{i=1}^{M} \sum_{j=1}^{N} w(i,j)^2} \times \sqrt{\sum_{i=1}^{M} \sum_{j=1}^{N} W'(i,j)^2}} \qquad (8.13)$$

where, $W(i,j)$ is the pixel value of the original watermark and $W'(i,j)$ is the pixel value of the extracted watermark; M, N are the width and height of the watermark, respectively.

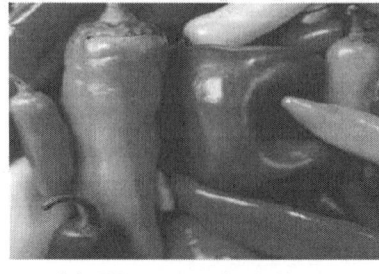 

(a) Watermarked image (b) Extracted watermark image

FIGURE 8.7: Host image after watermarking and the extracted watermark image

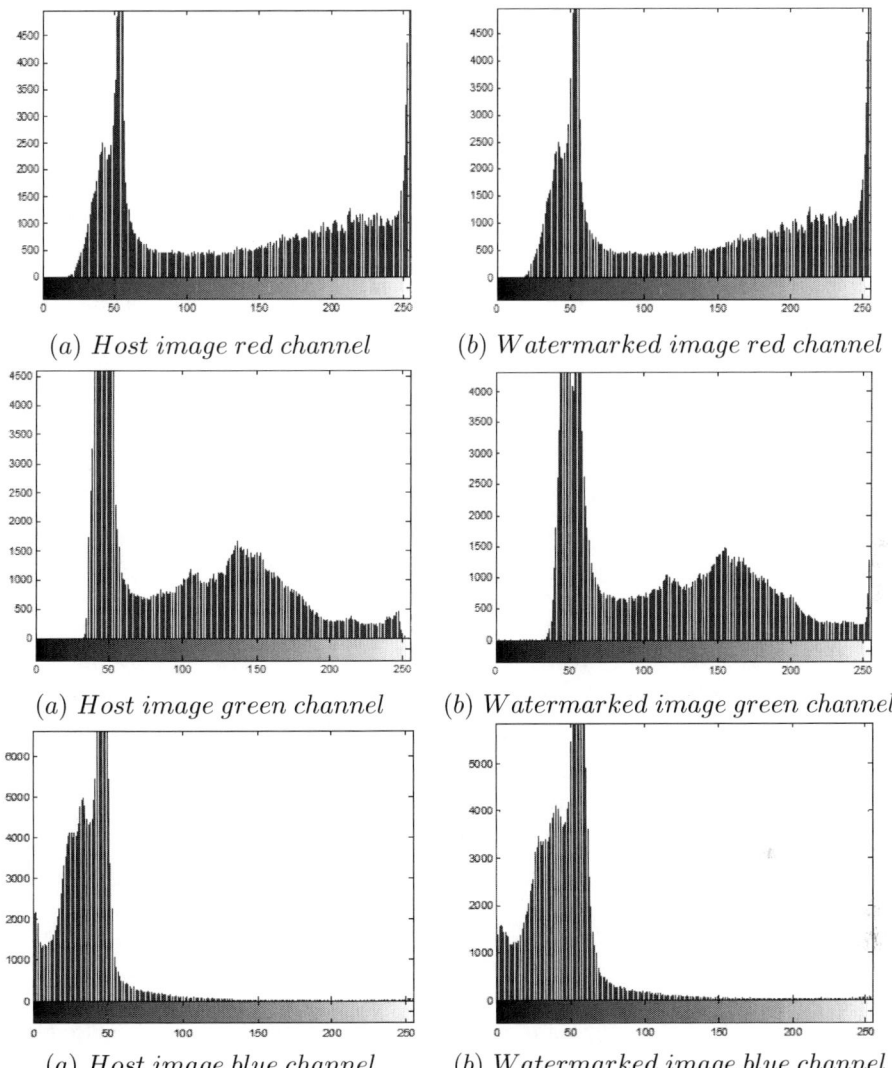

FIGURE 8.8: Histogram of host image and watermarked image R, G, and B channels

The PSNR value for the above original and extracted watermark images is 32.1. This is a good value for this higher watermarking capacity algorithm. Normalized correlation for the watermarked images is 0.94. For checking the visual quality of the host image before and watermarking after the histogram for the three channels before and after the watermark embedding, are taken. The histogram looks similar for all the three channels before and after the watermark embedding. The results are depicted in Figure 8.8.

8.7 Conclusion

A novel watermarking algorithm for digital color images using DWT-SVD and the cuckoo search algorithm is proposed in this chapter. By using the advantage of DWT and SVD, the amount of embedded watermark information is improved and the watermark information is loaded in the low-level band of the DWT and SVD section. The cuckoo search method is used to improve the visual information of the embedded watermarked host image. The simulation experiment results show that the digital watermarking algorithm proposed in this paper has characteristics such as good transparency and accurate results of watermarking detection.

Bibliography

[1] Ingemar J. Cox, Gwenaël Doërr, and Teddy Furon. Watermarking is not cryptography. In *International Workshop on Digital Watermarking*, pages 1–15. Springer, 2006.

[2] Ingemar J. Cox, Joe Kilian, F. Thomson Leighton, and Talal Shamoon. Secure spread spectrum watermarking for multimedia. *IEEE transactions on Image Processing*, 6(12):1673–1687, 1997.

[3] Claudia I. Gonzalez, J.R. Castro, Patricia Melin, and Oscar Castillo. Cuckoo search algorithm for the optimization of type-2 fuzzy image edge detection systems. In *2015 IEEE Congress on Evolutionary Computation (CEC)*, pages 449–455. IEEE, 2015.

[4] Prayoth Kumsawat, Kitti Attakitmongcol, and Arthit Srikaew. A new approach for optimization in image watermarking by using genetic algorithms. *IEEE Transactions on Signal Processing*, 53(12):4707–4719, 2005.

[5] Ki R. Kwon and Ahmed H. Tewfik. Adaptive watermarking using successive sub-band quantization and perceptual model based on multiwavelet transform. In *Electronic Imaging 2002*, pages 334–348. International Society for Optics and Photonics, 2002.

[6] Jun Jing Liu, Hua Jiang, and Gui Ying Liang. A watermarking algorithm for digital image based on logistic and svd. In *2010 International Conference on Intelligent Computing and Integrated Systems (ICISS)*, pages 140–144. IEEE, 2010.

[7] Wei Lu, Hong-Tao Lu, and Fu-Lai Chung. Chaos-based spread spectrum robust watermarking in dwt domain. In *Proceedings of 2005 International*

Conference on Machine Learning and Cybernetics, 2005., volume 9, pages 5308–5313. IEEE, 2005.

[8] Wei Lu and Hongtao Lu. Content dependent image watermarking using chaos and robust hash. In *Computational Intelligence and Security Workshops, 2007. CISW 2007. International Conference on*, pages 660–663. IEEE, 2007.

[9] V. Santhi and D.P. Acharjya. Improving the security of digital images in hadamard transform domain using digital watermarking. In *Improving Information Security Practices through Computational Intelligence*, pages 228–254. IGI Global, 2016.

[10] Xin-She Yang and Suash Deb. Cuckoo search via Lévy flights. In *World Congress on Nature & Biologically Inspired Computing, 2009. NaBIC 2009.*, pages 210–214. IEEE, 2009.

Chapter 9

Digital Image Watermarking Scheme in Transform Domain Using the Particle Swarm Optimization Technique

Sarthak Nandi
VIT University, Vellore, India

V. Santhi
VIT University, Vellore, India

9.1	Introduction	246
9.2	Foundations of Transformation Techniques	248
	9.2.1 Discrete Wavelet Transform	248
	9.2.2 Singular Value Decomposition	249
	9.2.3 Particle Swarm Optimization (PSO)	249
9.3	Proposed Watermarking Scheme	250
	9.3.1 Adaptive Calculation of Scaling and Embedding Factors Using PSO	250
	9.3.2 Watermark Embedding Process	251
	9.3.3 Watermark Extraction Process	253
9.4	Performance Analysis	254
9.5	Conclusion	261
	Bibliography	261

With the Internet, the chances of piracy of an image have increased many times over, and this has made digital watermarking of images very useful. This chapter proposes a novel watermarking scheme to embed a digital watermark in an image. The inserted watermark is invisible to the human eye. The embedding and scaling factors are adaptively calculated using a particle swarm optimization technique, while embedding and extraction are carried out using discrete wavelet transform and singular value decomposition. The proposed technique can be used against copyright infringement of images by inserting a digital watermark in an image. The major advantage is that it can be applied to both color and gray scale images. The robustness of this technique has been verified by making the images watermarked using this technique subject to various attacks that degrade the quality of the image. The results obtained show that the proposed technique is robust to various attacks and can be used effectively to tackle copyright infringement.

9.1 Introduction

The internet and the availability of affordable smartphones, laptops, etc., have enabled people to share images with millions of other people at the click of a button, as it is extremely easy to produce and distribute unlimited number of copies of any digital information. This has also increased the possibility of copyright infringement of images. Digital watermarking is a technique with which this illegal practice can be curbed.

Digital watermarking is a process through which some data, called a digital watermark, is embedded in a digital signal, which can be an image, audio, video, etc. [6, 9]. This digital watermark can then be used to identify the owner or the authorized users of the digital signal. Digital watermarking can be generally classified into two types: visible and invisible watermarking. Visible watermarking has an advantage over invisible watermarking in that visible watermarking can be used to identify the owner with minimal effort, and makes it easier to avoid copyright issues, as any potential violator can be discouraged on seeing the watermark. However, it might be possible to remove a visible watermark using some image editing applications. On the other hand, a person might not suspect that an image contains an invisible watermark, and if any copyright infringement is found, the violator can be prosecuted with relative ease. A typical watermarking scheme consists of a cover image, a watermark, an embedding algorithm and an extraction algorithm, [4, 15, 22].

A watermark can be inserted in either the frequency domain or the spatial domain. A frequency domain watermarked image is more robust towards noise than a spatial domain watermarked image [15]. The frequency components of an image are obtained by applying a transformation technique on the image. Some of the frequently used transformations are discrete Fourier transform (DFT), discrete cosine transformation (DCT), Walsh Hadamard transformation, and discrete wavelet transformation (DWT) [22]. The mathematical models for embedding and extracting a watermark into and from an image are defined in Equations (9.1) and (9.2), respectively.

$$Im = \beta Im + \alpha W \qquad (9.1)$$

where Im is the watermarked image; W is the watermark image, i.e., the data to be embedded; Im is the original or cover image; α is the scaling factor used to control the strength of the watermark, and β is the embedding factor used to create space for inserting the watermark.

$$W = frac(Im \beta Im) \alpha \qquad (9.2)$$

where W is the extracted watermark.

As the original image is required while extracting a watermark, the algorithm can be classified as a non-blind watermarking algorithm. An early watermarking scheme that hides an undetectable electronic watermark using least significant bit (LSB) manipulation in gray scale image is proposed in [21]. In this method, a pseudo random sequence is used as the electronic watermark, which is generated randomly through linear shift register. Manoharan presented an enhancement procedure proposed over the watermarking results on medical images, using singular value decomposition (SVD), contourlet transform, and DCT. The enhanced watermarked image is subjected to various attacks, and it is proved that the enhanced image does not disturb the watermark content embedded in the cover data.

A DCT domain-based spread spectrum approach has been proposed in [7]. In this, the authors have emphasized that the watermarks should be embedded in perceptually significant components, and N-highest-valued coefficients are used for inserting watermarks to make the watermark more robust to common attacks. In [11], the number of watermarks embedded is proportional to the energy contained in each band through the selection of different weights for different energy levels. An Rotation, Scaling, and Translation (RST) resilient watermarking method in which the most appropriate SIFT features are selected for inserting the watermark image (features called SIFT for Watermarking, or SIFTW) has been proposed in [14]. Color image watermarking in the multidimensional Fourier domain is proposed in [20]. Watermark casting is performed by estimating the just-noticeable distortion of the images, to ensure watermark invisibility.

Yinghui has proposed a particle swarm optimization (PSO) method based on multi-wavelet digital watermarking in [25], one that fully takes human visual system characteristics into account. In [13], a blind watermarking algorithm based on the significant difference of wavelet coefficient quantization is proposed. However, it pointed out in [17] that the method is potentially insecure. To overcome this security issue, an intelligent watermarking technique by invoking PSO in the wavelet domain is proposed in [23].

A hybrid watermarking scheme based on SVD and PSO is proposed in [19]. They embedded the watermark in the DCT domain of the DWT sub-band of the cover image. In [12] the authors have described a watermarking algorithm utilizing the genetic algorithm and particle swarm optimization to find the optimal values of perceptual loss ratio. Venkateswara et al. have proposed a watermarking scheme using SVD and PSO in the DWT domain in [18]. They have incorporated PSO in the scheme to find the optimum scaling factors.

Veysel et al. have proposed a DWT-SVD-based watermarking method that uses particle swarm optimization to calculate the optimal value for the modifications in the DWT-SVD domain in [3]. Chakraborty et al. have proposed a technique in [5] that uses PSO-based parameter optimization to hide medical information. They have embedded a hospital logo or electronic patient record in the DWT domain. Sivavenkateswara Rao et al. have proposed a multipurpose image watermarking algorithm using the SVD technique [20], in which the PSO optimized scaling factor is used to insert a watermark in singular values of decomposed image. The proposed watermarking is used for multi purposes including ownership verification and tamper detection.

In the above methods of watermarking using particle swarm optimization, the authors have utilized the watermarked image as well as the watermarked images subjected to various attacks to calculate the best scaling factor value using PSO. However, this makes the process more computationally complex, especially because the particle swarm optimization algorithm requires many iterations. The aim of this chapter is to propose a watermarking process utilizing PSO that is less computationally complex but also gives good results. Section 2 discusses the background of DWT and SVD. In Section 3 the proposed algorithm is presented in detail. The performance analysis is carried out in Section 4. Section 5 concludes the work.

9.2 Foundations of Transformation Techniques

A watermark inserted in an image must be robust in order to endure any geometric or signal distortions. These distortions may remove portions of the image, leading to irretrievable loss of image data. Embedding the watermark in the frequency domain spreads the watermark over the whole spatial extent of the image and is therefore less likely to be affected by distortions [7]. This section discusses few important transformation techniques.

9.2.1 Discrete Wavelet Transform

The DWT is a technique for multi-resolution decomposition of images. DWT is performed as a multi-stage transformation. An image is decomposed into three sub-bands of finest scale wavelet coefficients, i.e., the detailed image, and one sub-band of the coarse-level coefficients, i.e., the approximation image in the first stage. Let the three sub-bands of finest-scale wavelet coefficients be HL1, HL1, and HH1, and the sub-band of coarse-level coefficients be LL1. To obtain the next coarse level of wavelet coefficients, the sub-band, LL1 is further decomposed into four sub-bands LL2, HL2, LH2, and HH2. The decomposition process continues to obtain sub-bands LLn, HLn, LHn, and HHn where n is the stage of decomposition until a certain final requirement is reached according to the user's application [24]. The Haar wavelet transform functions are defined as:

$$har(0,0,\theta) = 1; 0 \leq \theta \leq 1$$

$$har(i,j,\theta) = \begin{cases} \sqrt{2^i}; & \frac{j-1}{2^i} \leq \theta < \frac{j-1/2}{2^i} \\ -\sqrt{2^i}; & \frac{j-1/2}{2^i} \leq \theta < \frac{j}{2^i}; i = 0,1,2,;j = 12^i \\ 0; & otherwise \end{cases} \quad (9.3)$$

9.2.2 Singular Value Decomposition

Singular value decomposition (SVD) is a linear algebraic technique that is used to decompose a given matrix into three matrices. An image consists of non-negative scalar entries. The SVD of an image Im of size $m \times n$ is defined as $Im = USV^T$, where U and V are orthogonal matrices, i.e., $U^T U = I$; $V^T V = I$'; S is a diagonal matrix and I is the identity matrix. The diagonal values of S are called the singular values of Im and they represent the luminance of Im. The matrices U and V are called the left and right singular vectors of A, respectively, and they preserve the geometrical properties of the image. UV together are called the SVD subspace of Im [8, 2]. The following are some of the properties of SVD.

1. The singular values of an image are very stable, i.e., when a small change is done to the singular values, there is a very small variation in the image.

2. Every real matrix and its transpose have the same non-zero singular values.

3. Singular values represent intrinsic properties.

9.2.3 Particle Swarm Optimization (PSO)

PSO is a computational algorithm proposed by Kennedy and Eberhart. It has proven both very effective and quick when applied to a diverse set of optimization problems. It simulates the behavior of birds or swarms. Additionally, it initializes a group of random particles (random solutions), and then obtains the optimal solution through iterations. In each iteration, each particle updates itself by tracking two optimum values: the first is the optimal solution found by the particle itself and it is called personal best, the other is the best current solution sought out by the whole population, called global best. When finding these two optimal values, the particles update their own speeds and arrive at new locations [1]. In PSO the velocity and position are represented as a vector as per the equation given below:

$$v_i(t+1) = wv_i(t) + c_1 r_1 (y_i(t) - x_i(t)) + c_2 r_2 (y_g(t) - x_i(t)) \quad (9.4)$$

x_i, is the current position of the particle, y_i denotes the personal best position, and y_g denotes the global best position found by the particles. The parameter w controls the influence of the previous velocity vector. Once the velocities of all particles have been updated, the particles move with their new velocities to their new positions according to the formula given in Equation 9.4.

$$x_i(t+1) = x_i(t) + v_i(t+1) \qquad (9.5)$$

After all particles have been moved to their new position, the function is evaluated in new positions and the corresponding personal best positions and the global best position are updated, i.e., the iterations continue. Typical termination criteria for PSO algorithms are to execute until a maximum number of iterations are completed or to terminate when the global best position has not been changed for a certain number of iterations, or if a function value has been found that is better than a required threshold value [10].

9.3 Proposed Watermarking Scheme

In this section an innovative method for invisible watermarking using particle swarm optimization is discussed. The watermarking is carried out in the DWT domain using the SVD technique. The proposed method adaptively calculates the embedding factor, taking into account only the content of the cover image. While inserting an invisible watermark, two properties need to be ensured: (i) the inserted watermark must not degrade the quality of the cover image, i.e., there should be no visible difference between the original and watermarked image, and (ii) the watermark itself must not be visible. In the subsequent section the procedure for adaptive calculation of scaling and embedding factors are discussed in detail.

9.3.1 Adaptive Calculation of Scaling and Embedding Factors Using PSO

In this section the adaptive calculation of scaling and embedding factors using particle swarm optimization is discussed. The scaling factor is denoted as α and the embedding factor is denoted as β. The proposed algorithm could be used to insert the watermark in both color images and gray scale images. If the color image is to be watermarked, then it needs to be transformed into YIQ domain in order to insert the watermark in the luminance components; otherwise gray scale values can be considered directly for inserting the same. In order to make the watermarking algorithm more robust, the luminance component Y is transformed into frequency components using DWT. The multi-resolution capability of DWT decomposes the signal into low frequency, middle frequency, and high frequency bands.

As low frequency components are more significant and robust to attacks, they are selected for inserting the secret data. The identified low frequency band is decomposed into singular matrix, left orthogonal, and right orthogonal matrices. To calculate the scaling factor using particle swarm optimization, 4 particles are taken. The maximum number of iterations is set at 100. Other digital watermarking techniques take many more particles and also have more iterations, but this needs much more processing. As can be seen in the results, even after taking a lesser number of particles and lesser iterations, the algorithm performs at par or better than other PSO-based watermarking techniques. The fitness function to be maximized is:

$$f = NCC_w + NCC_img \tag{9.6}$$

where NCC_w denotes the normalized correlation coefficient (NCC) of the extracted watermark and NCC_img denotes the NCC of the watermarked image. Once the scaling factor α is calculated, the embedding factor β is calculated as:

$$\beta = 1 - \alpha \tag{9.7}$$

The use of embedding factor β is to make room in the image for adding the watermark. Thus the strength of the image is reduced in proportion to the strength of the inserted watermark.

9.3.2 Watermark Embedding Process

In this section the watermark embedding process is discussed in detail. The pictorial representation of the watermark embedding process is shown in Figure 9.1. The image to be watermarked is called the cover image. The luminance component of the cover image is converted into frequency components. The low frequency band is decomposed into singular values. From these singular values the embedding and scaling factors are calculated. As the parameters

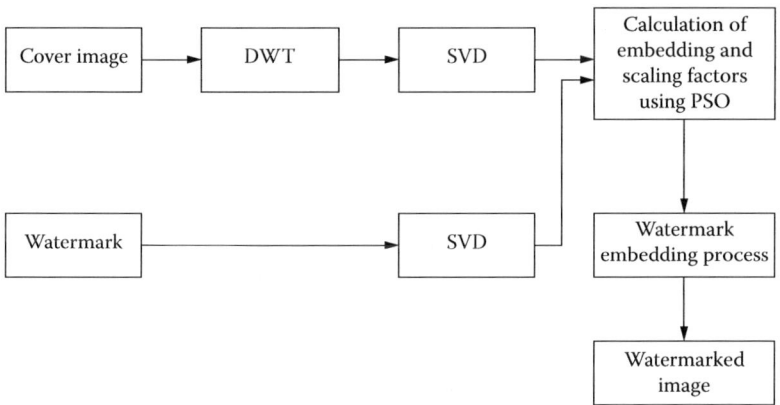

FIGURE 9.1: Watermark embedding process

are calculated adaptively from the content of the cover image, the inserted watermark does not distort the quality of the cover image. The pseudocode for the embedding process is given below:

Algorithm [Pseudocode of Embedding Process]

Im = Watermark Insertion (Im, W)
Input: Cover image (Im), Watermark (W)
Output: Watermarked image (Im)

1. Read Im, W (Read cover image and watermark image)

2. $[Y, I, Q] = RGB2YIQ(Im)$ (Conversion of cover image Im from RGB to YIQ domain)

3. $[LLY, LHY, HLY, HHY] = DWT(Y)$ (DWT on luminance component Y)

4. $[U, S, V] = SVD(LLY)$ (SVD on frequency component of Y)

5. $[Uw, Sw, Vw] = SVD(W)$ (SVD on watermark)

6. $\alpha = PSO(Im, W)$ (Calculation of scaling factor using PSO)

7. $\beta = 1 - \alpha$ (Calculation of embedding factor)

8. $S = \alpha Sw + \beta S$ (Watermark insertion)

9. $LLY = U*S*V^T$ (Inverse SVD to get modified frequency component)

10. $Y = IDWT(LLY, LHY, HLY, HHY)$ (Inverse DWT to get modified luminance component)

11. $[R, G, B] = YIQ2RGB(Y, I, Q)$ (Conversion of YIQ components to RGB components)

12. $Im = RECONST(R, G, B)$ (Reconstruction of watermarked image in RGB domain)

In the proposed watermarking algorithm, the watermark is inserted in the luminance component if the cover image considered is color in nature. The scaling factor denoted by controls the strength of the watermark image while the embedding factor denoted by α provides space for inserting the watermark. The calculated values of both parameters should range between 0 and 1 $(0 < \alpha < 1, 0 < \beta < 1)$.

9.3.3 Watermark Extraction Process

During the extraction process, the original image is used to obtain the embedding and scaling factors. The watermarked image is transformed into frequency components, and the band in which the watermark was embedded is transformed into the SVD domain. The S'_W matrix of the watermark is extracted from the SVD domain and the orthogonal components obtained from the original watermark are used to retrieve the watermark. The watermark extraction process is shown in Figure 9.2. The pseudocode for the extraction process is given below.

Algorithm [Pseudocode for Extraction Process]

Input: $W = Watermark_Extraction(Im, W, Im)$
Output: Extracted watermark (W)

1. Read Im, W, Im (Read cover image, watermark image and watermarked image)
2. $[Y, I, Q] = RGB2YIQ(Im)$ (Conversion of cover image Im from RGB to YIQ domain)
3. $[Y, I, Q] = RGB2YIQ(Im)$ (Conversion of watermarked image Im from RGB to YIQ domain)
4. $[LLY, LHY, HLY, HHY] = DWT(Y)$ (DWT on luminance component Y of cover image)
5. $[LLY, LHY, HLY, HHY] = DWT(Y)$ (DWT on luminance component Y of watermarked image)
6. $[U, S, V] = SVD(LLY)$ (SVD on frequency component of Y)
7. $[U, S, V] = SVD(LLY)$ (SVD on frequency component of Y)
8. $[Uw, Sw, Vw] = SVD(W)$ (SVD on watermark image)
9. $\alpha = PSO(Im, W)$ (Calculation of scaling factor)
10. $\beta = 1 - \alpha$ (Calculation of embedding factor)
11. $Sw = (S\beta S)/\alpha$ (Watermark extraction)
12. $W = Uw * Sw * VwT$ (Inverse SVD transform to recreate watermark)

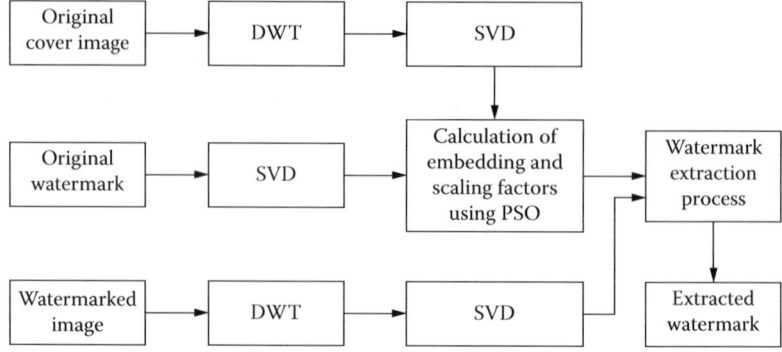

FIGURE 9.2: Watermark extraction process

9.4 Performance Analysis

The performance of the proposed watermarking algorithm is checked out in this section. The proposed work is implemented on an Intel dual core CPU with a 1 GB RAM configured system using MATLAB version R2009b. The test images used to test the performance of the proposed algorithm are shown in Figure 9.3. The original watermark used is shown in Figure 9.4. The watermarked images are shown Figure 9.5. The metrics used are explained below.

Normalized correlation coefficient (NCC): The similarity of the embedded and the extracted watermarks is checked by calculating the normalized correlation coefficient (NC) with Equation 9.8. For two identical images, NCC value is equivalent to 1, indicating the perfect similarity between them, whereas 0 indicates no similarity between them:

$$NC = \frac{\sum_{i=1}^{m}\sum_{j=1}^{n} W(i,j) * W_e(i,j)}{\sqrt{\sum_{i=1}^{m}\sum_{j=1}^{n} W^2(i,j)}\sqrt{\sum_{i=1}^{m}\sum_{j=1}^{n} W_e^2(i,j)}} \qquad (9.8)$$

where W and W_e are original and extracted watermarks of size $M \times N$. Peak signal-to-noise ratio (PSNR): This is an objective metric used to measure the quality of a watermarked image. The peak signal-to-noise ratio (PSNR) is a term for the ratio between the maximum possible power of a signal and the power of noise in the signal. PSNR is usually represented in the logarithmic

FIGURE 9.3: Original images

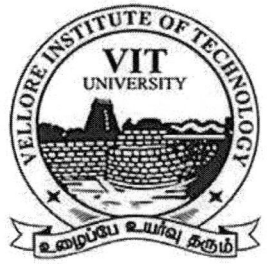

FIGURE 9.4: Original watermark

FIGURE 9.5: Watermarked images

TABLE 9.1: Calculated scaling and embedding parameters from test images.

S. No.	Type	Name of Image	Scaling Parameter	Embedding Parameter
1	Color	Lena	0.067301964	0.932698036
2	Color	Airplane	0.062420516	0.937579484
3	Color	Baboon	0.060040078	0.939959922
4	Color	Pepper	0.06004	0.93996
5	Gray scale	Barbara	0.061199138	0.938800862
6	Gray scale	Boat	0.068389376	0.931610624
7	Gray scale	Cameraman	0.064473887	0.935526113
8	Gray scale	Houses	0.062390292	0.937609708

scale. PSNR is calculated using the mean squared error (MSE). MSE and PSNR are calculated using Equations 9.9 and 9.10, respectively.

$$MSE = \frac{1}{mn} \sum_{i=1}^{m} \sum_{j=1}^{n} (W(i,j) - W_e(i,j))^2 \qquad (9.9)$$

$$PSNR = 10 log_{10} \frac{255^2}{MSE} \qquad (9.10)$$

where W and W_e are original and extracted watermarks of size $M \times N$.

In order to insert a watermark in a cover image, the scaling and embedding parameters are calculated adaptively. The adaptive calculation procedure is discussed in Section using swarm optimization. The adaptively calculated scaling and embedding factors for various test images are shown in Table 9.1. The watermark has been inserted using a calculated scaling and embedding factor. Similarly, the extracted watermark without any attack is compared with the original watermark, and it is observed that the similarity is more than 90 percent. The calculated NCC values without any attack for different images are shown in Table 9.2. The proposed algorithm is tested for both

TABLE 9.2: Similarity measure between original and extracted watermark.

S. No.	Type	Name of Image	NCC of extracted watermark
1	Color	Lena	0.99999
2	Color	Airplane	0.99999
3	Color	Baboon	0.99999
4	Color	Pepper	0.99999
5	Gray scale	Barbara	0.99999
6	Gray scale	Boat	0.99999
7	Gray scale	cameraman	0.99999
8	Gray scale	Houses	0.99999

color and gray scale images, and is presented in Table 9.3 and Table 9.4 respectively.

The above metrics show that this technique performs very well when the watermarked image is of perfect quality. However, this is not always possible in real-world scenarios where the quality of the watermarked image may be degraded. This may happen due to loss of information during transmission, when compressing the image for transmission, or it may be done on purpose as an attempt to destroy the embedded watermark. Thus, the watermarked image must be robust so that the watermark can be identified even if its quality is reduced. In order to test the robustness of the watermarked image, it is tested against various attacks. Attacks on the image basically mean simulations to reduce the quality of the image. Thus, if we are able to extract the watermark from an attacked image, the technique would be proven to be robust. The obtained results are tabulated below. The watermarking is carried out on both gray scale and color images. The obtained results for a gray scale image are given in Table 9.3. Similarly, color image results are shown in Table 9.4. The results and the analysis of each attack are given below.

1. Histogram equalization: In this attack, for all intensity values the number of pixels existing in an image are equal. It could be implemented only in gray scale images. The process of making the number of pixels equal for all intensity values is called histogram equalization. This is done for gray scale images only. The similarity of extracted watermarks has varied from 67 to 99% for the test images. The PSNR varied from 65 to 72 dB for all the gray scale images.

2. Translation: This is considered to be one of the geometric attacks. The image could be translated from one coordinate positon to another coordinate position in a two-dimensional Cartesian coordinate system. In this work, the image will be translated by (2,2). This translation process could be considered an attack. After implementing it, the similarity of the extracted watermark is measured, and it is observed that it is above 99%. The calculated PSNR varies between 62 and 69 dB for all images.

3. Resize: This is another important attack, which is frequently carried by users based on their requirements. Due to resizing, the intensity values of the pixels could undergo drastic changes. In this proposal, the watermarked images were resized by a factor of 0.5 and then resized again by a factor of 2. After implementing the resizing attack, the extracted watermark similarity values are measured, and it is observed that they vary from 86% to 99% for all the images. The calculated PSNR value for the extracted watermarks is above 71 dB in almost all cases.

4. Compression: It is considered to be a very serious attack as it removes redundant pixels' intensity values. It is important to test this attack, as almost all images are compressed using the JPEG algorithm, which

TABLE 9.3: Extracted watermark's calculated PSNR and NCC values for gray scale test image House.

Type of Attack	Attacked Watermarked Image	PSNR	Retrived Watermark	NCC
Histogram equalization		72.22		0.86
Translated by (2,2)		62.07		0.99
Resized to 1/2 of original		71.71		0.86
Compressed to 10% of original		77.55		0.99
Gaussian noise $M = 0$, $V = 0.001$		77.87		0.99

(continued)

Digital Image Watermarking in Transform Domain Using PSO 259

TABLE 9.3: (continued) Extracted watermark's calculated PSNR and NCC values for gray scale test image House.

Type of Attack	Attacked Watermarked Image	PSNR	Retrived Watermark	NCC
Salt and pepper noise $D = 0.005$		75.95		0.99
Gaussian filter hsize = 3 and $\sigma = 0.5$		80.02		0.95

removes high frequency components from the image signal. In this proposal, the watermarked bitmap images were compressed to between 3% and 14% of the original size. After implementing a compression attack, the similarity measure of the extracted watermark is observed. The similarity in all the cases was near or above 99%, and the PSNR value for almost all the cases was above 77 dB.

5. Noise addition: The noise attacks are implemented by increasing the density of Gaussian as well as salt and pepper noise. In the case of Gaussian noise, the similarity values are more than 99% for all the extracted watermarks, while the PSNR values vary between 76 dB and 78dB. In the case of salt and pepper noise, the calculated similarity values vary between 98% and 99%, while the PSNR values vary from 75 dB to 77dB.

6. Filter: In general, filtering is considered to be one image processing operation that could be used to enhance the quality of images. These could be low-pass, high-pass, band-pass, or band-reject filters. In this proposal, a Gaussian filter has been used as a low filter on all the watermarked images. The calculated similarity of all the watermarked images is observed, and it is more than 95%. The PSNR value for most of the cases has been observed, and it is above 80 dB.

TABLE 9.4: Extracted watermark calculated PSNR and NCC after various attacks from color test image Lena.

Type of Attack	Attacked Watermarked Image	PSNR	Retrived Watermark	NCC
Resized to 1/2 of original		83.27		0.98
Compressed to 10% of original		80.68		0.99
Gaussian noise $M = 0$, $V = 0.001$		77.92		0.99
Salt and pepper noise $D = 0.005$		75.75		0.99
Gaussian filter hsize $= 3$ and $\sigma = 0.5$		86.66		0.99

As per the results obtained, the proposed watermarking algorithm is robust against various attacks. Moreover, the hidden information is not recognizable, as it is inserted using scaling and embedding factors obtained using particle swarm optimization.

9.5 Conclusion

A novel adaptive invisible watermarking scheme has been proposed in this chapter. It particle swarm optimization to calculate the embedding and scaling factors adaptively. The adaptive calculation of the scaling and embedding factor is carried out in the frequency domain. The proposed methodology can be used to prevent copyright infringement of images in the public or private domain by securely inserting a watermark into the color or gray scale image. The efficiency of the proposed algorithm is verified by the results obtained through various attacks implemented on the watermarked images.

Bibliography

[1] Debi Prasanna Acharjya and Ahmed P Kauser. Swarm intelligence in solving bio-inspired computing problems: Reviews, perspectives, and challenges. *Handbook of Research on Swarm Intelligence in Engineering*, pages 74–98, 2015.

[2] Ali Al-Haj and Tuqa Manasrah. Non-invertible copyright protection of digital images using dwt and svd. In *2nd International Conference on Digital Information Management, 2007. ICDIM'07.*, volume 1, pages 448–453. IEEE, 2007.

[3] Veysel Aslantas, A. Latif Dogan, and Serkan Ozturk. DWT-SVD based image watermarking using particle swarm optimizer. In *2008 IEEE International Conference on Multimedia and Expo*, pages 241–244. IEEE, 2008.

[4] Hal Berghel and Lawrence O'Gorman. Protecting ownership rights through digital watermarking. *Computer*, 29(7):101–103, 1996.

[5] Sayan Chakraborty, Sourav Samanta, Debalina Biswas, Nilanjan Dey, and Sheli Sinha Chaudhuri. Particle swarm optimization based parameter optimization technique in medical information hiding. In *2013 IEEE*

International Conference on Computational Intelligence and Computing Research (ICCIC), pages 1–6. IEEE, 2013.

[6] Yongchang Chen, Weiyu Yu, and JiuChao Feng. A digital watermarking based on discrete fractional fourier transformation dwt and svd. In *Control and Decision Conference (CCDC), 2012 24th Chinese*, pages 1383–1386. IEEE, 2012.

[7] Ingemar J. Cox, Joe Kilian, F. Thomson Leighton, and Talal Shamoon. Secure spread spectrum watermarking for multimedia. *IEEE Transactions on Image Processing*, 6(12):1673–1687, 1997.

[8] Nagaraj V. Dharwadkar, B.B. Amberker, and Avijeet Gorai. Non-blind watermarking scheme for color images in rgb space using dwt-svd. In *2011 International Conference on Communications and Signal Processing (ICCSP)*, pages 489–493. IEEE, 2011.

[9] Mohan S. Kankanhalli, K.R. Ramakrishnan, et al. Adaptive visible watermarking of images. In *IEEE International Conference on Multimedia Computing and Systems, 1999.*, volume 1, pages 568–573. IEEE, 1999.

[10] J. Kennedy and R. Eberhart. Particle swarm optimization, IEEE International, First Conference on Neural Networks, 1995.

[11] Young-Sik Kim, O-Hyung Kwon, and Rae-Hong Park. Wavelet based watermarking method for digital images using the human visual system. *Electronics Letters*, 35(6):466–468, 1999.

[12] Zne-Jung Lee, Shih-Wei Lin, Shun-Feng Su, and Chun-Yen Lin. A hybrid watermarking technique applied to digital images. *Applied Soft Computing*, 8(1):798–808, 2008.

[13] Wei-Hung Lin, Shi-Jinn Horng, Tzong-Wann Kao, Pingzhi Fan, Cheng-Ling Lee, and Yi Pan. An efficient watermarking method based on significant difference of wavelet coefficient quantization. *IEEE Transactions on Multimedia*, 10(5):746–757, 2008.

[14] Y.-T. Lin, C.-Y. Huang, and Greg C. Lee. Rotation, scaling, and translation resilient watermarking for images. *IET Image Processing*, 5(4):328–340, 2011.

[15] J. Samuel Manoharan, Dr. Kezi C. Vijila, and A. Sathesh. Performance analysis of spatial and frequency domain multiple data embedding techniques towards geometric attacks. *International Journal of Security (IJS)*, 4(3):28–37, 2010.

[16] Samuel Manoharan. An efficient reversible data embedding approach in medical images for health care management. 2012.

[17] Peter Meerwald, Christian Koidl, and Andreas Uhl. Attack on watermarking method based on significant difference of wavelet coefficient quantization. *IEEE Transactions on Multimedia*, 11(5):1037–1041, 2009.

[18] V. Siva Venkateswara Rao, Rajendra S. Shekhawat, and V.K. Srivastava. A reliable digital image watermarking scheme based on svd and particle swarm optimization. In *2012 Students Conference on Engineering and Systems (SCES)*, pages 1–6. IEEE, 2012.

[19] V. Sivavenkateswara Rao, Rajendra S. Shekhawat, and V.K. Srivastava. A dwt-dct-svd based digital image watermarking scheme using particle swarm optimization. In *2012 IEEE Students' Conference on Electrical, Electronics and Computer Science (SCEECS)*, pages 1–4. IEEE, 2012.

[20] Tsz Kin Tsui, Xiao-Ping Zhang, and Dimitrios Androutsos. Color image watermarking using multidimensional Fourier transforms. *IEEE Transactions on Information Forensics and Security*, 3(1):16–28, 2008.

[21] Ron G. Van Schyndel, Andrew Z. Tirkel, and Charles F. Osborne. A digital watermark. In *IEEE International Conference on Image Processing, 1994. Proceedings. ICIP-94*, volume 2, pages 86–90. IEEE, 1994.

[22] Ben Wang, Jinkou Ding, Qiaoyan Wen, Xin Liao, and Cuixiang Liu. An image watermarking algorithm based on dwt, dct, and svd. In *IEEE International Conference on Network Infrastructure and Digital Content, 2009. IC-NIDC 2009.*, pages 1034–1038. IEEE, 2009.

[23] Yuh-Rau Wang, Wei-Hung Lin, and Ling Yang. An intelligent watermarking method based on particle swarm optimization. *Expert Systems with Applications*, 38(7):8024–8029, 2011.

[24] Feng Wen-ge and Liu Lei. SVD and DWT zero-bit watermarking algorithm. In *2010 2nd International Asia Conference on Informatics in Control, Automation and Robotics (CAR)*, volume 3, pages 361–364. IEEE, 2010.

[25] Pan Yinghui. Digital watermarking particle swarm optimization based on multi-wavelet. *Journal of Convergence Information Technology*, 5(3), 2010.

Part III

Bio-Inspired Computing Applications to Image and Video Processing

Chapter 10

Evolutionary Algorithms for the Efficient Design of Multiplier-Less Image Filter

Abhijit Chandra

Jadavpur University, India

10.1	Introduction	268
10.2	Theoretical Background of Evolutionary Algorithms	270
	10.2.1 Comparison Among State-of-the-Art Evolutionary Algorithms	270
	10.2.2 Recent Evolutionary Algorithms based on GA	272
10.3	Proposed Models	274
10.4	Results and Observations	276
10.5	Conclusion	291
	Bibliography	292

This chapter makes an attempt to throw sufficient light on the impact of evolutionary algorithms in some recent areas of image processing, and involves the design of a hardware efficient image filter. Reducing the hardware complexity of digital systems has received reasonable attention from the research community over the last few decades. Hardware costs of digital filters may have been minimized by encoding the individual coefficients as the sum of signed powers of two (SPT), which can substitute the operation of multiplication by means of simple addition or subtraction and shifting. Design of such a multiplier-less low-pass image filter with the aid of evolutionary computational algorithms is explicitly demonstrated in this chapter. Two such popularly employed, recent evolutionary techniques, namely differential evolution (DE) and self-organizing random immigrants genetic algorithm (SORIGA), are elaborately discussed in this regard. Characteristics of the image filters have been analyzed from different perspectives and subsequently compared with some other standard two-dimensional filters. Finally, the filters that were designed have been employed to minimize the effect of Gaussian noise from a few benchmark test images. Performance of these image denoising filters has been analyzed with respect to a few performance parameters, such as peak signal to noise ratio (PSNR), structural similarity index measure (SSIM), image enhancement factor (IEF), image quality index (IQI), and so on. A number of image filters, as available from the literature, have also been included for the purpose of making a fair comparison.

10.1 Introduction

Design of two-dimensional image filters has undergone through serious revolutions over the past few years, with a constant aim to improve the performance of the designed filter. In order to improve the hardware efficiency, mask coefficients of those filters are represented in the form of SPT, and these were subsequently named multiplier-less filters. The first written article on the design of two-dimensional (2D) multiplier-less filters appeared in 1987 when Pei and Jaw [25] took a pioneering initiative using a special class of multiplier-less 1D filter with coefficients in the form of SPT, and incorporating McClellan transformation to map the one-dimensional filter into a two-dimensional one. These filters are efficient and reliable for high speed computation, which had made them attractive in terms of hardware usage.

Because the structure in [25] is valid only for original first-order McClellan transformation, one new analytical approach for the determination of the coefficients of the first-order McClellan transformation was subsequently presented by Kwan and Chan [19]. It has also been established that the analytical approach results in further improvement over the original first-order McClellan transformation. The design problem of 2D linear-phase filter was later targeted by a generalized McClellan transformation with order more than one, and was formulated as a linear programming (LP) optimization problem to maximize the transition width of a 1D FIR filter subject to the inequality constraints in the 2D frequency domain [10], followed by a local search method for efficiently finding the appropriate powers-of-two coefficients. This optimization algorithm eliminates the drawbacks of high computational cost and huge memory storage, as encountered in conventional LP-based algorithms.

A new algorithm was introduced in [31] for the synthesis of a 2D linear-phase filter with finite precision coefficients, in which the minimax strategy has been employed. Linear programming and a branch and bound technique have been jointly incorporated into this algorithm, for which two strategies are compared. These two search techniques are a depth-first-search and hybrid strategy consisting of depth-first-search and breadth-first-search. Efficiency of the proposed method in [31] for the design of 2D filters with different specifications and sizes has been validated with the help of a large number of design examples. Minimax design of 2D multiplier-less FIR filters with SPT coefficients was later addressed by one simulated annealing (SA)-based design technique in [3], which has proved to be intrinsically very flexible. These filters have proved to be extremely useful in the context of video signal processing, as demonstrated by a number of filter design examples. Design of 2D linear-phase FIR filters with continuous and discrete coefficients was later described in [20]. It had initially been formulated as an LP problem with inequality constraints. Based on the obtained continuous coefficients, an efficient method was proposed for designing a 2D filter with powers-of-two coefficients in the spatial domain.

A number of artificially intelligent optimization techniques have been successfully employed in the design of 2D multiplier-less filters with SPT coefficients. Sriranganathan and his co-researchers have taken a pioneering initiative in this regard by designing circularly symmetric and diamond-shaped, low-pass linear-phase powers-of-two FIR filters with the aid of the genetic algorithm (GA) [32]. An objective function of the optimization technique has been formulated by minimizing the weighted ripple in both pass-band and stop-band. The authors had also established that the filters designed with the aid of GA yield better or comparable performance than the filters designed with the aid of LP and SA. Another efficient design method of multiplier-less 2D state-space digital filters (SSDF) with the help of GA has been proposed in [43]; it seems to be attractive for high speed operation and simple implementation. The design problem is described by Roesser's local state-space model, and formulated subject to the stability of the resultant filter. Two different types of GA, namely binary-GA and integer-GA, were later employed [36] to determine the periodically shift variant (PSV) coefficients of a 2D filter. Impulse response of the 2D PSV filter in closed form is initially obtained followed by the application of GA to search for the filter coefficients. GA-based design strategy of 2D FIR filters with complex-valued frequency response was proposed in [39], extending the concept of 1D filter design. As an attempt to realize the filter coefficients with an evolutionary algorithm, the proposed technique generates real-valued chromosomes by minimizing the quadratic measure of error in the frequency band. It has also been shown that some of the coefficients in the designed filters are inherent to zero, and therefore result in significant savings in design time. Design of 2D filters has also been attempted by means of an advanced GA in [5]. It can adapt the genetic operators during the genetic life cycle and maintain the simplicity and ease of implementation as well. Adaptive GA has synthesised filters with good response characteristics along with a significant reduction in the error criteria and CPU time. GA combined with singular value decomposition (SVD) has been employed in [13] for the design of 2D filters in which the role of GA was to optimize the design of a 1D filter. Design of a 2D SPT linear-phase filter has recently been accomplished by developing a 1D multiplier-less linear-phase FRM FIR filter followed by multiplier-less transformation [23, 22]. The resulting 1D filter is converted to the CSD space using a new discrete optimization based on a modified gravitational search algorithm (GSA) [23] and modified harmony search algorithm (HSA) [22], which are endowed with the features of reduced computational complexity and time. During the course of optimization, GSA and HSA have been adapted in such a way that candidate solutions turn out to be integers and efficient exploration and exploitation of the search space are realized. One comparative study among various evolutionary algorithms applied for the design of 2D FIR filters has been elaborated in [4]. Finally, a new type of GA has been proposed with the introduction of some innovative concepts, to optimize the trade-off between diversity and elitism in the genetic population.

10.2 Theoretical Background of Evolutionary Algorithms

A method of optimization can be found inside many natural processes, including Darwinian evolution. It has been well observed by researchers that there exists a fine correlation between the theory of optimization and biological evolution. This was finally given a recognizable shape in the form of the genetic algorithm (GA) in 1992 by Holland [16]. Since then, the field of evolutionary optimization techniques has undergone substantial growth and has come up with a number of computational tools of practical interest. Each of these techniques has its own advantages and limitations over others. This issue is described in detail in this section, followed by a discussion of the development of recent evolutionary algorithms based on GA.

10.2.1 Comparison Among State-of-the-Art Evolutionary Algorithms

Different types of non-linear and non-differentiable functions have found a wide application in a number of branches of engineering. In order to successfully optimize these non-linear and non-differentiable functions, innovation of a new optimization technique was necessary. In the year 1995, one such new floating-point-encoded evolutionary algorithm for global optimization was proposed [34, 35, 26, 27, 11, 12]. As this new optimization algorithm involves a new type of differential operator the method was termed differential evolution (DE). It incorporates a very few control parameters with negligible parameter tuning, and can be implemented very easily. This has made the algorithm very popular since its inception [11]. This particular optimization process can be utilized in an effective way in various engineering problems, such as design of digital filters [33, 7, 18, 21], electromagnetic inverse problems [28, 24], antennas [42], composite materials [29], and so on.

DE has opened a new door of evolutionary computation and has been successfully confirming its supremacy to date. Acceptability of DE is being enhanced day by day because of its powerful features of common interest.

Complexity of computational methods is one of the reasons why researches have avoided them for so long. However, the debut of DE has changed things a lot and makes people think about applying an evolutionary approach in their own field of interest. The main body of the algorithm takes very few lines to code in any programming language, which makes DE uncomplicated. Although particle swarm optimization (PSO) is implemented by means of minimum complexity, its performance is immensely affected over a wide range of problems. As pointed out by a number of studies [11, 12], performance of DE is significantly superior to its antecedents, such as stochastic approach, biased roulette wheel (MA-S2), adaptive Lévy evolutionary programming (ALEP), and early cooperative particle swarm optimization (CPSO-H) in problems having unimodal or multimodal search spaces with separable and non-separable variables. The second important observation about DE is the presence of a very few control parameters associated with it, namely weighting factor, recombination probability, and population size [34, 35, 26, 12]. In order to ensure proper functioning of DE, the program developer needs to take care only with those limited values, to guarantee higher accuracy, better convergence speed, and robustness. Finally, the space complexity is another important issue that has helped in extending DE for handling large-scale and expensive optimization problems.

GA, PSO, and DE are evolutionary or swarm computation paradigms that have shown their superior performance on complex non-linear function optimization problems. Their general applicability to numerical optimization problems has been extensively carried out in the literature on a suite of 34 numerical benchmark problems. There are very few functions for which a consistent performance pattern is observed for all algorithms [27]. Both PSO and DE converge exponentially to the optimum fitness value. However, GA, being a simple evolutionary algorithm, is many times slower than PSO and DE. More specifically, it needs several hours to find the solution whereas this can be achieved within a few seconds for PSO and DE [12]. For any particular function, DE converges towards the optimum value with an exponential progressing rate. Under a similar scenario, PSO performs moderately better than GA. For a test function consisting of plateaus, PSO becomes incompetent in optimizing them while GA and DE can easily find the optimum solution.

Multimodal functions have urged a special interest in the field of optimization theory. DE prominently outperforms other optimization techniques and is capable of finding out the optimum solution for such functions. On the other hand, neither GA nor PSO can find the global optimum solution for these functions in any of the trial runs. As far as the other relevant functions are concerned, DE yields an optimum or near-optimum solution much more quickly than any other algorithms [12]. However, there exist a very few noisy problems for which performance of DE becomes the worst as compared to other techniques. DE has also been found suitable for multi-objective optimization problems [2].

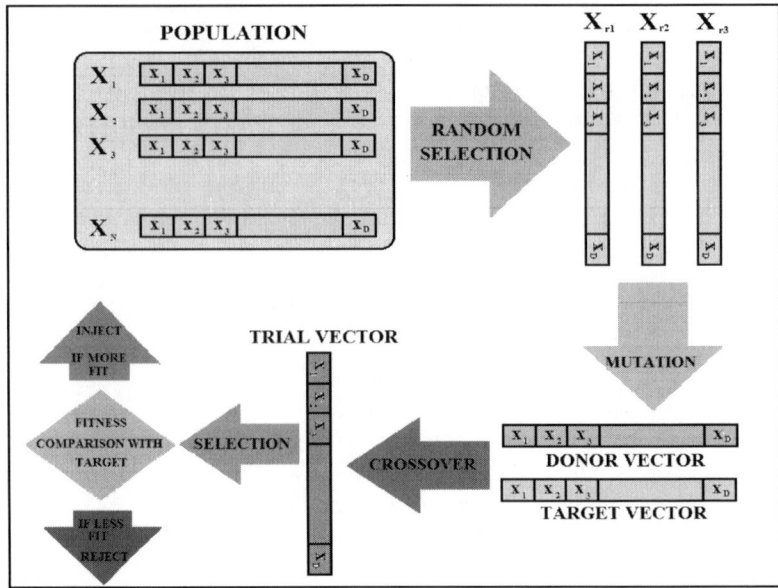

FIGURE 10.1: Schematic diagram of differential evolution algorithm

In brief, DE emerges as the best performing algorithm as far as the research outcome is concerned, except for one or two functions where DE either shows a slower response than its competitors or stagnates on a suboptimal value. The most remarkable feature associated with DE is its capability for producing the same results consistently for a number of trial runs [11, 12]. In contrast to this, performance of PSO is highly dependent on the randomized initial value of the individuals. This difference becomes reflective for 100 dimensional benchmark problems. Hence, PSO must be executed several times in order to obtain the best result. Since PSO is more sensitive to parameter changes than other optimization mechanisms, different settings of parameter are crucial for PSO to optimize the wide range of problems at hand [11]. GA and DE can be operated using the same settings as the initial population, and they continue to produce the same convergence speed, which may not be achieved by PSO. Except for only a few functions, DE also shows a great fine-tuning ability, which has produced sufficient interest among researchers. A complete schematic diagram of the DE algorithm is depicted in Figure 10.1.

10.2.2 Recent Evolutionary Algorithms based on GA

The genetic algorithm (GA) was the first computational model proposed by Holland and his co-researchers that truly reflected the idea of genetic evolution. The most attractive feature of GA has been its ability of parallel computation, which yields a sufficient number of candidate solutions for any given problem of interest [16]. However, GA suffers from premature convergence, where the optimum solution is generally confined to the local optimum point. As a matter of fact, a number of new variants of the conventional genetic algorithm (CGA) have been proposed in recent times, and they have attempted to overcome the limitations of CGA.

One such modification takes place in the form of a micro-genetic algorithm (μGA), which is popular mostly because of its small size of population. μGA starts with a population of five randomly chosen individuals, among which the individual with the largest fitness value is copied to the next generation, and the other four individuals are selected by means of the tournament selection strategy [1]. Moreover, unlike the rates for CGA, cross-over and mutation rate for μGA are set to 1 and 0, respectively. Due to the presence of a limited number of potential solutions, convergence speed of μGA is relatively faster than that of CGA. However, because of the presence of small population size, there is a likelihood that μGA may be trapped into a local optimum point. Although the performance of μGA seems to be satisfactory for a medium chromosome, it gets affected when the size of the chromosome is large enough. Subsequently, μGA has been modified to improve the convergence characteristics by varying the probabilities of cross-over and mutation in a different convergence stage, instead of having fixed probabilities for the whole evolution process [6]. This has prevented the premature convergence of μGA and improved the convergence speed for long chromosome optimization scenarios.

Concept of flux of immigrants in a biological population was exploited very recently to develop a new computational model, known as random immigrants GA (RIGA) [15]. RIGA helps the population members to escape from the local minima point during occasional environmental changes. The flux of immigrants generally enhances the genetic diversity of the entire population during genetic searching, and hence allows the search mechanism to find out its global optimal point with relative ease. RIGA differs from CGA in the sense that it necessitates the replacement of some of the population members with randomly generated vectors, which has given rise to a new terminology called replacement rate, signifying the number of newly entered individuals in the current population set.

Occurrence of self-organizing criticality (SOC) in natural evolution processes has specially drawn the attention of the researchers, in particular as SOC helps the systems to self-organize themselves into a critical state even without any external perturbations. The Bak-Sneppen simulation model [30] has already described a strong relation between the natural evolution and SOC, which demonstrates how the replacement of the worst individual along with its next two neighbors originates a chain reaction in a critical state, and thus affects all the individual members in the population. Incorporation of SOC into a natural evolution process protects an individual element from getting trapped in a local optimum point in the fitness landscape. The idea of SOC has recently been adopted in RIGA in the form of a new evolutionary algorithm, called self-organizing RIGA (SORIGA) [37]. However, uncertainty in ensuring self-organizing behavior always exist in this scheme, since the new random immigrant individuals with lower fitness value may be substituted for an individual with higher fitness during the process of selection. In order to address this issue properly, two major modifications were adopted in the execution of SORIGA.

The new immigrants who enter the population set during the time of replacement in SORIGA are preserved in another set, called the subpopulation. Members of the subpopulation precisely identify those elements that replace the worst individuals along with their neighbors. Moreover, each member of the entire population set is associated with a bit that identifies whether or not that particular element has been replaced at least once in the current chain reaction (replacement event). In the case that the individual element is not replaced by any random immigrant, the new individual is selected from the main population; otherwise, the selection mechanism involves the members from the subpopulation. In SORIGA, the individuals in the current population not belonging to the subpopulation are not allowed to replace members from the subpopulation [37]. In brief, SORIGA can be interesting in dynamic optimization problems where diversity is a key factor. SORIGA controls the diversity of the population using self-organization, i.e., without an explicit control of the diversity of the population.

10.3 Proposed Models

Design strategy of a two-dimensional multiplier-less low-pass image filter with the aid of self-organizing random immigrants genetic algorithm (SORIGA) [37] has been illustrated in this section. Fitness function of the optimization process has been constructed as the inverse of the squared error between the frequency samples of the ideal and designed low-pass filters. It has been mathematically illustrated as follows:

$$\mathfrak{F}_i = \left\{ \sum_{u=1}^{N} \sum_{v=1}^{N} |W(\omega_\mu, \omega_\nu) - H_i(\omega_\mu, \omega_\nu)|^2 \right\}^{-1} \quad (10.1)$$

where N is the number of frequency samples in the vertical and horizontal directions, $H_i(\omega_\mu, \omega_\nu)$ is the normalized frequency response of the ith mask coefficient $h_i(m,n)$, and $W(\omega_\mu, \omega_\nu)$ is the frequency response of ideal 2-D low-pass filter, defined as below, where z signifies the corresponding cut-off frequency.

$$W(\omega_\mu, \omega_\nu) = \begin{cases} 1 & for \sqrt{(u-\frac{N}{2})^2 + (v-\frac{N}{2})^2} \leq z \\ 0 & for \sqrt{(u-\frac{N}{2})^2 + (v-\frac{N}{2})^2} > z \end{cases} \quad (10.2)$$

Mask coefficients of the 2-D filter have been encoded as chromosomes, which are allowed to take part in subsequent genetic operations, such as replacement, selection, cross-over, and mutation, in accordance with SORIGA. It can be established that one $(2M+1) \times (2M+1)$ filter mask is in need of at most $\mathbb{M} = (M+1)^2$ number of distinct coefficients, each of which is encoded by at most Δ bits, where Δ is the word length of the mask coefficient. As a matter of fact, the entire set of potential solutions is populated by P number of chromosomes, each of length $\mathbb{M}.\Delta$, and stored in the set $\mathbb{I} = \mathcal{C}_i, i \in \mathbb{P}_{main} = \{1, 2, 3, \cdots, \mathcal{P}\}$, where an individual chromosome $\mathcal{C}_i$ may be represented as:

$$\mathcal{C}_i = [c_{i,0} c_{i,1} c_{i,2} \ldots c_{i,L-1}] \forall i \in \mathbb{P}_{main} \quad (10.3)$$

The proposed design strategy starts with identifying the worst chromosome $\mathcal{C}_{worst}$ for which $\mathfrak{F}_{worst} < \mathfrak{F}_i \ \forall \ i \in \mathbb{P}_{main}$. Chromosome $\mathcal{C}_{worst}$ along with its neighbors are subsequently replaced by random individuals and stored in a sub-population $\mathbb{P}_{sub}$. The method of selection identifies the most promising chromosomes from the population set $\mathbb{P}_{CR} \subset \{\mathbb{P}_{main} \cup \mathbb{P}_{sub}\}$ in accordance with Darwin's theory of "survival of the fittest." During the selection of each individual from the new population $\mathbb{P}_{CR}$, each of the members are chosen either from the main population $\mathbb{P}_{main}$ if its index was not involved in the current replacement event, or from the sub-population $\mathbb{P}_{sub}$ in the case where its index is found to be involved in the current replacement event.

Chromosomes from the set $\mathbb{P}_{CR}$ are allowed to take part in subsequent genetic operations, such as cross-over and mutation. Let, for example, $\mathcal{C}_i$ and $\mathcal{C}_{i+1}$ be two arbitrary chromosomes that exchange their body parts at a point during the cross-over procedure of SORIGA, and consequently generate two new chromosomes, namely $\mathfrak{R}_i$ and $\mathfrak{R}_{i+1}$, to take an active part in the mutation operation to follow. The scheme has been depicted mathematically as follows:

$$\mathfrak{R}_i = [\mathcal{C}_{i,1}, \mathcal{C}_{i,2}, \cdots, \mathcal{C}_{i,\mu}, \mathcal{C}_{i+1,\mu+1}, \mathcal{C}_{i+1,\mu+2}, \cdots, \mathcal{C}_{i+1,L}] \forall i \in \mathbb{P}_{CR} \quad (10.4)$$

$$\mathfrak{R}_{i+1} = [\mathcal{C}_{i+1,1}, \cdots, \mathcal{C}_{i+1,\mu}, \mathcal{C}_{i,\mu+1}, \mathcal{C}_{i,\mu+2}, \cdots, \mathcal{C}_{i,L}] \forall i \in \mathbb{P}_{CR} \quad (10.5)$$

The operation of bit-reversal, formally known as mutation, enhances the potential diversity of the recombined chromosomes to a significant extent. The mutated chromosome $\mathfrak{W}_i$ is related to the recombined chromosome $\mathfrak{R}_i$ in accordance with the following equation:

$$\mathfrak{W}_i = \begin{cases} \mathfrak{R}_{i,j} & for \ j \neq rand(1,L) \\ 1 - \mathfrak{R}_{i,j} & for \ j = rand(1,L) \end{cases} \quad (10.6)$$

Being a population-based algorithm, SORIGA yields a number of solutions with different fitness functions at the end of every iteration. The process of replacement, selection, cross-over, and mutation is executed several times until and unless one optimum chromosome $\mathcal{C}_{opt}$ is obtained for which $\mathfrak{F}(\mathcal{C}_{opt}) >$

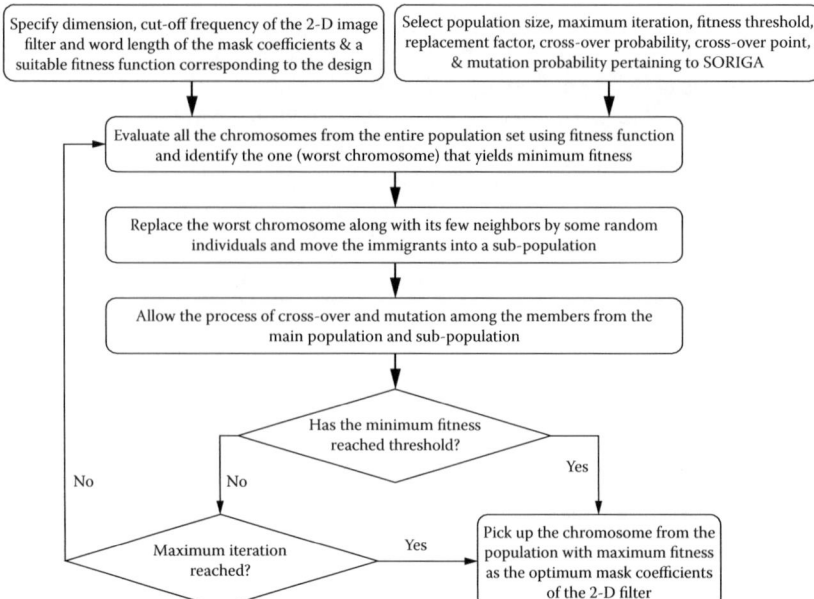

FIGURE 10.2: Flow chart for the design of the SORIGAML2D filter

$\mathfrak{F}(\mathcal{C}_i) \ \forall i \in \{\mathbb{P}_{main} \cup \mathbb{P}_{CR}\}$. Optimum mask coefficients $\mathbb{H}^{opt}$ of the 2-D low-pass filter can simply be obtained from $\mathcal{C}_{opt}$ using the function Θ as:

$$\mathbb{H}^{opt}_{(2M+1)\times(2M+1)} = \Theta_{(2M+1)}\{\xi_{opt,0}, \xi_{opt,1}, \cdots, \xi_{opt,(M-1)}\} \quad (10.7)$$

where

$$\xi_{i,l} = \sum_{j=0}^{\Delta-1} c_{i,k} 2^{-j} \forall i \in \{\mathbb{P}_{main} \cup \mathbb{P}_{CR}\} and \forall l = \{0, 1, 2, \cdots, \mathbb{M}-1\} \quad (10.8)$$

with $j = k \bmod(\Delta)$ and $l = (k-j)/\Delta$

The entire design process of the SORIGA-optimized multiplier-less two-dimensional (SORIGAML2D) low-pass filter has been explicitly illustrated by means of a flow chart, as shown in Figure 10.2. A similar strategy for designing a multiplier-less image filter using the DE algorithm (DEML2D) is presented in Figure 10.3 for the purpose of illustration.

10.4 Results and Observations

This section clearly demonstrates the performance of a 2D multiplier-less low-pass filter designed with the aid of DE and SORIGA, respectively. Characteristics of the designed filter in the frequency domain have been elaborately

Evolutionary Algorithms for Design of Multiplier-Less Image Filter 277

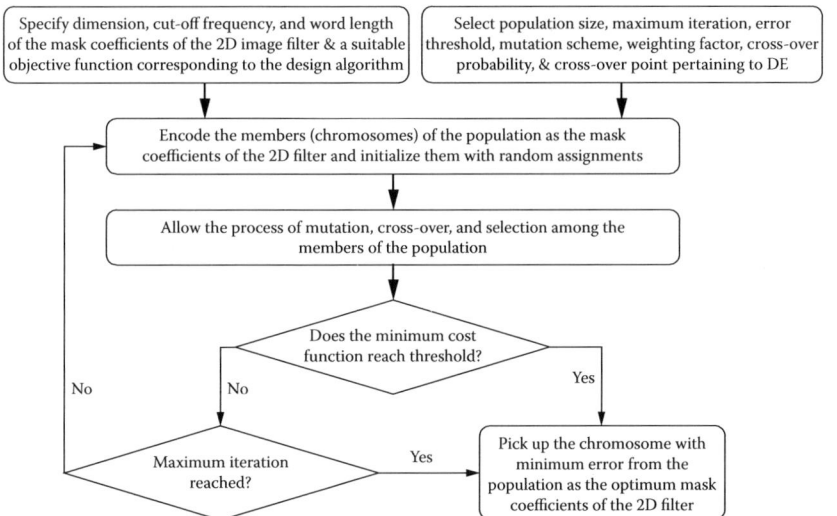

FIGURE 10.3: Flow chart for the design of the DEML2D filter

presented by means of 2D and 3D views of the frequency response. Finally, the designed filter has been utilized for reducing the effect of zero-mean Gaussian noise from a number of test images. In connection with this, a comparative analysis has been carried out between the proposed and other state-of-the-art filters in terms of relevant performance parameters.

The quality of the 2D low-pass filter in eliminating the high frequency components from the input image can best be evaluated by observing the behavior of the filter concerned in the transformed domain. Frequency response of the DE-optimized multiplier-less 2D (DEML2D) and SORIGA-optimized multiplier-less 2D (SORIGAML2D) filter of size 3×3 has been depicted in Figure 10.4, along with that of a conventional low-pass linear 2D filter of similar order. Figure 10.5 demonstrates the nature of the frequency response for the DEML2D and SORIGAML2D filter of higher order along with some of the state-of-the-art 2D filters of recent interest.

The above figures unambiguously establish the effectiveness of DEML2D and SORIGAML2D filters, which allow the low frequency components of an image while suppressing its high frequency substituent (like noise) to a significant extent. It can be clearly seen from the above illustrations that the frequency response of the newly proposed filters is distinctively divided into two segments corresponding to the pass-band and stop-band of the filter. This has consequently enhanced the sharpness of the designed filter, as opposed to the other conventional low-pass 2D filter. However, as observed from Figure 10.4, frequency response of mean, circular, and Gaussian filters in particular consists of grey variations, which results in considerable blurring of the processed image. This has been demonstrated further by displaying the three-

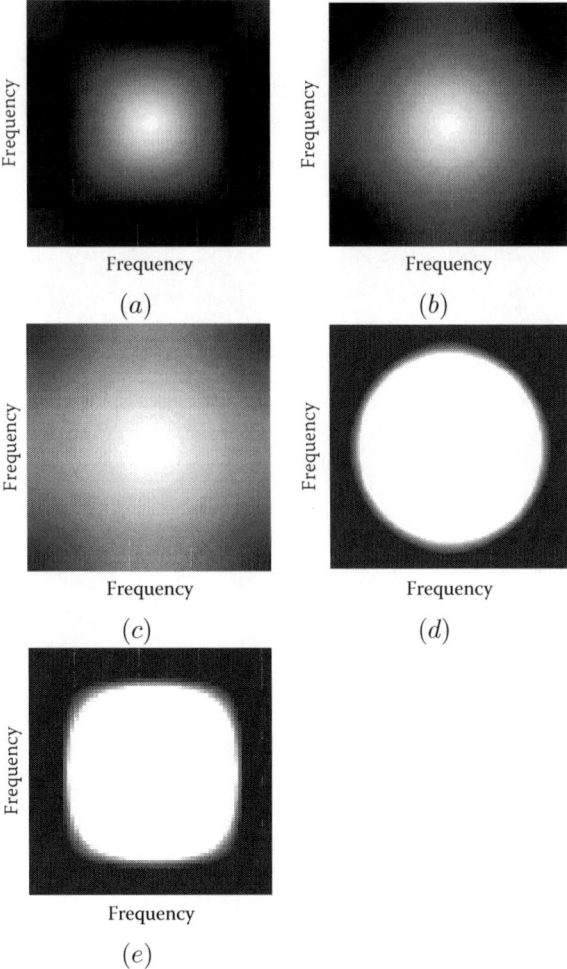

FIGURE 10.4: 2D view of frequency response corresponding to (a) mean, (b) circular averaging, (c) Gaussian, (d) DEML2D [9], (e) SORIGAML2D [8] filter of size 3 × 3

dimensional (3D) view of each of these low-pass image filters as shown in Figure 10.6, Figure 10.7, and Figure 10.8.

Frequency characteristics of different 2D low-pass filters can provide qualitative information regarding the efficiency of the filtering process. However, a more realistic analysis may be carried out in terms of various performance parameters that are directly related to the quality of the filtered image. Denoising performance of the filter is generally evaluated in terms of a few parameters, namely peak signal-to-noise ratio (PSNR), structural similarity index measure (SSIM), image enhancement factor (IEF) and image quality index (IQI), which

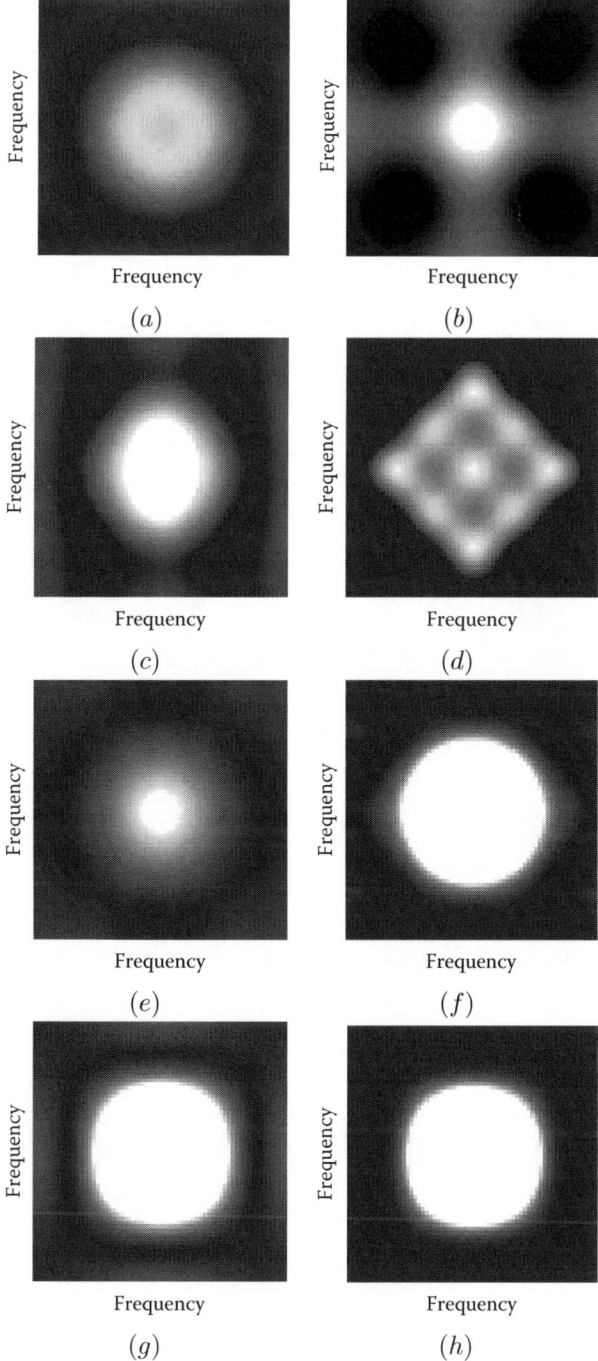

FIGURE 10.5: 2D view of frequency response for the filters in (a) Sriranganathan [32], (b) Siohan [31], (c) Lee [20], (d) Chen [10], (e) DEML2D5X5, (f) DEML2D7X7, (g) SORIGAML2D5X5, (h) SORIGAML2D7X7

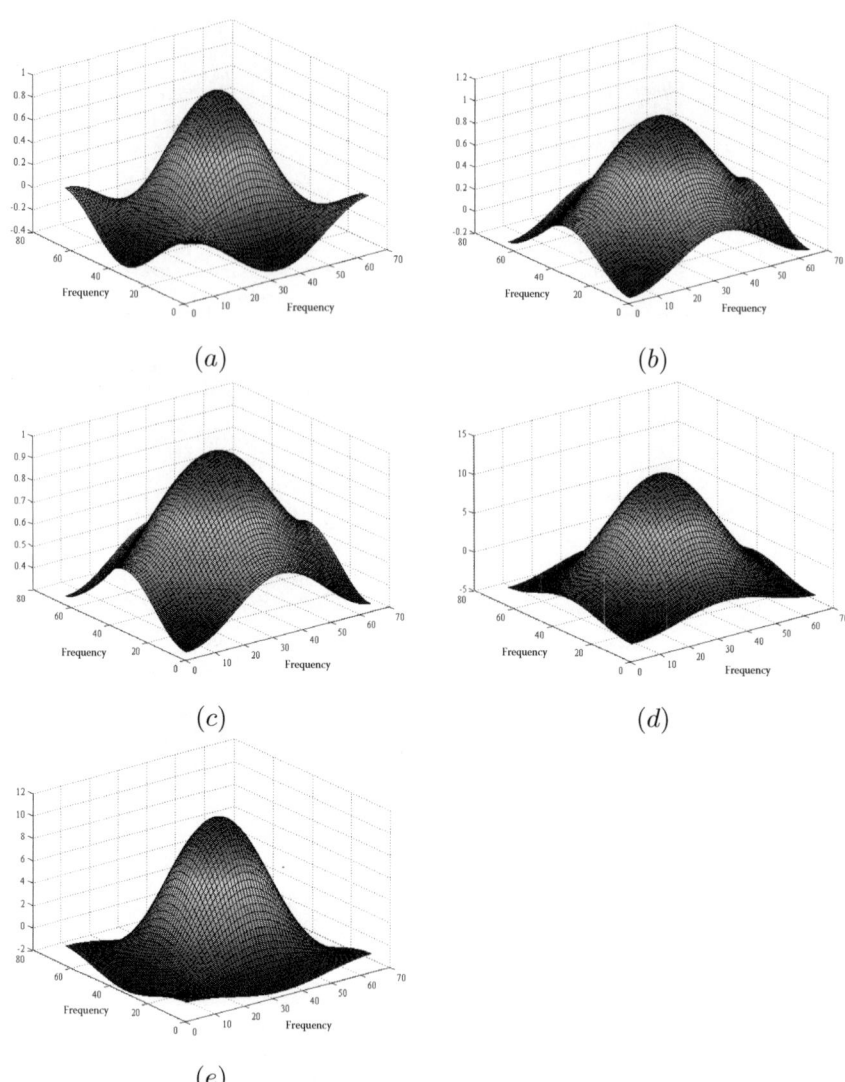

FIGURE 10.6: 3D view of frequency response corresponding to (a) mean, (b) circular averaging, (c) Gaussian, (d) DEML2D [9], (e) SORIGAML2D [8] filter of size 3×3

are defined as follows [41, 40].

$$PSNR = 20\log_{10}\left(\frac{\max_{i,j}\{I(i,j)\}}{\sqrt{\frac{1}{RC}\sum_{i=1}^{R}\sum_{j=1}^{C}\{I(i,j)-F(i,j)\}^2}}\right) \quad (10.9)$$

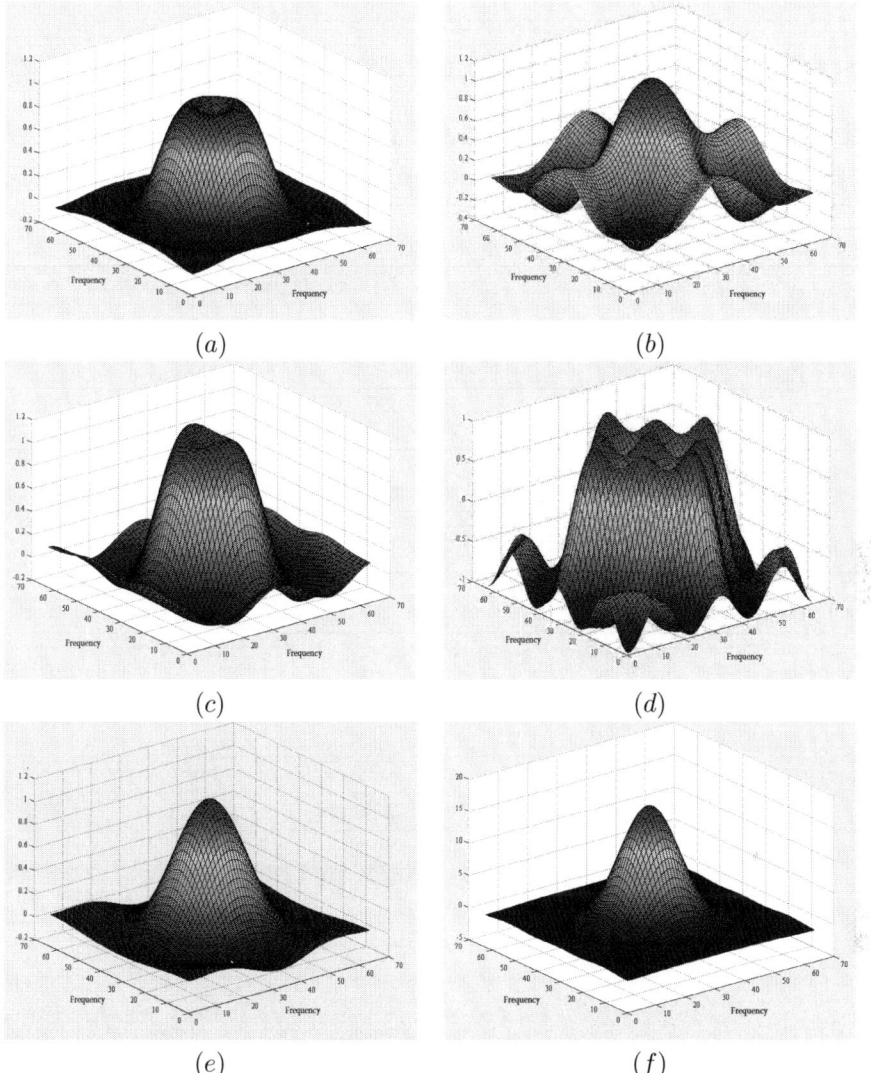

FIGURE 10.7: 3D view of frequency response for the filters in (a) Sriranganathan [32], (b) Siohan [31], (c) Lee [20], (d) Chen [10], (e) DEML2$D_{5\times5}$ (f) DEML2$D_{7\times7}$

$$SSIM(I,F) = \frac{2\mu_I\mu_F(2\sigma_{IF} + K_2)}{(\mu_I^2 + \mu_F^2 + K_1)(\mu_I^2 + \mu_F^2 + K_2)} \quad (10.10)$$

$$IEF = \frac{\sum_{i=1}^{R}\sum_{j=1}^{C}\{N(i,j) - I(i,j)\}^2}{\sum_{i=1}^{R}\sum_{j=1}^{C}\{F(i,j) - I(i,j)\}^2} \quad (10.11)$$

$$IQI = corr(I,F).lum(I,F).cont(I,F) \quad (10.12)$$

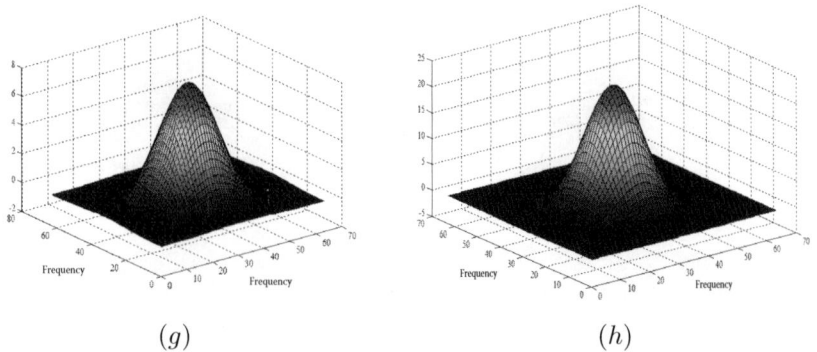

(g)   (h)

FIGURE 10.8: 3D view of frequency response for the filters in (g) SORIGAML2D$_{5\times5}$ and (h) SORIGAML2D$_{7\times7}$

where, $corr(I,F) = \dfrac{\sigma_{IF}}{\sigma_I \sigma_F}$, $lum(I,F) = \dfrac{2\mu_I \mu_F}{(\mu_I^2 + \mu_F^2)}$, and $cont(I,F) = \dfrac{2\sigma_I \sigma_F}{\sigma_I^2 + \sigma_F^2}$.

$I(i,j)$, $N(i,j)$, and $F(i,j)$ represent the input image, noisy image, and filtered image of dimension $R \times C$, respectively. Parameters μ and σ indicate the mean intensity and standard deviation from the mean intensity, respectively, and are defined as:

$$\mu_X = \frac{1}{RC}\sum_{i=1}^{R}\sum_{j=1}^{C} X(i,j) \text{ where } X = I \text{ or } F \qquad (10.13)$$

$$\sigma_X = \frac{1}{RC}\sum_{i=1}^{R}\sum_{j=1}^{C} \{X(i,j) - \mu_X\} \text{ where } X = I \text{ or } F \qquad (10.14)$$

$$\sigma_{IF} = \frac{1}{RC}\sum_{i=1}^{R}\sum_{j=1}^{C} \{I(i,j) - \mu_I\}\{F(i,j) - \mu_F\} \qquad (10.15)$$

Based on the above parameters, performance of the DEML2D and SORIGAML2D filters in eliminating Gaussian noise from four different test images, namely Lena, Moon, MRI, and Cameraman, have been analyzed in this work. In connection to this, original, noisy, and filtered images are displayed in Figure 10.9, Figure 10.10, Figure 10.11, Figure 10.12, Figure 10.13, and Figure 10.14, respectively, for visualizing the improvement in image quality. However, due to space constraints, detailed numerical analysis corresponding to the Moon image alone has been presented in Table 10.1, Table 10.2, Table 10.3, and Table 10.4.

Evolutionary Algorithms for Design of Multiplier-Less Image Filter   283

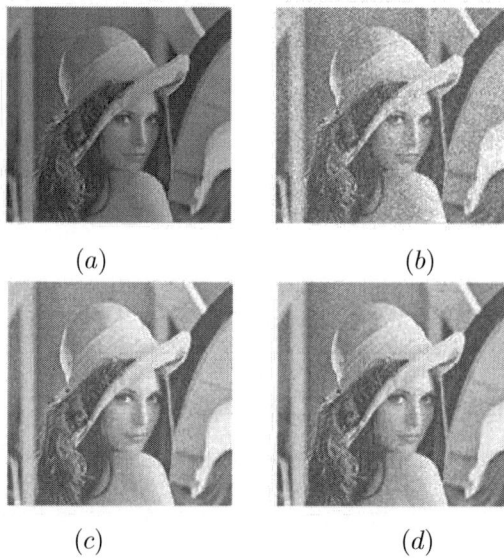

FIGURE 10.9: (a) Original, (b) noisy (Gaussian noise with $\sigma = 0.25$), (c) mean-filtered, (d) circular-filtered Lena image

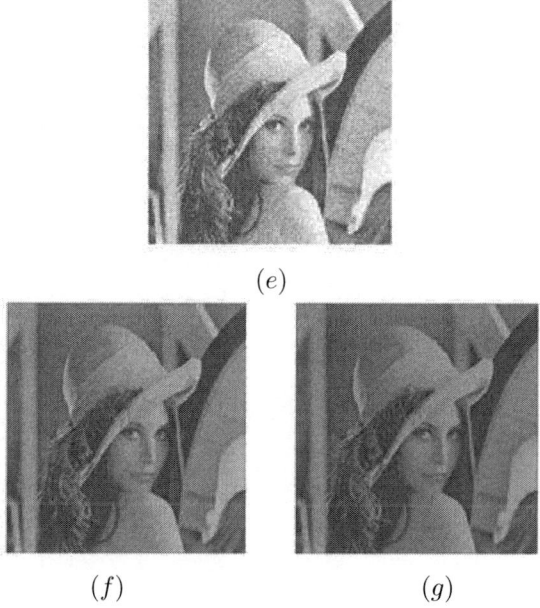

FIGURE 10.10: (e) Gaussian-filtered, (f) DEML2D-filtered, (g) SORIGAML2D-filtered Lena image

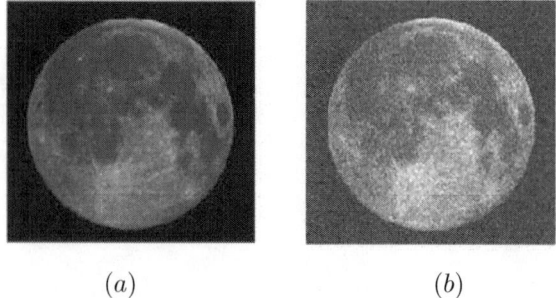

(a) (b)

FIGURE 10.11: (a) Original (b) noisy (Gaussian noise with $\sigma = 0.25$) Moon image

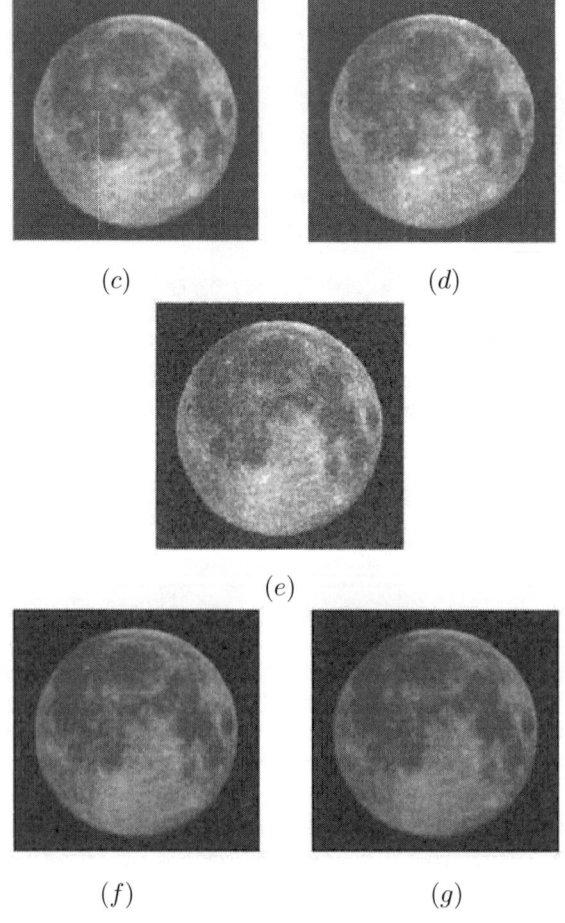

(c) (d)

(e)

(f) (g)

FIGURE 10.12: (c) Mean-filtered, (d) circular-filtered, (e) Gaussian-filtered, (f) DEML2D-filtered, (g) SORIGAML2D-filtered Moon image

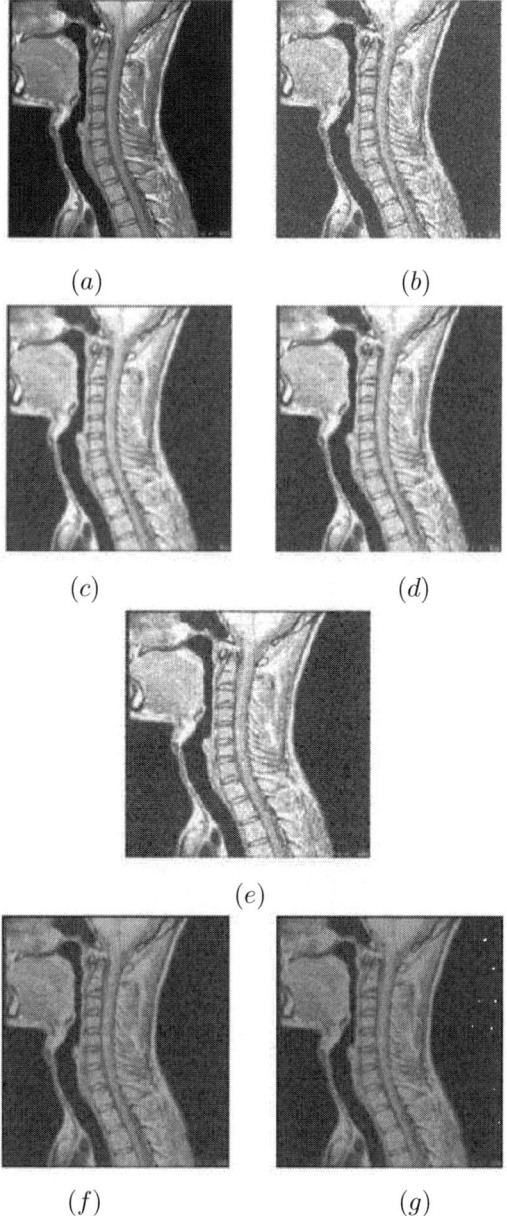

FIGURE 10.13: (a) Original, (b) noisy (Gaussian noise with $\sigma = 0.25$), (c) mean-filtered, (d) circular-filtered, (e) Gaussian-filtered, (f) DEML2D-filtered, (g) SORIGAML2D-filtered MRI image

FIGURE 10.14: (a) Original, (b) noisy (Gaussian noise with $\sigma = 0.25$), (c) mean-filtered, (d) circular-filtered, (e) Gaussian-filtered, (f) DEML2D-filtered, (g) SORIGAML2D-filtered Cameraman image

TABLE 10.1: Comparison in terms of PSNR (DB) resulting from various filters at different noise intensities for Moon image.

Filter type	Standard deviation of Gaussian noise				
	0.1	0.15	0.2	0.25	0.3
Mean (3×3)	19.05	16.11	13.8	11.97	10.44
Mean (7×7)	18.89	16.08	13.82	11.99	10.49
Disk (3×3)	18.96	16.04	13.77	11.92	10.41
Disk (7×7)	19.14	16.18	13.87	12.02	10.48
Gaussian (3×3)	18.39	15.72	13.56	11.78	10.3
Gaussian (7×7)	18.37	15.72	13.57	11.79	10.3
Siohan	16.16	13.63	11.61	9.96	8.56
Lee	17.62	14.91	12.75	11.02	9.55
DEML2D (3×3)	23.07	20.96	18.56	16.43	14.65
SORIGAML2D (3×3)	22.65	21.47	19.45	17.42	15.63
DEML2D (7×7)	19.41	20.01	19.97	19.34	18.29
SORIGAML2D (7×7)	22.6	21.45	19.46	17.45	15.65
Filter type	Standard deviation of Gaussian noise				
	0.35	0.4	0.45	0.5	
Mean (3×3)	9.14	8.03	7.07	6.24	
Mean (7×7)	9.19	8.09	7.13	6.29	
Disk (3×3)	9.11	8	7.05	6.21	
Disk (7×7)	9.18	8.08	7.12	6.28	
Gaussian (3×3)	9.03	7.95	6.99	6.17	
Gaussian (7×7)	9.04	7.95	7	6.18	
Siohan	7.37	6.33	5.44	4.65	
Lee	8.31	7.24	6.31	5.5	
DEML2D (3×3)	13.14	11.83	10.74	9.77	
SORIGAML2D (3×3)	14.09	12.77	11.64	10.65	
DEML2D (7×7)	17.1	15.9	14.78	13.76	
SORIGAML2D (7×7)	14.13	12.81	11.66	10.68	

The supremacy of DEML2D and SORIGAML2D filters of two different dimensions in minimizing the effect of Gaussian noise from a standard test image is well established in Table 10.1 to Table 10.4. It can be well observed that, except for SSIM, $DEML2D_{3\times 3}$ produces better results at lower noise density while $DEML2D_{7\times 7}$ yields the best result for higher noise density. The SORIGAML2D filter, on the other hand, exhibits comparable or better results even with the increase in the mask size. Moreover, except for the performance of the $DEML2D_{7\times 7}$ filter at very low noise density, recently proposed filters yield superior solutions as far as the value of IEF is concerned. On the other hand, DEML2D and SORIGAML2D filters yield better SSIM and IQI values, particularly at a higher noise density, than the other 2D filters considered in

TABLE 10.2: Comparison in terms of SSIM resulting from various filters at different noise intensities for Moon image.

Filter type	Standard deviation of Gaussian noise				
	0.1	0.15	0.2	0.25	0.3
Mean (3×3)	0.906	0.8507	0.7935	0.7401	0.6898
Mean (7×7)	0.9025	0.8484	0.7925	0.7372	0.6867
Disk (3×3)	0.9019	0.8479	0.79	0.7363	0.6862
Disk (7×7)	0.9062	0.8515	0.7946	0.7403	0.69
Gaussian (3×3)	0.8837	0.8303	0.7735	0.7188	0.6713
Gaussian (7×7)	0.8851	0.8299	0.7733	0.7194	0.6701
Siohan	0.8515	0.7881	0.7271	0.6699	0.6205
Lee	0.8798	0.8206	0.7611	0.7055	0.6567
DEML2D (3×3)	0.9316	0.9106	0.8749	0.8337	0.7917
SORIGAML2D (3×3)	0.9185	0.905	0.8787	0.8434	0.8026
DEML2D (7×7)	0.8701	0.8076	0.7468	0.6915	0.6412
SORIGAML2D (7×7)	0.9149	0.9026	0.8766	0.8404	0.8031
Filter type	Standard deviation of Gaussian noise				
	0.35	0.4	0.45	0.5	
Mean (3×3)	0.6435	0.6029	0.5668	0.5315	
Mean (7×7)	0.6404	0.5995	0.5622	0.5277	
Disk (3×3)	0.6409	0.6007	0.5648	0.5307	
Disk (7×7)	0.6446	0.6025	0.5652	0.531	
Gaussian (3×3)	0.6272	0.5882	0.5534	0.5209	
Gaussian (7×7)	0.6278	0.5873	0.553	0.5202	
Siohan	0.5782	0.5402	0.5075	0.4779	
Lee	0.6122	0.5735	0.5377	0.5069	
DEML2D (3×3)	0.7482	0.7062	0.666	0.6233	
SORIGAML2D (3×3)	0.7622	0.7216	0.6807	0.6388	
DEML2D (7×7)	0.5966	0.5575	0.5231	0.4926	
SORIGAML2D (7×7)	0.761	0.7208	0.6805	0.6393	

this work. As far as the performance comparison in terms of PSNR and IEF is concerned, SORIGAML2$D_{3\times3}$ yields comparable or better performance, particularly at higher noise densities, than DEML2$D_{3\times3}$. It may also be clearly inferred from these numerical entries that the SORIGAML2$D_{7\times7}$ filter produces better PSNR and IEF than the equivalent DEML2D filter at lower noise density.

Performances of the DEML2D and SORIGAML2D filters have also been compared with respect to some recent algorithms pertaining to the design of a two-dimensional filter. Impact of different 2D filters in eliminating the effect of Gaussian noise from standard test images has been measured in terms of

TABLE 10.3: Comparison in terms of IEF resulting from various filters at different noise intensities for Moon image.

Filter type	Standard deviation of Gaussian noise				
	0.1	0.15	0.2	0.25	0.3
Mean (3×3)	1.581	1.319	1.197	1.134	1.096
Mean (7×7)	1.535	1.313	1.203	1.144	1.108
Disk (3×3)	1.547	1..302	1.187	1.125	1.089
Disk (7×7)	1.611	1.342	1.215	1.147	1.109
Gaussian (3×3)	1.359	1.207	1.131	1.089	1.064
Gaussian (7×7)	1.362	1.209	1.131	1.089	1.064
Siohan	0.8136	0.7477	0.7239	0.7151	0.7129
Lee	1.1372	1.003	0.9422	0.912	0.8959
DEML2D (3×3)	3.976	4.052	3.583	3.175	2.893
SORIGAML2D (3×3)	3.629	4.547	4.395	3.996	3.642
DEML2D (7×7)	1.711	3.241	4.97	6.188	6.687
SORIGAML2D (7×7)	3.588	4.531	4.393	4.009	3.648
Filter type	Standard deviation of Gaussian noise				
	0.35	0.4	0.45	0.5	
Mean (3×3)	1.073	1.058	1.046	1.039	
Mean (7×7)	1.086	1.071	1.061	1.054	
Disk (3×3)	1.066	1.051	1.041	1.034	
Disk (7×7)	1.084	1.068	1.057	1.049	
Gaussian (3×3)	1.048	1.037	1.029	1.023	
Gaussian (7×7)	1.048	1.037	1.029	1.024	
Siohan	0.7135	0.7156	0.7181	0.7207	
Lee	0.8869	0.8818	0.8785	0.8769	
DEML2D (3×3)	2.688	2.542	2.43	2.342	
SORIGAML2D (3×3)	3.363	3.155	2.995	2.867	
DEML2D (7×7)	6.704	6.476	6.171	5.865	
SORIGAML2D (7×7)	3.38	3.172	3.009	2.881	

SSIM. Experimental observations have been listed in Table 10.5, Table 10.6, and Table 10.7.

Careful observation of the above entries reveals the fact that DEML2D and SORIGAMl2D outperform other state-of-the-art filters by a large margin. Moreover, performance of these filters improves with the intensity of the standard deviation for noise. Among the rest, an image filter designed with the aid of a DEPSO-based approach produces the best result. To be specific, the DEML2D filter yields 7.35%, 16.85%, and 22.86% higher SSIM than the DEPSO-based design for the Lena image at noise standard deviations of 0.1, 0.2, and 0.3, respectively. Corresponding improvements in the Moon and Peppers images have been calculated as 7.03%, 16.05% and 17.46% and 5.59%,

TABLE 10.4: Comparison in terms of IQI resulting from various filters at different noise intensities for Moon image.

Filter type	Standard deviation of Gaussian noise				
	0.1	0.15	0.2	0.25	0.3
Mean (3×3)	0.9048	0.8499	0.7933	0.7392	0.6893
Mean (7×7)	0.8984	0.8449	0.7893	0.7354	0.6854
Disk (3×3)	0.9024	0.8475	0.7912	0.7912	0.7367
Disk (7×7)	0.9062	0.8515	0.7949	0.7401	0.6894
Gaussian (3×3)	0.8844	0.8294	0.7735	0.7197	0.6708
Gaussian (7×7)	0.8839	0.8293	0.7731	0.7198	0.6711
Siohan	0.8505	0.7866	0.7258	0.6707	0.6213
Lee	0.8793	0.8201	0.7608	0.7066	0.6564
DEML2D (3×3)	0.9293	0.9074	0.8728	0.8728	0.8323
SORIGAML2D (3×3)	0.915	0.902	0.875	0.84	0.801
DEML2D (7×7)	0.8024	0.8219	0.822	0.81	0.7896
SORIGAML2D (7×7)	0.9136	0.9017	0.8743	0.8394	0.8009
Filter type	Standard deviation of Gaussian noise				
	0.35	0.4	0.45	0.5	
Mean (3×3)	0.644	0.6031	0.5663	0.5322	
Mean (7×7)	0.6398	0.5988	0.5614	0.5271	
Disk (3×3)	0.6424	0.6019	0.5654	0.5316	
Disk (7×7)	0.6441	0.6027	0.5653	0.5307	
Gaussian (3×3)	0.6287	0.5895	0.5538	0.5212	
Gaussian (7×7)	0.6271	0.588	0.5526	0.5199	
Siohan	0.5785	0.5406	0.5079	0.4786	
Lee	0.6124	0.5735	0.5385	0.5069	
DEML2D (3×3)	0.7476	0.7057	0.665	0.624	
SORIGAML2D (3×3)	0.761	0.721	0.68	0.638	
DEML2D (7×7)	0.7634	0.7322	0.6958	0.6556	
SORIGAML2D (7×7)	0.7606	0.7199	0.6789	0.6369	

TABLE 10.5: Comparison among various filters in terms of SSIM at different noise intensities for Lena image.

Type of filter	Standard deviation of Gaussian noise		
Bilateral filtering (BF) [38]	0.798	0.7552	0.5877
Stochastic denoising (SD) [14]	0.852	0.534	0.3532
DEPSO-based design [17]	0.8692	0.7702	0.7004
DEML2D [9]	0.9331	0.9	0.8605
SORIGAML2D [8]	0.8857	0.8875	0.855

11.6% and 14.24%, respectively. The SORIGAML2D filter, on the other hand, produces 1.89%, 15.23%, and 22.07% more SSIM values for the Lena image at noise standard deviations of 0.1, 0.2, and 0.3, respectively. In regard to this, the Moon and Peppers images had respectively exhibited 5.47%, 16.42%

TABLE 10.6: Comparison among various filters in terms of SSIM at different noise intensities for Moon image.

Type of filter	Standard deviation of Gaussian noise		
Bilateral filtering (BF) [38]	0.6306	0.6453	0.5594
Stochastic denoising (SD) [14]	0.8567	0.5873	0.3843
DEPSO-based design [17]	0.8545	0.7435	0.6654
DEML2D [9]	0.9146	0.8628	0.7816
SORIGAML2D [8]	0.9012	0.8656	0.7942

TABLE 10.7: Comparison among various filters in terms of SSIM at different noise intensities for Peppers image.

Type of filter	Standard deviation of Gaussian noise		
Bilateral filtering (BF) [38]	0.8429	0.7851	0.5948
Stochastic denoising (SD) [14]	0.8562	0.5196	0.3372
DEPSO-based design [17]	0.8913	0.8087	0.7422
DEML2D [9]	0.9411	0.9025	0.8479
SORIGAML2D [8]	0.8342	0.8372	0.8021

and 19.36% and -6.41% (indicating deterioration), 3.52%, and 8.07% improvements under similar circumstances.

10.5 Conclusion

This chapter throws sufficient light on the design strategy of multiplier-less 2D filters with the aid of evolutionary computation techniques. Design of 2D filters has already received sufficient attention from researchers over the last few years. In connection to this, the impact of two recently proposed evolutionary optimization techniques has been exclusively studied in this work. More specifically, algorithms like DE and SORIGA had been employed for the design of such filters whose mask coefficients are represented in the form of the sum of signed powers-of-two. These filters have consequently been used to reduce the effect of Gaussian noise of different intensities from a few standard test images. In regard to this, performance of the DEML2D and SORIGAML2D filters had been studied with respect to some relevant parameters like PSNR, SSIM, IEF, and IQI, and subsequently the supremacy of these filters has been established over a few conventional image filters and some other state-of-the-art image denoising algorithms. A comparative study between DEML2D and SORIGAML2D filter has also been included for this purpose.

Bibliography

[1] Complexity reduction of high-speed FIR filters using micro-genetic algorithm. In *First International Symposium on Control, Communications and Signal Processing, 2004.*, pages 419–422. IEEE, 2004.

[2] B.V. Babu and M. Mathew Leenus Jehan. Differential evolution for multi-objective optimization. In *The 2003 Congress on Evolutionary Computation, 2003. CEC'03.*, volume 4, pages 2696–2703. IEEE, 2003.

[3] Luca Banzato, Nevio Benvenuto, and Guido M. Cortelazzo. A design technique for two-dimensional multiplierless fir filters for video applications. *IEEE Transactions on Circuits and Systems for Video Technology*, 2(3):273–284, 1992.

[4] Kamal Boudjelaba, Djamel Chikouche, and Frédéric Ros. Evolutionary techniques for the synthesis of 2-d fir filters. In *Statistical Signal Processing Workshop (SSP), 2011 IEEE*, pages 601–604. IEEE, 2011.

[5] Kamal Boudjelaba, Frédéric Ros, and Djamel Chikouche. An advanced genetic algorithm for designing 2-d fir filters. In *2011 IEEE Pacific Rim Conference on Communications, Computers and Signal Processing (PacRim)*, pages 60–65. IEEE, 2011.

[6] Ling Cen and Yong Lian. A modified micro-genetic algorithm for the design of multiplierless digital fir filters. In *TENCON 2004. 2004 IEEE Region 10 Conference*, pages 52–55. IEEE, 2004.

[7] Abhijit Chandra and Sudipta Chattopadhyay. A novel approach for coefficient quantization of low-pass finite impulse response filter using differential evolution algorithm. *Signal, Image and Video Processing*, 8(7):1307–1321, 2014.

[8] Abhijit Chandra and Sudipta Chattopadhyay. Design of powers-of-two image filter using evolutionary optimization technique. *6th International Conference on Computers and Devices for Communication (CODEC 2015)*, 2015.

[9] Abhijit Chandra and Sudipta Chattopadhyay. A new strategy of image denoising using multiplier-less fir filter designed with the aid of differential evolution algorithm. *Multimedia Tools and Applications*, 75(2):1079–1098, 2016.

[10] Charng-Kann Chen and Ju-Hong Lee. Mcclellan transform based design techniques for two-dimensional linear-phase fir filters. *IEEE Transactions on Circuits and Systems I: Fundamental Theory and Applications*, 41(8):505–517, 1994.

[11] Swagatam Das, Ajith Abraham, and Amit Konar. Particle swarm optimization and differential evolution algorithms: technical analysis, applications and hybridization perspectives. In *Advances of Computational Intelligence in Industrial Systems*, pages 1–38. Springer, 2008.

[12] Swagatam Das and Ponnuthurai Nagaratnam Suganthan. Differential evolution: A survey of the state-of-the-art. *IEEE Transactions on Evolutionary Computation*, 15(1):4–31, 2011.

[13] Bashier Elkarami and Majid Ahmadi. An efficient design of 2-d fir digital filters by using singular value decomposition and genetic algorithm with canonical signed digit (csd) coefficients. In *2011 IEEE 54th International Midwest Symposium on Circuits and Systems (MWSCAS)*, pages 1–4. IEEE, 2011.

[14] Francisco J. Estrada, David J. Fleet, and Allan D. Jepson. Stochastic image denoising. In *BMVC*, pages 1–11, 2009.

[15] John J. Grefenstette et al. Genetic algorithms for changing environments. In *PPSN*, volume 2, pages 137–144, 1992.

[16] John H. Holland. *Adaptation in Natural and Artificial Systems: an Introductory Analysis with Applications to Biology, Control, and Artificial Intelligence*. MIT Press, 1992.

[17] Jingyu Hua, Wangkun Kuang, Zheng Gao, Limin Meng, and Zhijiang Xu. Image denoising using 2-d fir filters designed with depso. *Multimedia tools and applications*, 69(1):157–169, 2014.

[18] Nurhan Karaboga. Digital iir filter design using differential evolution algorithm. *EURASIP Journal on Applied Signal Processing*, 2005:1269–1276, 2005.

[19] H.K. Kwan and C.L. Chan. Circularly symmetric two-dimensional multiplierless fir digital filter design using an enhanced mcclellan transformation. *IEE Proceedings G (Circuits, Devices and Systems)*, 136(3):129–134, 1989.

[20] Ju-Hong Lee, Shih-Jen Yang, and Ding-Chiang Tang. Minimax design of 2-d linear-phase fir filters with continuous and powers-of-two coefficients. *Signal Processing*, 80(8):1435–1444, 2000.

[21] Bipul Luitel and Ganesh K. Venayagamoorthy. Differential evolution particle swarm optimization for digital filter design. In *IEEE Congress on Evolutionary Computation, 2008. CEC 2008.(IEEE World Congress on Computational Intelligence)*, pages 3954–3961. IEEE, 2008.

[22] Manju Manuel and Elizabeth Elias. Design of sharp 2d multiplier-less circularly symmetric fir filter using harmony search algorithm and frequency transformation. 2012.

[23] Remya Krishnan Manuel and Manju Elizabeth Elias. Design of multiplierless 2-d sharp wideband filters using frm and gsa. *Global Journal of Research In Engineering*, 12(5-F), 2012.

[24] Krzysztof Arkadiusz Michalski. Electromagnetic imaging of elliptical–cylindrical conductors and tunnels using a differential evolution algorithm. *Microwave and Optical Technology Letters*, 28(3):164–169, 2001.

[25] Soo-Chang Pei and Sy-Been Jaw. Efficient design of 2-d multiplierless fir filters by transformation. *IEEE Transactions on Circuits and Systems*, 34(4):436–438, 1987.

[26] Kenneth Price and Rainer Storn. Differential evolution: A simple evolution strategy for fast optimization. *Dr. Dobb's Journal*, 22(4):18–24, 1997.

[27] Kenneth Price, Rainer M. Storn, and Jouni A. Lampinen. *Differential Evolution: a Practical Approach to Global Optimization*. Springer Science & Business Media, 2006.

[28] Anyong Qing. Electromagnetic inverse scattering of multiple two-dimensional perfectly conducting objects by the differential evolution strategy. *IEEE Transactions on Antennas and Propagation*, 51(6):1251–1262, 2003.

[29] Anyong Qing, Xin Xu, and Yeow Beng Gan. Anisotropy of composite materials with inclusion with orientation preference. *IEEE Transactions on Antennas and Propagation*, 53(2):737–744, 2005.

[30] David M. Raup. How nature works: The science of self-organized criticality. *Complexity*, 2(6):30–33, 1997.

[31] Pierre Siohan and Acyl Benslimane. Finite precision design of optimal linear phase 2-d fir digital filters. *IEEE Transactions on Circuits and Systems*, 36(1):11–22, 1989.

[32] S. Sriranganathan, D.R. Bull, and D.W. Redmill. Design of 2-d multiplierless fir filters using genetic algorithms. 1995.

[33] Rainer Storn. Designing nonstandard filters with differential evolution. *IEEE Signal Processing Magazine*, 22(1):103–106, 2005.

[34] Rainer Storn and Kenneth Price. Differential evolution-a simple and efficient adaptive scheme for global optimization over continuous spaces. *International Computer Science Institute*, TR-095-012, 1995.

[35] Rainer Storn and Kenneth Price. Differential evolution–a simple and efficient heuristic for global optimization over continuous spaces. *Journal of Global Optimization*, 11(4):341–359, 1997.

[36] R. Thamvichai, Tamal Bose, and Randy L. Haupt. Design of 2-d multiplierless filters using the genetic algorithm. In *Conference Record of the Thirty-Fifth Asilomar Conference on Signals, Systems and Computers, 2001*, volume 1, pages 588–591. IEEE, 2001.

[37] Renato Tinós and Shengxiang Yang. A self-organizing random immigrants genetic algorithm for dynamic optimization problems. *Genetic Programming and Evolvable Machines*, 8(3):255–286, 2007.

[38] Carlo Tomasi and Roberto Manduchi. Bilateral filtering for gray and color images. In *Sixth International Conference on Computer Vision, 1998*, pages 839–846. IEEE, 1998.

[39] Shian-Tang Tzeng. Design of 2-d fir digital filters with specified magnitude and group delay responses by ga approach. *Signal Processing*, 87(9):2036–2044, 2007.

[40] Zhou Wang and Alan C. Bovik. A universal image quality index. *IEEE Signal Processing Letters*, 9(3):81–84, 2002.

[41] Zhou Wang, Alan C. Bovik, Hamid R. Sheikh, and Eero P. Simoncelli. Image quality assessment: from error visibility to structural similarity. *IEEE Transactions on Image Processing*, 13(4):600–612, 2004.

[42] Shiwen Yang, Yeow Beng Gan, and Anyong Qing. Sideband suppression in time-modulated linear arrays by the differential evolution algorithm. *IEEE Antennas and Wireless Propagation Letters*, 1(1):173–175, 2002.

[43] Lee Young-Ho, Masayuki Kawamata, and Tatsuo Higuchi. Design of multiplierless 2-d state-space digital filters over a powers-of-two coefficient space. *IEICE TRANSACTIONS on Fundamentals of Electronics, Communications and Computer Sciences*, 79(3):374–377, 1996.

Chapter 11

Fusion of Texture and Shape-Based Statistical Features for MRI Image Retrieval System

N. Kumaran

Annamalai University, Chidamvaram, India

R. Bhavani

Annamalai University, Chidamvaram, India

11.1	Introduction	298
11.2	Problem Description and Motivations of CBMIR	299
	11.2.1 Challenges of CBMIR	300
	11.2.2 Objectives and Scope	300
11.3	Visual Image Content Descriptors	301
	11.3.1 Texture	301
	11.3.2 Shape	301
	11.3.3 Color	302
11.4	Proposed Approach	302
11.5	Texture Based Feature Extraction	303
	11.5.1 Gabor Wavelet Transform	306
11.6	Shape-Based Feature Extraction	308
11.7	Feature Selection	312
11.8	K-NN Classification	314
11.9	Image Retrieval	316
	11.9.1 Progressive Retrieval Strategy	316
	11.9.2 Euclidean Distance Measure	317
11.10	Results and Discussion	317
11.11	Conclusion and Future Work	318
	Bibliography	320

Content-based image retrieval (CBIR) is one of the largest and most influential research fields. It is an area that has grown in the last two decades and is extensively employed at the present day, with applications in many fields. Content-based medical image retrieval (CBMIR) is one of the most essential applications of CBIR. Today a great number of medical images are being produced in hospitals, as a consequence of new technologies. The traditional concept-based image retrieval systems have more limitations for the retrieval of medical images from this immense database. This will direct various new systems for storage, organization, indexing, and retrieval of medical images using low-level contents. An efficient multi-feature fusion-based hybrid framework for magnetic resonance imaging (MRI) medical image retrieval is discussed in this chapter.

11.1 Introduction

In past years, medical database systems only made available textual information about patients in treatment; soon after, this information was stored in huge databases where queries could be answered by searching for the text information. At the present day, there is a tremendous number of medical images being generated in hospitals, and it is expected that the quantity of such images will further increase exponentially in the time to come. The significance of new medical technologies, such as picture archiving and communication systems (PACS), ultrasound, magnetic resonance imaging (MRI), computed tomography (CT), and X-ray radiography, has resulted in a volatile growth in the number of images stored in the databases.

The growth of medical imaging has attracted the awareness of the medical and other academic communities. This directs the research for innovative computer vision techniques, and permits us to develop tools for assisting the medical experts taking advantage of visual information in medical images. This also has led to various new frameworks in CBMIR [14] for storage, organization, indexing, and retrieval of the medical images in various domains, such as a medical diagnosis, research, and teaching. It is essential to extend the abilities of such application domains by developing image database systems supporting the automated archiving and retrieval of medical images by content. The primary objective of the CBMIR system is to retrieve the images for the given query from the vast pile of medical databases, with high accuracy, by doing feature extraction, categorization, and the similarity measure process. Generally, medical images have individual characteristics and frequently need different retrieval algorithms compared to non-medical images. Some of the basic principles of CBMIR systems are:

1. Requiring capable storing and accurate retrieval systems for the growth in creation and use of medical images.

2. Comparing query and database medical images in a manner that reveals a doctor's similarity judgments.

3. Retrieving database images automatically from the visual content of the various parts of a human body query image.

4. Improving accuracy for the best medical treatment and to reduce the time.

CBMIR systems are search engines for medical image databases. This system is used to index and retrieve medical images according to their low-level and high-level image contents [20]. A typical CBMIR system is divided into an off-line feature extraction process and an on-line image retrieval process. In the off-line process, the CBMIR system automatically extracts the visual content of the image features, like texture, shape, color, and spatial information of each image in the database, based on its pixel values, and stores them within the system as image signatures or feature vectors.

In the online image retrieval process, a radiologist can submit the query image to the CBMIR system in search of preferred images. The system extracts the feature vectors of the query image, and using the preferred model and the distance measure formulas, the distance between the feature vectors of the query image and the database images are computed and ranked.

11.2 Problem Description and Motivations of CBMIR

Some years ago, analog images (hard-copy format) were the early support for the medical field, but they are getting rarer. These medical images are categorized by human experts and retrieved through textual keywords, file names, and patient identification numbers. In recent years, medical images have been placed in digital format (soft-copy format). These digital images have more complexity and less quality compared to analog images. Hence, traditional medical image retrieval methods are not able to recognize the visual content inside the image. For that reason, major limitations are created for medical image retrieval and treatment. To overcome these limitations, various content-based medical image retrieval systems have been developed.

The enormous quantity of digital images created in health care centers and hospitals motivates the requirement for automatic storage and retrieval systems. The power to search medical images by content is becoming more and more substantial, mainly with the present technology towards medical diagnosis. In addition, medical image data have been improved rapidly in size and dimension due to an enormous range of imaging modalities available, and various clinical tests carried out in digital form. This will lead to an increased need for a well-organized CBMIR system. Medical applications are the main concern where CBIR can achieve more outside the experimental sphere, because of an aging population in developed nations. In spite of the progress already accomplished in the few systems available, a great deal of work must be performed in order to produce a commercial CBMIR system.

11.2.1 Challenges of CBMIR

Medical images acquired from different scanning devices with various technologies may exhibit a variety of features and formats. Low resolution and strong noise are the two common characteristics in most medical images. In medical imaging, the extraction of meaningful gray-level features is quite straightforward because the color of each pixel in the image is restricted to a gray-level's intensity. Medical images are more visually similar and therefore will have similar visual characteristics. These discriminations among similar medical images are difficult, which will decrease retrieval performance.

Medical images normally contain a rich set of semantics about the description of images, such as modality, orientation, part of the organ, etc. The deep gap between visual features and high-level semantics concepts is a major obstacle to more effective image retrieval. An efficient hybrid framework [15] is needed to improve the retrieval's accuracy and decrease the semantic gap between the original visual characteristics and the extracted feature vectors of the medical images. Among the various medical CBIR systems available, it is found that in most systems, there is no proper framework available to combine the suitable visual features. To improve the retrieval performance of a content-based medical image retrieval system, effective feature representation methods are needed to extract the discriminant salient features from the medical images.

11.2.2 Objectives and Scope

The objective of this work is to develop an efficient content-based medical image retrieval system, using low-level statistical features for the various organs in human body MRI scan images. In order to achieve these objectives, we must:

- Study and compare the existing texture and shape feature extraction techniques available in CBMIR systems.

- Propose a well-organized multi-feature fusion-based hybrid framework for MRI medical image retrieval, using Gabor wavelet transform (GWT) and texture spectrum (TS)-based texture features, and Prewitt edge region-based shape features.

- Develop an efficient image retrieval framework by combining the promising low-level statistical features, using a K-nearest neighbor (K-NN) classifier with the forward greedy feature selection (FGFS) algorithm.

- Utilize the decisions of both the progressive retrieval strategy and the Euclidean distance (ED) measure for quick indexing and accurate image retrieval.

11.3 Visual Image Content Descriptors

Normally, image substance may consist of both semantic and visual content. The image-specific visual content can be general or domain explicit. The general visual content of an image comprises texture, shape, and color [18]. The domain-explicit visual content of an image is an application support and may require domain knowledge. The semantic content of an image is obtained either by textual footnote or by composite inference procedures using visual content. This work pays attention to the visual contents of MRI images.

11.3.1 Texture

Texture is a dominant visual content that has commonly been applied in computer vision and pattern recognition, for recognizing visual patterns with properties of homogeneity that cannot result merely from the occurrence of a particular color or intensity [7]. The texture patterns are characterized by the interrelated elements such as the number of distinguishable gray-level primitives, the spatial interaction among these primitives, and the size of the image area. Three common ways of analyzing texture are statistical approaches, structural approaches, and spectral approaches.

11.3.2 Shape

Shape is an important visual and a primitive feature for image content description [2]. In many CBIR systems, the shape features have been used for accurate image recognition, classification, and retrieval. The techniques used for shape description and representation can be classified into (i) region-based methods, and (ii) contour-based (or) boundary-based methods. All the pixels are considered for region-based shape feature extraction methods [11], since they extract only the global properties. A region-based method frequently causes shorter feature vectors and simpler matching algorithms. However, they are unsuccessful in creating an accurate similarity retrieval. These include geometric moments, invariant moments, Zernike moments, Legendre moments, grid method, etc.

11.3.3 Color

In image retrieval, color is one of the most extensively used visual features. The color is comparatively strong to background difficulty and self-sufficient in terms of image orientation and size. Each pixel in an image has joint probability of the intensities of the three color values that build its discrimination potentiality better than single-dimensional gray values. The regularly used color feature representations are color histogram [17], color moments [4], and color sets [13].

11.4 Proposed Approach

In this work, the nonlinear median filter using (3×3) square kernel is applied for MRI scan image pre-processing. Median filter is preferred since it is less sensitive to extreme values and capable to eliminate outliers without dropping sharpness of the image.

In the off-line process, the proposed CBMIR system automatically extracts the visual content of the image features, such as texture and shape, from each pre-processed image in the database based on its pixel values, and stores them within the system as image signatures or feature vectors. The two sets of texture features are extracted using the TS and GWT method. The shape features are extracted from the Prewitt edge region. For optimal feature selection, the FGFS algorithm is used. Then, K-NN is used to training the various parts of human body MRI medical images.

In the online image retrieval process, the user can submit the query image to the CBMIR system in search of preferred images. The system extracts and selects the feature vectors of the query image. With the FGFS feature selection algorithm, the K-NN model, and the combination of progressive retrieval strategy and ED measure formulas, the distance between the feature vectors of the query image and the database images is computed and ranked. Finally, the best retrieved images are displayed according to the ranking order. The efficiency of this system has been measured using accuracy, precision, recall, and F-measure. The block diagram for the multi-feature fusion-based hybrid framework is shown in Figure 11.1. The hybrid fusion of TS, GWT, and Prewitt edge region-based low-level statistical features have the highest accuracy, precision, recall, and F-measure, of 99.86%, 99.7%, 99.8%, and 99.75%, respectively.

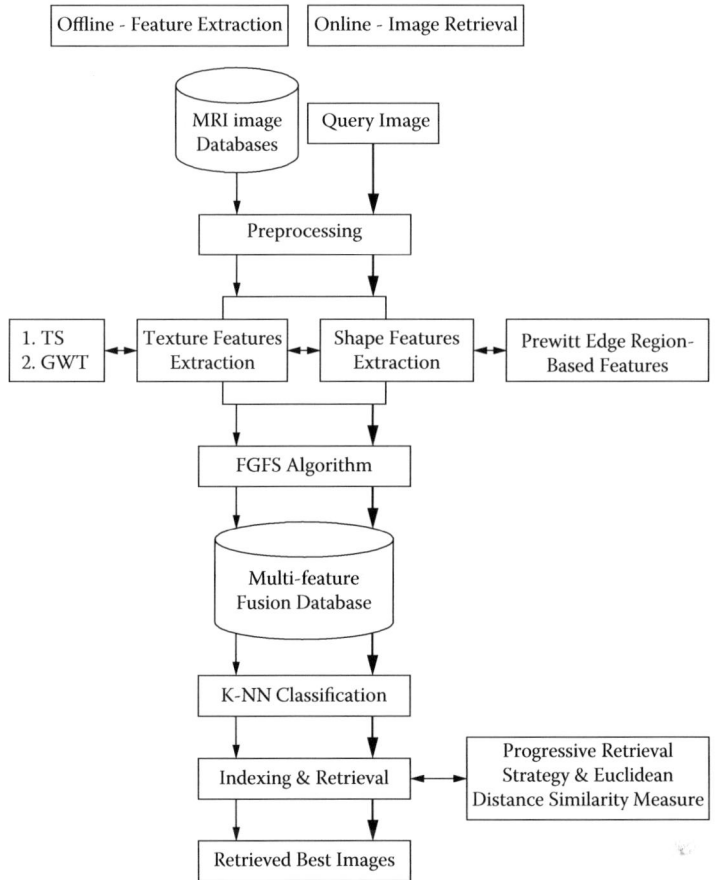

FIGURE 11.1: Block diagram of the proposed work

11.5 Texture Based Feature Extraction

Texture spectrum [10] is applied to extract the other set of texture features. Based on texture characteristics, the texture information [8] is extracted in all eight directions of an image using the TS approach. TS is defined as the strength of texture unit (TU) and texture unit number (NTU) values. Texture units describe the local texture information of a pixel and its nearest pixels, and all texture units in the whole image, exposing the global texture information of the image. In this work, the local texture information for a pixel is extracted from the neighboring pixels of (3×3) matrix, which characterize the smallest texture unit. Figure 11.2 shows the neighborhood of pixels in a

P_1	P_2	P_3
P_8	P_0	P_4
P_7	P_6	P_5

FIGURE 11.2: (3 × 3) pixel matrix

(3 × 3) matrix, which will be represented by a set containing nine elements,

$$P = \{P_0, P_1, P_2, P_3, P_4, P_5, P_6, P_7, P_8\} \tag{11.1}$$

where P_0 denotes the intensity value of the central pixel and $P_i; \{i = 1, 2, \cdots, 8\}$ is the intensity value of the neighboring pixel i. Then, the corresponding TU is determined by a set containing eight elements,

$$TU = \{T_1, T_2, T_3, T_4, T_5, T_6, T_7, T_8\} \tag{11.2}$$

where $T_i\{i = 1, 2, \cdots, 8\}$ is determined by the following formula:

$$T_i = \begin{cases} 0 & \text{if } P_i < P_0 \\ 1 & \text{if } P_i = P_0 \\ 2 & \text{if } P_i > P_0 \end{cases} \tag{11.3}$$

Since every single element of TU has any one of three possible values, the group of the entire eight elements results in 38, giving 6561 feasible texture units. There is no individual method to tag and arrange the 6561 texture units. In this work, the total 6561 texture units are tagged by applying the following equation, where N_{TU} represents the texture unit number. P_i is the ith element of the texture unit set TU.

$$N_{TU} = \sum_{i=1}^{8} P_i * 3^{i-1} \tag{11.4}$$

Figure 11.3 shows an example of texture unit transformation by applying Equation 11.3. Based on the concepts of TS, the following eight texture features are extracted.

192	190	208
195	195	215
190	205	230

→

0	0	2
1	P_0	2
0	2	2

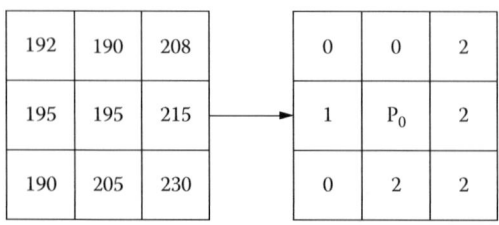

FIGURE 11.3: Example of texture unit transformation

1. Black-white symmetry (BWS)

$$BWS = \left[1 - \frac{\sum_{k=0}^{3279} |F(k) - F(3281+k)|}{\sum_{i=0}^{6560} F(k)}\right] * 100 \quad (11.5)$$

where $F(k)$ denotes the occurrence frequency of the texture unit number k. BWS values are normalized from 0 to 100, and measure the symmetry between the left part (0–3279) and right part (3281–6560) in the texture spectrum, with the axis of symmetry at position 3280.

2. Geometric symmetry (GS)

$$GS = \left[1 - \frac{1}{4}\sum_{l=1}^{4} \frac{\sum_{k=0}^{6560} |F_l(k) + F_l + 4(k)|}{2 * \sum_{k=0}^{6560} F_l(k)}\right] * 100 \quad (11.6)$$

where, $F_l(k)$ is the occurrence frequency of the texture unit numbered k in the texture spectrum under the ordering method l, where $k = 0, 1, \cdots, 6560$ and $l = 1, 2, \cdots, 8$. GS values are normalized from 0 to 100 and measure the symmetry between the spectra under the ordering methods P_1 and P_5, P_2 and P_6, P_3 and P_7, P_4 and P_8, as shown in Figure 11.2.

3. Degree of direction (DD)

$$DD = \left[1 - \frac{1}{6}\sum_{m=1}^{3}\sum_{n=m+1}^{4} \frac{\sum_{k=0}^{6560} |F_m(k) - F_n(k)|}{2 * \sum_{k=0}^{6560} F_m(k)}\right] * 100 \quad (11.7)$$

where $F_m(k)$ & $F_n(k)$ is the occurrence frequency of the texture unit numbered k in the texture spectrum under the ordering method m & n. DD values are normalized from 0 to 100 and measure the degree within an image.

4. Central symmetry (CS)

$$CS = \sum_{k=0}^{6560} F(k) * [G(k)]^2 \quad (11.8)$$

where $G(k)$ denotes the number of pairs having the same value as the elements $(P1, P8)$, $(P2, P7)$, $(P3, P6)$ and $(P4, P5)$, as shown in Figure 11.2.

5. Orientational Features (OF): For the eight elements of a pixel matrix shown in Figure 11.2, if $P1 = P2 = P3$ and $P5 = P6 = P7$, then the original image has a micro-structure of horizontal lineament. Let $S(k)$ denote the occurrence frequency of the texture unit number k, in the texture spectrum, while $V(k, l, m)$ represents the number of elements

having the same value in P_k, P_l, and P_m. $HS(k), VS(k), DS1(k)$ and $DS2(k)$ denote the horizontal, vertical, diagonal-1, and diagonal-2 measures of the texture unit numbered k; then the following features can be extracted.

(a) Micro-horizontal structure (MHS): If $HS(k)$ is defined as $HS(k) = V(P1, P2, P3) \times V(P5, P6, P7)$, then MHS is given by

$$MHS = \sum_{k=0}^{6560} S(k) * HS(k) \tag{11.9}$$

(b) Micro-vertical structure (MVS): If $VS(k)$ is defined as $VS(k) = V(P1, P8, P7) \times V(P3, P4, P5)$, then MVS is given by

$$MVS = \sum_{k=0}^{6560} S(k) * VS(k) \tag{11.10}$$

(c) Micro-diagonal structure-1 (MDS1): If $DS1(k)$ is defined as $DS1(k) = V(P1, P2, P8) \times V(P4, P5, P6)$, then $MDS1$ is given by

$$MDS_1 = \sum_{k=0}^{6560} S(k) * DS_1(k) \tag{11.11}$$

(d) Micro-diagonal-structure-2 (MDS2): If $DS2(k)$ is defined as $DS2(k) = V(P2, P3, P4) \times V(P6, P7, P8)$, then $MDS2$ is given by

$$MDS_2 = \sum_{k=0}^{6560} S(k) * DS_2(k) \tag{11.12}$$

11.5.1 Gabor Wavelet Transform

Gabor wavelets-based texture feature extraction has been broadly applied to efficient CBMIR systems. Hence, in this work the Gabor wavelets [12] are used to extract the other set of texture features. The two-dimensional Gabor function can be defined as:

$$w(p, q) = \frac{1}{2\pi\sigma_p\sigma_q} exp\left(-\frac{1}{2}\left(\frac{p^2}{\sigma_p^2} + \frac{q^2}{\sigma_q^2}\right) + j2\pi F_p\right) \tag{11.13}$$

where $w(p, q)$ is the mother Gabor wavelet; (p, q) is the total number of rows and columns of the mother Gabor wavelet; F is the modulation frequency; and σ_p & σ_q is the standard deviation of the Gaussian kernel. The self-similar filter dictionary is attained by suitable scaling and rotation of the mother

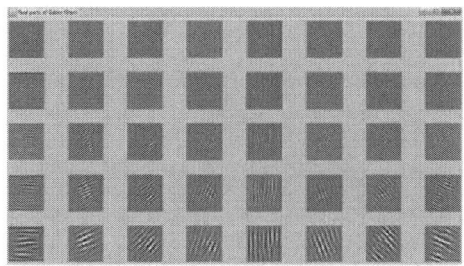

(b) Real parts of the input image

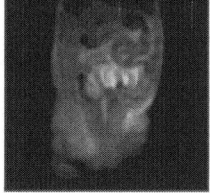

(a) Input image

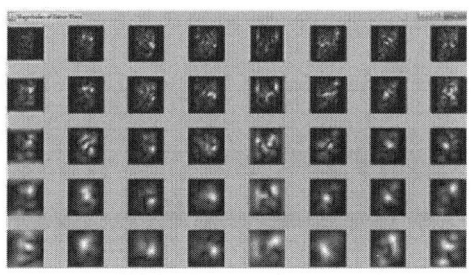

(c) Magnitudes of the input image

FIGURE 11.4: GWT of an input image

Gabor wavelet by Equation 11.14, where s and d are scale and orientation of the wavelets, respectively with $s = 0, 1, 2, 4$ and $d = 0, 1, 2, 7$.

$$w_{sd}(p, q) = z^{-s} w(\tilde{p}, \tilde{q}) \quad (11.14)$$

$$\tilde{p} = z^{-s}(p * cos\theta + q * sin\theta) \quad (11.15)$$

$$\tilde{q} = z^{-s}(-p * sin\theta + q * cos\theta) \quad (11.16)$$

where $z > 1$ and $\theta = \frac{d\pi}{8}$

The discrete Gabor wavelet transform for a given input image $I(p, q)$ with size $(M \times N)$ is defined by the convolution of that input image together with a Gabor kernel. It is defined as below, where, u and v are filter mask size variables, ranging from 0 to 60; w_{sd}^* is a complex conjugate of w_{sd}.

$$W_{sd}(p, q) = \sum_{u=1}^{60} \sum_{v=1}^{60} I(p - u, q - v) w_{sd}^*(u, v) \quad (11.17)$$

The real parts and the magnitudes for five scales and eight orientations of the given input image are shown in Figure 11.4.

The group of magnitudes $D(s, d)$ is obtained from the input image using the Gabor filters with five scales and eight orientations. It is defined as

$$D(s, d) = \sum_p \sum_q |W_{sd}(p, q)| \quad (11.18)$$

From the energy content of the magnitudes, the following texture features are extracted.

1. Mean (μ_{sd})

$$\mu_{sd} = \frac{D(s, d)}{M * N} \quad (11.19)$$

2. Standard deviation (σ_{sd})

$$\sigma_{sd} = \frac{\sqrt{\sum_{p=1}^{M} \sum_{q=1}^{N} (W_{sd}(p, q) - \mu_{sd})^2}}{M * N} \quad (11.20)$$

3. Skewness (S_{sd})

$$S_{sd} = \frac{1}{M * N} \sum_{p=1}^{M} \sum_{q=1}^{N} \frac{(|W_{sd}(p, q) - \mu_{sd}|)^2}{\sigma_{sd}^2} \quad (11.21)$$

4. Kurtosis (K_{sd})

$$K_{sd} = \frac{1}{M * N} \sum_{p=1}^{M} \sum_{q=1}^{N} \frac{(|W_{sd}(p, q) - \mu_{sd}|)^4}{\sigma_{sd}^4} - 3 \quad (11.22)$$

By using Equation 11.19 to Equation 11.22, 160 (5×8×4) texture feature vectors (GW_{sd}) are extracted based on the Gabor wavelet transform. This is given as

$$GW_{sd} = \{\mu_{00}, \sigma_{00}, K_{00}, S_{00}, ..., \mu_{47}, \sigma_{47}, K_{47}, S_{47}\} \quad (11.23)$$

11.6 Shape-Based Feature Extraction

Prewitt operator is a method of edge detection that estimates the maximum response of a set of convolution kernels to locate the local edge orientation for every pixel element. Normally it is applied to calculate the approximate gradient magnitude and orientation of an edge within the gray scale input image. The Prewitt operator is a discrete derivative operator by means of calculating the gradient components of the image intensity function. The Prewitt edge

1	1	1
0	0	0
−1	−1	−1

(a) P_x

−1	0	1
−1	0	1
−1	0	1

(b) P_y

FIGURE 11.5: (3×3) convolution kernels for Prewitt operator

detector is composed of a pair of 2D convolution masks or kernels of 3x3, one to compute the gradient components in the vertical direction, and another one to compute the gradient components in the horizontal direction. These 2D convolution masks $(P_x \& P_y)$ are shown in Figure 11.5.

The vertical and horizontal edge gradient magnitudes of an input image are merged together to calculate the absolute magnitude of the gradient. The gradient magnitude is defined as

$$|P| = \sqrt{P_x^2 + P_y^2} \qquad (11.24)$$

Usually, the approximate gradient magnitude is calculated as

$$|P| = |P_x| + |P_y| \qquad (11.25)$$

The gradient orientation is defined by

$$\theta = arctan\left(\frac{P_y}{P_x}\right) \qquad (11.26)$$

Generally, edge gradient magnitude of an input image yields better response compared to edge gradient orientation. So, only the edge gradient magnitude of an input image is considered in this work. Example outputs of Prewitt operator-based edge-detected MRI scan images are shown in Figure 11.6. After the Prewitt operator edges have been detected, the edge region is selected using best-fit ellipse with the help of the ImageJ software tool. The results of edge region selection are shown in Figure 11.7. From the selected region, the following sixteen shape features are extracted.

1. Area: Total number of pixels in the selected region.

2. Mean (μ_g): It is defined as below, where g_K is gray value in the selected edge region, and N is number of gray values.

$$\mu_g = \frac{\sum_{k=1}^{N} g_k}{N} \qquad (11.27)$$

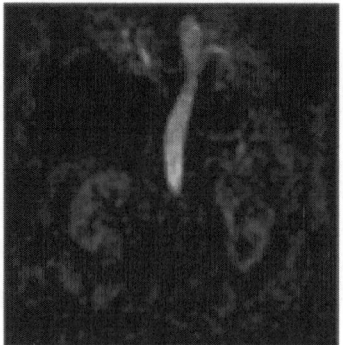
(a) MRI lower limb image

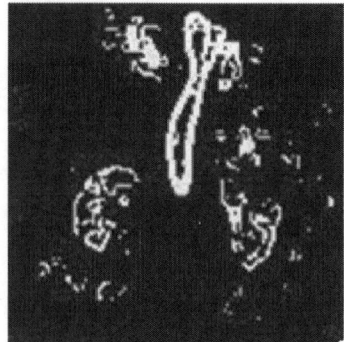
(b) Prewitt edge detected MRI lower limb image

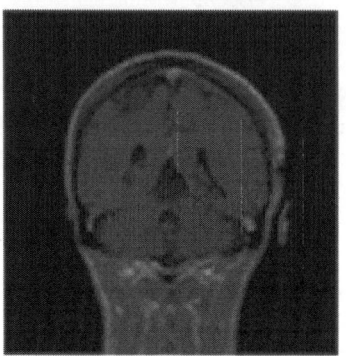
(c) MRI brain image

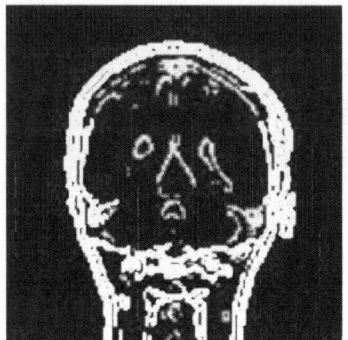
(b) Prewitt edge detected MRI brain image

FIGURE 11.6: Results of Prewitt operator-based edge detection

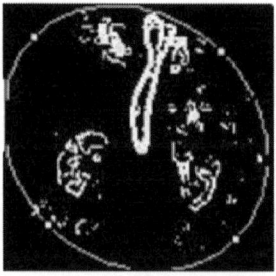
(a) Selected region for MRI lower limb image

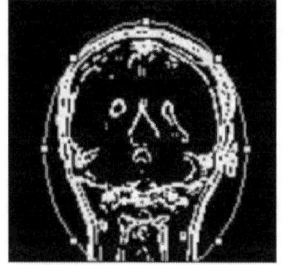
(b) Selected region for MRI brain image

FIGURE 11.7: Results of region selection for Prewitt operator-based edges

3. Standard deviation (σ_g): It is defined as below.

$$\mu_g = \frac{\sum_{k=1}^{N} g_k}{N} \tag{11.28}$$

4. Centroid: The average value of the x and y coordinates of all the pixels in the selected edge region ($X\&Y$).

5. Center of mass: The brightness weighted average value of x and y coordinates of all pixels in the selected edge region ($MX\&MY$).

6. Perimeter: The length of the outside boundary of the edge region.

7. Major and minor axis: The primary and secondary axis of the best-fitting ellipse.

8. Angle: The angle between the major axis of the best-fitting ellipse and a line parallel to the x-axis of the image.

9. Eccentricity (E): It is defined as below, where a, b are semi-major and semi-minor axis, respectively.

$$E = \frac{\sqrt{a^2 + b^2}}{a} \tag{11.29}$$

10. Aspect ratio (AR)

$$AR = \frac{Primary\ axis}{Secondary\ axis} \tag{11.30}$$

11. Roundness (R)

$$R = 4 * \frac{Area}{\pi (Primary\ axis)^2} \tag{11.31}$$

12. Maximum and minimum caliper: The maximum and minimum distance between any two points along the selected edge region.

13. Integrated density (ID)

$$ID = Area * \mu_g \tag{11.32}$$

14. Skewness S

$$S = \frac{\sum_{k=1}^{N} (g_k - \mu_g)^3}{N * \sigma_g^3} \tag{11.33}$$

15. Kurtosis K

$$K = N * \frac{\sum_{k=1}^{N} (g_k - \mu_g)^4}{\left(\sum_{k=1}^{N} (g_k - \mu_g)^2\right)^2} - 3 \tag{11.34}$$

16. Area fraction (AF): the percentage of pixels in the selected edge region.

11.7 Feature Selection

A greedy search algorithm [5] is a mathematical procedure that appears to be uncomplicated, making it easy to find solutions for complex and multi-step optimization problems by deciding which subsequent step will turn over the most obvious solution. In general, the greedy approach does not look at the larger problem as a whole, and also, it does not re-evaluate the solutions once the decision has been accepted. A greedy algorithm is a step-by-step process with the intention of the problem-solving heuristic of producing the nearly best possible solution at each stage, with the expectation of determining an entire optimum solution. Usually, the greedy search algorithm has five elements:

- Candidate set: It is used to find solutions.
- Selection criteria: It is used to select a best candidate set.
- Possibility function: It determines whether a candidate set can be used to subscribe to a solution.
- Objective function: It allots a value to a complete solution or an incomplete solution.
- Solution operation: It specifies at what time a complete solution will come.

In some situations, a greedy search approach may not generate the best possible solution, but on the other hand, it may give a locally optimal solution that is fairly correct to the global optimal solution. In this work, the greedy search algorithm is employed for the optimum feature selection procedure. The greedy feature selection algorithm [9] is executed by building modification to the set of original features and only keeps the absolute best possible set of features, which causes an improvement in classification accuracy. Normally it is classified into two types, such as the forward greedy feature selection (FGFS) approach and the backward greedy feature elimination (BGFE) approach.

The FGFS approach performs by functioning with just one feature first and then adding all the other features one by one towards the optimal solution. As each feature is appended in, the classifier accuracy is calculated with the feature set. If there is an improvement in accuracy, then only newly added features are maintained. The BGFE approach performs by starting with an entire set of features and removing features one by one towards the optimal solution. As each feature is eliminated the classifier accuracy is calculated with the new feature set. If there is an improvement in accuracy, then only the feature is eliminated. Normally, FGFS has less computational cost compared to BGFE. Actually, executing the classifier on a data set with few features is faster than executing with the entire data set of features. The FGFS approach favors only less features that are predicted to be relevant, and the BGFE approach favors only less features that are predicted to be irrelevant. Hence, in this work only the FGFS-based optimal feature selection process is employed.

The algorithm for the FGFS approach is stated below. The FGFS algorithm begins with an empty feature set ($FS(0) = \phi$) and the n original feature set ($F(0) = f1, f2, \cdots, fn$). At each iteration $'i'$, remove one feature from the original feature set and evaluate the performance score. If the performance score is high, the feature will be added to the empty feature set. If the performance score is low, the next feature will be chosen from the original feature set. This will continue until it finds the optimal solution or the original feature set is exhausted. At last, the best performance $score(opt)$, corresponding iteration $index(iter)$, and selected optimal feature set ($FS^{(iter)}$) are displayed and used for further processing.

Algorithm for FGFS

1. Initialize $FS(0) = \phi, F^{(0)} = f_1, f_2, \cdots, f_n, i = 0, opt = 0$ and $iter = 0$.

2. For each iteration, remove one feature from $F(0)$ and evaluate the performance score (opt).

3. If $opt > max$, then the removed feature will be added to $FS^{(0)}$.

4. Else the next feature will be chosen from $F^{(0)}$.

5. The steps 3 and 4 are repeated until the algorithm finds the optimal solution (or) $FS^{(0)}$ is exhausted.

6. Finally, opt, iter, and $FS^{(iter)}$ are displayed.

In this work, although eight texture spectrum features are extracted from the MRI scan image to get accurate classification results, only four FGFS-based optimum features are employed in the K-NN classification algorithm. The total number of features extracted and the total number of features selected using the FGFS-based feature selection technique for various feature extraction methods and the K-NN classification algorithm are shown in Table 11.1.

According to TS, eight features are extracted. Using the FGFS technique, four best features such as BSW, GS, MVS, and MHS are selected for K-NN classification, where K=1, 3, and 5. According to GWT, 160 features are extracted in the form of mean, standard deviation, skewness, and kurtosis for five scales and eight orientations. In that, using the FGFS technique, 34 best features for K-NN classification, where K=1, 3, and 5, are randomly selected.

TABLE 11.1: Total number of optimum features selected using FGFS.

Sl. No.	Feature extraction method	Number of features extracted	Optimum features by FGFS K-NN		
			K=1	K = 3	K=5
1	TS	8	4	4	4
2	GWT	160	34	34	34
3	Prewitt edge Region-based features	20	6	6	6

According to Prewitt edge region, 20 features are extracted. Using the FGFS technique, six best features such as area, mean, standard deviation, major axis length, angle, and integrated density are selected for K-NN classification, where K=1, 3, and 5.

11.8 K-NN Classification

K-nearest neighbor [1, 6] is a simple supervised learning technique that stores the features of all available images and classifies them into new classes based on a distance function. K-NN is non-parametric and a lazy learning algorithm because it does not make any assumptions on the feature vectors, and there is no explicit training phase. Mostly it is used in statistical estimation and pattern recognition. K-NN is a technique for classifying images based on nearest training examples in the feature space, as shown in Figure 11.8. Here, the test point "x" would be labeled as the category of black points, where k = 5.

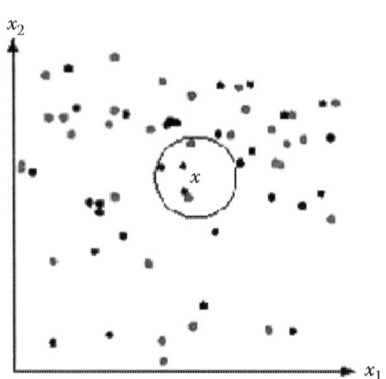

FIGURE 11.8: K-NN illustration (k = 5)

K-NN is a simplest classification algorithm for predicting the class of a test example. The algorithm is stated below. If $K = 1$, then the image is merely grouped to the class of its nearest neighbor. The best selection of K value depends upon the database; normally, higher values of K reduce the outcome of noise on the classification, but formulate the boundaries between classes having similar densities. The K value can be selected by various heuristic techniques; for example, cross validation. The special case where the class is predicted to be the class of the closest training sample (i.e., $K = 1$) is called the nearest neighbor algorithm. The proposed work is carried out for three different values of K (i.e., $K = 1, 3,$ and 5).

Algorithm for K-NN Classifier

1. Store all input data in the training set.
2. For each pattern in the test set, do the following step.
   (a) Search for the K nearest patterns to the input pattern using a Euclidean distance measure.
3. For classification, compute the confidence for each class as C_i/K, where C_i is the number of patterns among the K nearest patterns belonging to class i.
4. The classification for the input pattern is the class with the highest confidence score.

In the training phase, all training images are stored with their labels. To establish a prediction for a test image, initially calculate its distance to every training image. Then, keep the K closest training images and check the label that is most similar among these images. This label is the prediction for the test image. The two key issues involved in the training phase of the K-NN model are: (i) setting the value of K and (ii) the type of distance metric to be used.

To remove ties between two different classes, the general preference for K is a small odd integer. If there are more than two classes, then the ties will be even when K is odd. The ties can also happen when two distance values are equal. An execution of K-NN requires a reasonable technique to avoid ties. To calculate distances of all training vectors to the test vector, ED measure is adopted. The accuracy of the K-NN classification algorithm can be severely affected by the occurrence of noise or irrelevant features, or the consistent feature scales and their importance.

11.9 Image Retrieval

Successful extraction and categorization of image content-based features are needed to provide a meaningful indexing and retrieval of similar images [16]. In this work, both the progressive retrieval strategy and ED measure-based indexing and retrieval techniques are proposed to match and compute the similarity distance between the query image and each image in the classified group.

11.9.1 Progressive Retrieval Strategy

After the query image is tested and categorized, the progressive retrieval strategy is employed. To balance between the indexing difficulties and retrieval performance, the progressive retrieval strategy is adopted in this work. Consider the feature vectors of a query image and a particular class image as $(Q1, Q2, \cdots, Qn)$ and $(C1, C2, \cdots, Cn)$, respectively. Then the particular class images all are roughly filtered using Equation 11.35, where the rough filtering constant is used to adjust the filtered class images.

$$F = \left(\beta Q_1 < c_1 < \frac{Q_1}{\beta}\right) \& \left(\beta Q_2 < c_2 < \frac{Q_2}{\beta}\right) \& \cdots \& \left(\beta Q_n < c_n < \frac{Q_n}{\beta}\right) \tag{11.35}$$

If "F" is false, then the classified image can be identified as being far apart from the query image, and therefore it is discarded; or else, the classified image be kept for further matching. Then the precise filtering threshold is applied as a matching criterion to find more precise images from the rough filtered images. Consider the feature vectors of a query image and a filtered image as $(Q_1, Q_2, \cdots Q_n)$ and $(F_1, F_2, \cdots, F_n)$, respectively. Then the matching criteria for the ith feature vector is defined as:

$$\alpha_i = \begin{cases} \frac{Min(Q_i, F_i)}{Max(Q_i, F_i)} & Q_i \neq 0 \quad or\, F_i \neq 0 \\ 1 & Q_i = 0 \quad or\, F_i = 0 \end{cases} \tag{11.36}$$

Thus, the matching criteria of the query image and the filtered image are

$$\alpha = \sum_{i=1}^{n} D_i \alpha_i \tag{11.37}$$

where

$$D_i = \frac{2i - 1}{n^2} \text{ and } \sum_{i=1}^{n} D_i = 1 \tag{11.38}$$

Obviously, the similarity criterion lies between them. If exceeds a given threshold, it means that mismatch occurs and that image should be discarded; or else, it will be kept as a precise image for further similarity measure.

11.9.2 Euclidean Distance Measure

The general technique for comparing a query image and the database image in content-based medical image retrieval is various distance measures [3]. An image distance measure compares the similarity of two images in various feature dimensions such as color, texture, and shape. The small difference in distance between the feature vectors of a query image and the database image indicates the high similarity of the image. Then the search results of the database images can be ranked based on their similarity distance to the query image.

ED metric, city block or Manhattan distance metric, Minkowsiki distance metric, and Canberra distance metric are the most frequently used similarity measurements in image retrieval. In this work, the ED metric [19] is used as a similarity measure because of its efficiency and effectiveness. It measures the distance between the feature vectors of a query image and precise image databases by calculating the square root of the sum of the squared absolute differences. It can be defined as below, where Q_k is the kth feature vector of the query image and P_k is the kth feature vector of the precise image.

$$E.D(Q,P) = \sqrt{\sum_{K=1}^{n}(|Q_K - P_K|)^2} \qquad (11.39)$$

Equation 11.39 is applied between the query image and all the precise image databases obtained from the progressive retrieval strategy. Then the similarity values are ranked and the top 20 related images are retrieved and displayed.

11.10 Results and Discussion

From the inferences of the individual results, the three feature extraction methods, namely GWT, TS, and Prewitt, and the K-NN classification technique along with the FGFS algorithm are chosen in this proposed efficient multi-feature fusion-based hybrid framework for an MRI medical image retrieval system. From these three methods, three different feature fusions (GWT-Prewitt, TS-Prewitt, and GWT-TS-Prewitt) are formed for a hybrid CBMIR framework. The retrieval performance of K-NN classification for the proposed eight different feature fusion methods along with the FGFS algorithm are evaluated in terms of accuracy, precision, recall, and F-measure.

Figure 11.9 shows the accuracy of the proposed feature fusion methods. The GWT, TS, and Prewitt edge region-based feature fusion method gave the highest accuracy, of 99.86% for $K = 5$ in K-NN classification among all other feature fusion methods. Table 11.2 shows the precision rate of K-NN classification for the proposed feature fusion methods along with FGFS algorithms. The GWT, TS, and Prewitt edge region-based feature fusion method with

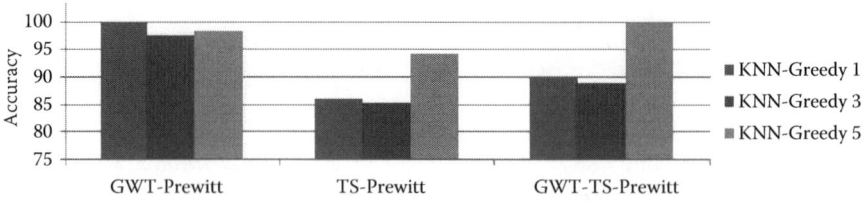

FIGURE 11.9: Accuracy of proposed system

FGFS algorithm gave the highest precision rate of 99.7% for $K = 5$ in K-NN classification among all other feature fusion methods.

Table 11.3 shows the recall rate of K-NN classification for the proposed feature fusion methods along with FGFS algorithms. The GWT, TS, and Prewitt edge region-based feature fusion method with FGFS algorithm gave the highest recall rate, of 99.8% for $K = 5$ in K-NN classification among all other feature fusion methods.

Figure 11.10 shows the F-measure of K-NN classification for the proposed feature fusion methods along with FGFS algorithms. The GWT, TS, and Prewitt edge region-based feature fusion method with FGFS algorithm gave the highest F-measure, of 99.75% for K = 5 in K-NN classification among all other feature fusion methods.

11.11 Conclusion and Future Work

Content-based image retrieval is one of the largest and most powerful research fields. CBIR is extensively employed at the present time, with applications in many fields. Content-based medical image retrieval is one of the most essential applications of CBIR. Today there is a great number of medical images being produced in hospitals as the consequence of new technologies. The traditional concept-based image retrieval systems have more limitations for the retrieval of medical images from this immense database. This will direct various new systems for storage, organization, indexing, and retrieval of medical images using low-level contents. The aim of this work is to develop an efficient multi-feature fusion-based hybrid image retrieval system using texture and

TABLE 11.2: Precision rate of proposed system.

	K	GWT-Prewitt	TS-Prewitt	GWT-TS-Prewitt
	1	99.61	85.73	89.61
KNN - FGFS	3	97.61	84.63	88.54
	5	98.27	93.65	99.7

Fusion of Texture and Shape for MRI Image Retrieval System

TABLE 11.3: Recall rate of proposed system.

	K	GWT-Prewitt	TS-Prewitt	GWT-TS-Prewitt
	1	99.67	85.53	89.69
KNN - FGFS	3	97.53	84.93	88.73
	5	98.27	93.67	99.8

shape-based low-level statistical features for the human body MRI scan images. This system has several stages, including image pre-processing, feature extraction, feature selection, classification, and image retrieval. In this work, two different texture feature techniques and a shape feature technique have been investigated individually for K-NN classification with FGFS feature selection algorithms. The combination of progressive retrieval strategy and ED measure has been proposed for indexing and retrieval of images. The GWT, TS, and Prewitt edge region-based hybrid multi-feature fusion method for real time MRI image retrieval system with FGFS algorithm gave the highest retrieval performance for K = 5 in K-NN classification among all other proposed methods. The GWT, TS, and Prewitt edge region-based hybrid multi-feature fusion method gave the highest accuracy, precision, recall, and F-measure of 99.86%, 99.7%, 99.8%, and 99.75%, respectively.

From the work carried out, it was found that for the medical image retrieval system, the selection of a good visual feature representation technique is essential. These visual features should be combined with a proper framework, so that the efficiency of each feature can be utilized. The combination of visual and semantic features will yield better retrieval performance. Further research is needed to organize the feature indices into an efficient data structure, to facilitate fast retrieval. Relevant feedback is one way to satisfy doctors and other users. This can be incorporated into future work. This method can be further extended to other types of medical images (X-ray, CT, PET, MRS, and fMRI) with a few modifications.

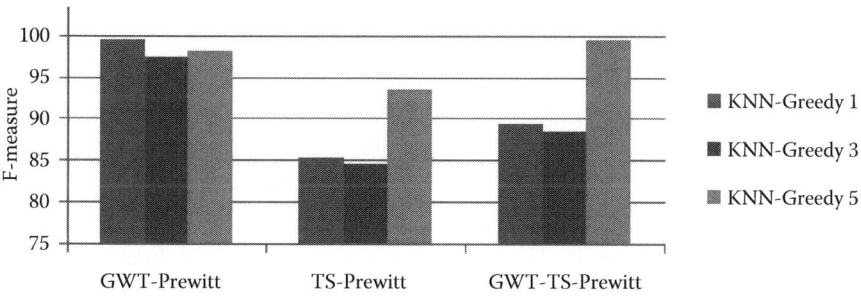

FIGURE 11.10: F-measure of proposed system

Bibliography

[1] M.M. Abdulrazzaq, Shahrul Azman Mohd, and Muayad A. Fadhil. Medical image annotation and retrieval by using classification techniques. In *Advanced Computer Science Applications and Technologies (ACSAT), 2014 3rd International Conference on*, pages 32–36. IEEE, 2014.

[2] A. Amanatiadis, V.G. Kaburlasos, A. Gasteratos, and S.E. Papadakis. Evaluation of shape descriptors for shape-based image retrieval. *IET Image Processing*, 5(5):493–499, 2011.

[3] Swarnambiga Ayyachamy and Vasuki S. Manivannan. Distance measures for medical image retrieval. *International Journal of Imaging Systems and Technology*, 23(1):9–21, 2013.

[4] Hariharapavan Kumar Bhuravarjula and V.N.S Vijaya Kumar. A novel content based image retrieval using variance color moment. *International Journal of Computational Engineering Research*, 1(3):93–98, 2012.

[5] Rich Caruana and Dayne Freitag. Greedy attribute selection. In *Proceedings of the Eleventh International Conference on Machine Learning*, pages 28–36. Morgan Kaufmann, 1994.

[6] P.A. Charde and S.D. Lokhande. Classification using k nearest neighbor for brain image retrieval. *International Journal of Scientific & Engineering Research*, 4(8):760–765, 2013.

[7] Asmita Deshmukh, Leena Ragha, and February Gargi Phadke. An effective cbir using texture. In *International Journal of Computer Applications Proceedings on International Conference and Workshop on Emerging Trends in Tchnology (ICWET)*, 2012.

[8] Ali Douik, Mehrez Abdellaoui, and Leila Kabbai. Content based image retrieval using local and global features descriptor. In *Advanced Technologies for Signal and Image Processing (ATSIP), 2016 2nd International Conference on*, pages 151–154. IEEE, 2016.

[9] Ahmed K. Farahat, Ali Ghodsi, and Mohamed S. Kamel. Efficient greedy feature selection for unsupervised learning. *Knowledge and Information Systems*, 35(2):285–310, 2013.

[10] Dong-Chen He and Li Wang. Texture unit, texture spectrum, and texture analysis. *IEEE Transactions on Geoscience and Remote Sensing*, 28(4):509–512, 1990.

[11] Hae-Kwang Kim and Jong-Deuk Kim. Region-based shape descriptor invariant to rotation, scale and translation. *Signal Processing: Image Communication*, 16(1):87–93, 2000.

[12] Bangalore S. Manjunath and Wei-Ying Ma. Texture features for browsing and retrieval of image data. *IEEE Transactions on Pattern Analysis and Machine Intelligence*, 18(8):837–842, 1996.

[13] Phillip A. Mlsna and Jeffrey J. Rodríguez. Efficient indexing of multi-color sets for content-based image retrieval. In *Image Analysis and Interpretation, 2000. Proceedings. 4th IEEE Southwest Symposium*, pages 116–120. IEEE, 2000.

[14] Henning Müller, Nicolas Michoux, David Bandon, and Antoine Geissbuhler. A review of content-based image retrieval systems in medical applications-clinical benefits and future directions. *International Journal of Medical Informatics*, 73(1):1–23, 2004.

[15] B. Ramamurthy and K.R. Chandran. CBMIR: Content based medical image retrieval using multilevel hybrid approach. *International Journal of Computers Communications & Control*, 10(3):382–389, 2015.

[16] M. Srinivas, R. Ramu Naidu, C.S. Sastry, and C. Krishna Mohan. Content based medical image retrieval using dictionary learning. *Neurocomputing*, 168:880–895, 2015.

[17] P.S. Suhasini, K. Krishna, and I.V. Murali Krishna. CBIR using color histogram processing. *Journal of Theoretical & Applied Information Technology*, 6(1), 2009.

[18] Dong Ping Tian et al. A review on image feature extraction and representation techniques. *International Journal of Multimedia and Ubiquitous Engineering*, 8(4):385–396, 2013.

[19] Liwei Wang, Yan Zhang, and Jufu Feng. On the euclidean distance of images. *IEEE Transactions on Pattern Analysis and Machine Intelligence*, 27(8):1334–1339, 2005.

[20] Qing-zhu Wang, Ke Wang, and Xin-zhu Wang. Medical image retrieval based on low level feature and high level semantic feature. In *Computer Engineering and Technology (ICCET), 2010 2nd International Conference on*, volume 7, pages V7–430. IEEE, 2010.

Chapter 12

Singular Value Decomposition–Principal Component Analysis-Based Object Recognition Approach

Chiranji Lal Chowdhary
VIT University, Vellore, India

D.P. Acharjya
VIT University, Vellore, India

12.1	Introduction	324
12.2	Methodology and Experimental Setup	326
	12.2.1 Implementation of ICA Algorithm	326
	12.2.2 ICA Data Presentation to the Neural Network	326
	12.2.3 Singular Value Decomposition	328
	12.2.4 Singular Value Decomposition—Pricipal Component Analysis	328
	12.2.5 Database	330
12.3	Results	331
	12.3.1 Result with Position Angle Sampling at Every 50^0 Using PCA	331
	12.3.2 Result with Position Angle Sampling at Every 25^0 Using PCA	333
	12.3.3 Result with Position Angle Sampling at Every 10^0 Using PCA	333
	12.3.4 Results with Position Angle Sampling at Every 50^0 Using ICA	334
	12.3.5 Results with Position Angle Sampling at Every 10^0 Using ICA	334
	12.3.6 Comparision between PCA, ICA, and SVD-PCA	334
	12.3.7 Results with Position Estimation	337
12.4	Conclusion and Future Enhencement	338
	Bibliography	339

A new feature extraction method for object recognition tasks based on singular value decomposition (SVA) and principal component analysis (PCA) has been proposed. Although PCA is intrinsically very close, SVD differs in important aspects that can affect the performance of the classifier used in the recognition system. The result is compared with other standard transforms, like PCA and ICA. In the proposed work, the appearance-based 3DOR techniques is studied and compared in equal working conditions regarding preprocessing and algorithm implementation, using a COIL-100 dataset, with its standard sets. The potential of SVD-PCA on a database of 7200 images of 100 different color objects is illustrated. Overall, it is observed that SVD-PCA performs significantly better than conventional and subsequent eigenvalue decompositions. Experimental results in appearance-based object recognition confirm that SVD-PCA offers better recognition rates over ICA and PCA. The excellent recognition rates achieved in all of the experiments performed indicates that the proposed method is well-suited for 3D object recognition in applications like surveillance, robot vision, biometrics, and security tasks, etc.

12.1 Introduction

An essential behavior of animals is the visual recognition of objects, which is vital for their survival. Human activity, for example, relies heavily on the classification or identification of a large variety of visual objects. Recently, object recognition has been found in a great range of applications. Object recognition is a difficult problem in computer vision. Earlier, artificial vision experiments tended to center around toy problems in which the world being observed was carefully controlled and constructed. Perhaps boxes in the shapes of regular polygons were identified, or simple objects such as a pair of scissors were used. In most cases the background of the image was carefully controlled to provide excellent contrast between the objects being analyzed and the surrounding world. Clearly, object recognition does not fall into this category of problems.

In computer vision, the recognition system typically involves some sort of a sensor and the use of a model database in which all the object representations are saved for decision making ability. When an object is viewed by a sensor, say a charge-coupled device (CCD) camera, the digitized image is processed so as to represent it in the same way as the models are represented in the database. Then a recognition algorithm tries to find the model to which the object best matches. For view-based recognition, the representations take into account the appearance of the object [16]. To achieve three-dimensional object recognition (3DOR), the positions of the object are also saved in the database. Then the objective of the algorithm is not only to recognize the object correctly, but also to identify its position as viewed by the camera.

SVD–PCA-Based Object Recognition Approach

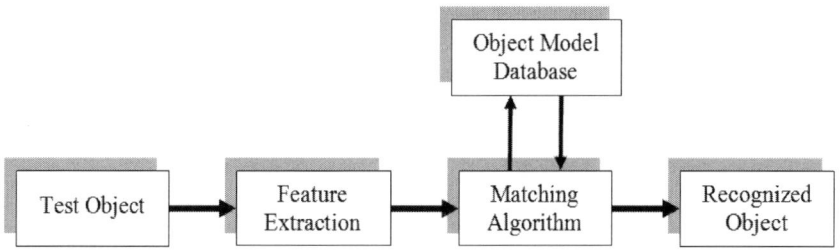

FIGURE 12.1: Typical object recognition system

Many approaches have been introduced for three-dimensional object recognition (3DOR) systems [21, 18, 1, 22]. Edges contain a great amount of information in a scene. It is also known that the eye pays more attention to edges in an image. This has prompted the use of edge information in the development of 3DOR algorithms. Traditionally, robot vision systems have also utilized the shape of objects for recognition. Many of these methods explicitly exploit the features extracted from only the shape of the objects, i.e., lines, curves, and vertices, which are called geometric features. But many factors that do not affect the intrinsic shape of an object do significantly determine how the object is viewed. The approaches that explicitly take these factors into account for use in object recognition have been categorized as view-based or appearance-based object recognition methods [11, 6, 9]. Thus, in contrast to the geometric features, the appearance of an object is the combined effect of its shape, reflectance properties, pose, and the illumination. As the database takes into account all of these attributes, appearance-based recognition tends to take into account the combined effect of the above factors [5, 15]. An overview of a typical object recognition system is depicted in Figure 12.1.

Three-dimensional object recognition (3DOR) [21, 18, 1, 22] is a rather nebulous term. Some schemes handle single segmented objects, whereas others interpret multiple object scenes. However, some of these other schemes are really performing two-dimensional (2D) processing using the three-dimensional (3D) information. There are systems that require objects to be placed on a turntable during the recognition process. Most efforts have limited the class of recognizable objects to polyhedrons, spheres, cylinders, cones, generalized cones, or a combination of these. Many projects fail to mention how well the proposed method can recognize objects from a large set of objects (e.g., at least 100 objects). Therefore, initially it is essential to discuss human visual capabilities and then describe how these relate to computer vision. The real world that we see and touch is primarily composed of solid objects. When people are given an object they have never seen before, they are typically able to gather information about that object from many different viewpoints. The process of gathering detailed object information and storing that information is referred to as model formation. Once familiar with many objects, we can normally identify them from an arbitrary viewpoint without further investigation.

Two of the most powerful subspace projection techniques are principal component analysis (PCA) [20] and the InfoMAX algorithm [13] for independent component analysis (ICA) [7, 17, 12, 13] in face or object recognition. It is evident that both PCA and ICA approaches to object recognition are depedent and computationally expensive. A common objective in $3D$ object recognition [2] is to devise a technique that improves the recognition rate. This chapter presents a new technique for object recognition using singular value decomposition and principal component analysis (SVD-PCA).

12.2 Methodology and Experimental Setup

Before presenting the results of the application of PCA, ICA, and SVD-PCA and comparisons for appearance-based $3D$ object recognition, the design aspects are explained. This section explains the data used in simulations, how to extract features from the image selected, and how the eigenspace associated with the database is constructed. Additionally, the presentation of data to the ICA and SVD-PCA techniques with their algorithms and how the SVD tool works are discussed.

12.2.1 Implementation of ICA Algorithm

The implementation of InfoMAX is publicly available MATLAB code written by Bell and Sejnowski [3], and is used to generate the results. InfoMAX has several parameters for this study: the block size was 50, the initial learning rate is 0.001, and after 1000 iterations, this rate is reduced by every 200 epochs to 0.005, 0.002, and then 0.0001. In this chapter, InfoMAX is trained for 1600 iterations. The file database contains 7200 images compiled from 100 color objects with 72 images in different positions. The object file database includes 100 different objects with different positions starting from 0 degree to 360 degree rotations. The original resolution of the images is 128×128 RGB images. In order to speed up the classification and to reduce the amount of data needed, this resolution is changed from RGB to graylevel image and reduced to 64×64 by simply resizing the image.

12.2.2 ICA Data Presentation to the Neural Network

In the InfoMAX algorithm, the weight matrix is a square matrix. Hence, the network attempts to extract many components as the number of mixtures being given to it. Various methods can be adopted to estimate the number of independent components in the given data, including a heuristic and trial and error method. In this proposed approach, PCA is used as a first step to make such an estimation [14]. This is motivated by the fact that the energy content corresponding to the eigenvectors obtained by using PCA provide an approximate indication of the number of components that one has to look for. For 64 × 64 images, most of the eigenvectors corresponding to the lower eigenvalues can be considered non-dominant, since most of the energy is concerntrated only in the first few eigenvectors. Neglecting these non-dominant components naturally leads to loss of some information. In the experiments considered, eigenvectors numbering from the first five up to the first 100 are used to demonstrate the performance of the algorithm.

First, PCA is performed on the training database. The first n eigenvectors are selected and given as the row of the input matrix to the ICA network [8]. This is justified by the fact that the eigenvectors are also another linear mixture of the input images. Note that the input data is assumed to be some linear mixture of the underlying independent components, and therefore, it can be replaced by any other linear mixture. Another important and related advantage of using PCA is that it reduces the dimensionality of the input data. An image is represented by a column vector of length N where N is the total number of pixels in each image. The database is assumed to consist of M images, so that it can be represented by a matrix X of size $N \times M$. To compute the singular values, a matrix XX^T is obtained, which possesses N eigenvectors. The first dominant principal component eigenvector is the input to the ICA neural network denotd by Q_n^T. But higher-order statistical characteristics still exist in the eigenvectors, whereas PCA removes only the second-order characteristics [4].

Also, before giving the data to the ICA network, it is sphered with the whitening filter W_z [4, 22]. With the input of the ICA as Q_n^T, i.e., the first eigenvectors as the rows of the input matrix, the ICA mapping expresses the row of the input matrix and is defined as below, where w_L is the square weight matrix learned by the network and U has independent images in its row.

$$W_L W_z Q_n^T = U \qquad (12.1)$$

The learned weight matrix and the whitening filter can be combined into one filter by the following expression, where W given in Equation 12.2 yields the combined filter. Thus Equation 12.1 reduces to Equation 12.3.

$$W = W_L W_Z \qquad (12.2)$$
$$W Q_n^T = U \qquad (12.3)$$

Reconstruct the data using PCA as given in Equation 12.4, where Γ_n^T is the representation of the database based on the first n dominant eigenvectors. Using Equation 12.3 and Equation 12.4, the reconstructed data from the ICA output U is presented in Equation 12.5.

$$\Gamma_n^T Q_n^T = X^T \qquad (12.4)$$
$$X^T = \Gamma_n^T W^1 U \qquad (12.5)$$

By using the coefficients of the training and the testing data, we match the object that is in the database by using the matching algorithm. The coefficient of the test and the training image is given below in Equation 12.6.

$$\Gamma_n^T W^{-1} = \rho \tag{12.6}$$

12.2.3 Singular Value Decomposition

Singular value decomposition (SVD) [21] is often used to solve the eigen problem for the covariance matrix. It is also used in robust statistics, to solve overdetermined systems of equations. As the underlying data represents gray-value images, the discussion is reduced to real-valued matrices. However, it works for square and non-singular matrices. Most of the modern SVD algorithms are based on the method of Golub and Reinsch [10]. SVD is one of the most important tools of numerical signal processing, and is employed in a variety of system and image processing applications, such as spectrum analysis, filter design, system identification, object recognition, etc.

If A is a real $m \times n$-matrix, then there exist two orthogonal matrices $U \in I\!R^{m \times m} = (u_1, u_2, \cdots, u_m)$ and $V \in I\!R^{n \times n}$ such that $A = U\Sigma V^T$, where $S \in I\!R^{m \times n} = diag(\sigma_1, \ldots, \sigma_p)$, $p = min(m, n)$ with $\sigma_1, \sigma_2, \cdots, \sigma_p \geq 0$.

$$\Sigma \in I\!R^{m \times n} = \begin{bmatrix} S & 0 \\ 0 & 0 \end{bmatrix}$$

We use the orthogonal matrix U as the projection vectors (instead of eigenvectors of covariance matrix) for feature extraction while implementing eigenimages. In contrast to eigenvector calculations involved in conventional algorithms, the SVD has several advantages, such as (i) it is computationally efficient and (ii) robust. The SVD are known to be more robust than the usual eigenvectors of covariance matrix. This is because the robustness is determined by the directional vectors rather than more scalar quantities like magnitudes (singular values stored in Σ). Since U and V matrices are inherently orthogonal in nature, these directions are encoded in U and V matrices.

12.2.4 Singular Value Decomposition—Pricipal Component Analysis

For symmetric and positive semi-defined matrices, sigular value decomposition (SVD) and eigenvalue decomposition (EVD) become equivalent. As the covariance matrix is a positive and semi-defined matrix, SVD may be applied to compute the EVD. The matrix multiplications $\hat{X}\hat{X}^T$ or $\hat{X}^T\hat{X}$ have to be performed, respectively. To even avoid these matrix multiplications, we applied SVD directly on the mean normalized data matrix $\hat{X}$ to compute the eigenvectors $u_i \in \mathcal{IR}^{m \times m}$ of the sample covariance matrix C.

Consider the SVD of the mean normalized sample matrix $\hat{X} \in \mathcal{IR}^{m \times m}$:

$$\hat{X} = U\Sigma V^T \tag{12.7}$$

Then, the SVD of $\hat{X}\hat{X}^T$ is given by

$$\hat{X}\hat{X}^T = U\Sigma \underbrace{V^T V}_{I_{m\times m}} \Sigma^T U^T = U\Sigma\Sigma^T U^T \qquad (12.8)$$

From Equations 12.7 and 12.8, it is clear that the left singular vectors of $\hat{X}$ and $\hat{X}\hat{X}^T$ are identical. The left singular vectors of $\hat{X}\hat{X}^T$ are also the eigenvectors of $\hat{X}\hat{X}^T$, and the squared singular values σ_j^2 of $\hat{X}$ are the eigenvalues Λ_j of $\hat{X}\hat{X}^T$. We apply SVD on $\hat{X}$ to estimate the eigenvalues and the eigenvectors of $\hat{X}\hat{X}^T$. The steps involved in the computation of the PCA projection matrix using SVD is summarized are Algorithm 1.

Algorithm 1 [Computation of PCA Projection Matrix Using SVD]

---

1. The input is taken as the data matrix X and the sample mean vector of all the columns are computed as

$$\bar{x} = \frac{1}{2}\sum_{j=1}^{n} x_j \qquad (12.9)$$

2. Normalize the input image using Equations 12.10 and 12.11

$$\hat{x}_j = x_j - \bar{x} \qquad (12.10)$$

$$\hat{X} = [\hat{x}_1 \cdots, \hat{x}_n] \qquad (12.11)$$

3. Compute left singular vectors $\breve{u}_j$ and singular value σ_j using normalized image.

4. By the singular vectors and singular values, compute the eigenvalues λ_j of C as

$$\lambda_j = \frac{\sigma_j^2}{n-1} \qquad (12.12)$$

5. Finally, compute eigenvectors u_j of C using the following equations.

$$u_j = \breve{u}_j \qquad (12.13)$$

$$U = [u_1, \ldots, u_{n-1}] \qquad (12.14)$$

12.2.5 Database

The database used is $COIL-100$ by Columbia Object Image Library, Columbia University, and is available online. The same database was used earlier for recognition tests with both ICA and PCA, but $COIL-100$ has yet been tested only with PCA. Therefore, the same database is used for both PCA, ICA, and SVD-PCA recognition tests and comparison with equal working conditions. In $COIL-100$, 100 color objects are used. Each object's image was taken by placing it on a turntable. The images were taken at every 5^0 of position angle, i.e., the object was rotated and an image was taken after every 5^0 of rotation, from 0^0 to 360^0 degrees. The gallery image set consisted of 7200 images of 100 objects, with 72 views of each of the objects and a picture roughly of size 128×128, involving a single object. Thus, we have chosen the training image set to be subset of the gallery of the image set. The probe image set consist of 3600 images for the 10^0 interval, 5700 images for the 25^0 interval, and the other probe image set consists of 6400 images for the 50^0, other than those objects selected for the training set. Thus, the interval view angle was the manifold parameter and the set of images gave the complete image manifold.

The size of the images was changed to 64×64 due to memory constraints. The 0^0 view angles of the 100 color objects are shown in Figure 12.2. Note

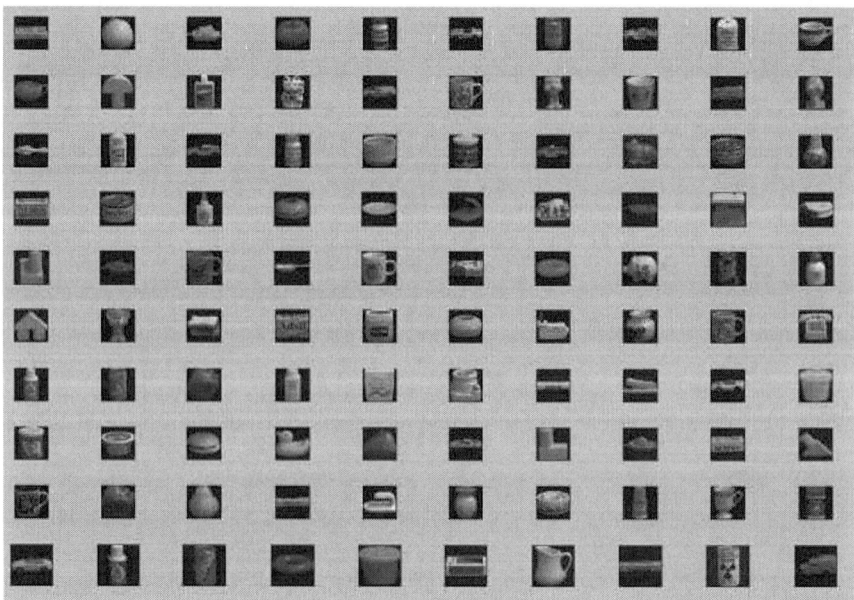

FIGURE 12.2: Columbia object image library ($COIL-100$)

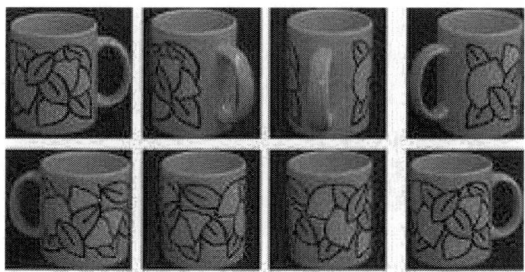

FIGURE 12.3: Training set of an object sampled at every 50^0

that 64×64 is a big enough size to demonstrate the effectiveness of the object recognition algorithms. All the images are changed from RGB to gray scale for matrix dimension.

12.3 Results

To construct the eigenspace of the object, a few of the images are chosen as the training images. Recall that the representation of the images makes a manifold with the position angle as the parameter in the high-dimensional space. Thus, the image manifold is sampled at regular intervals of position angle to make the training images. In the first experiment, images separated by 50^0 in position angle are chosen to construct the representative eigenspace of the database. That is to say, there are 8 training images for each object, making a total of 800 training images.

The training views of the 59th object are shown in Figure 12.3. In another experiment, images separated by 25^0 in pose angle are chosen to construct the representative eigenspace of the database. That is to say, there are 15 training images for each object, making a total of 1500 training images. The training views of the object are shown in Figure 12.4. In the third experiment, images separated by 10^0 in position angle are chosen to construct the representative eigenspace of the database. That is to say, there are 36 training images for each object, making a total of 3600 training images.

12.3.1 Result with Position Angle Sampling at Every 50^0 Using PCA

Training images are sampled at every 50^0. That is to say, there are 8 training images for each object, making a total of 800 training images. The images that are not in the training set are considered to be test images, thus making

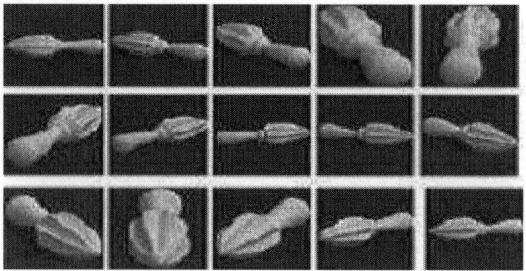

FIGURE 12.4: Training set of an object sampled at every 25^0

a total of 6400 test images. In the following experiments, the numbers of eigenvalue turned parameters are $q = 18, 24, 36, 48$. Figure 12.5 shows the eigenspace manifold of objects in the database set with the position angle manifolds being sampled at 50^0.

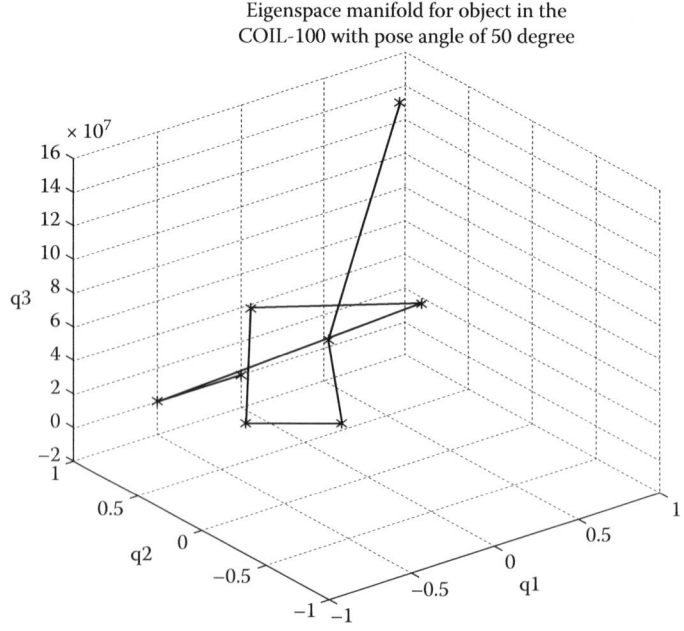

FIGURE 12.5: Eigenspace manifolds for an object in the database with the position angle manifold being sampled at 50^0

SVD–PCA-Based Object Recognition Approach

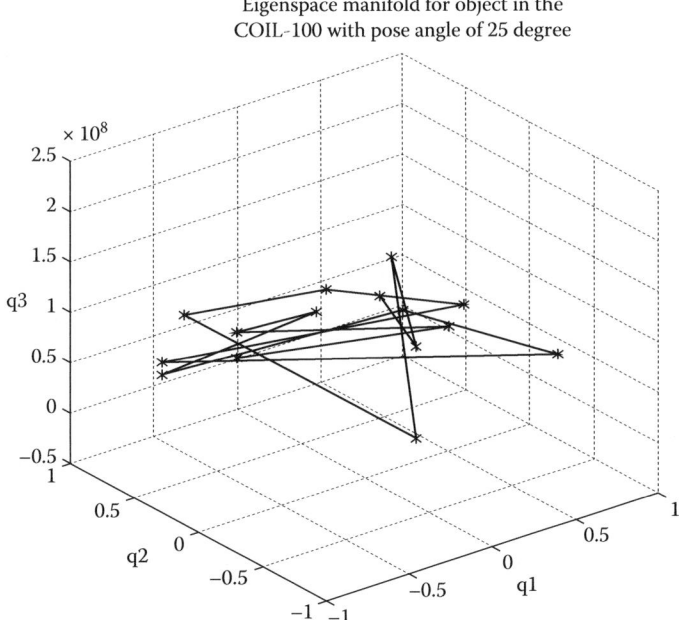

FIGURE 12.6: Eigenspace manifolds for an object in the database with the position angle manifold being sampled at 25^0

12.3.2 Result with Position Angle Sampling at Every 25^0 Using PCA

Training images are sampled at every 25^0. That is to say, there are 15 training images for each object, making a total of 1500 training images. The images that are not in the training set are considered to be test images, thus making a total of 5700 test images. In the following experiments, the numbers of eigenvalue turned parameters are $q = 18, 24, 36, 48$. Figure 12.6 shows the eigenspace manifold for objects in the database set with the position angle manifolds being sampled at 25^0.

12.3.3 Result with Position Angle Sampling at Every 10^0 Using PCA

Training images are sampled at every 10^0. That is to say, there are 36 training images for each object, making a total of 3600 training images. The images that are not in the training set are considered to be test images, thus making a total of 3600 test images. In the following experiments, the numbers of eigenvalue turned parameters are $q = 18, 24, 36, 48$. Figure 12.7 shows the

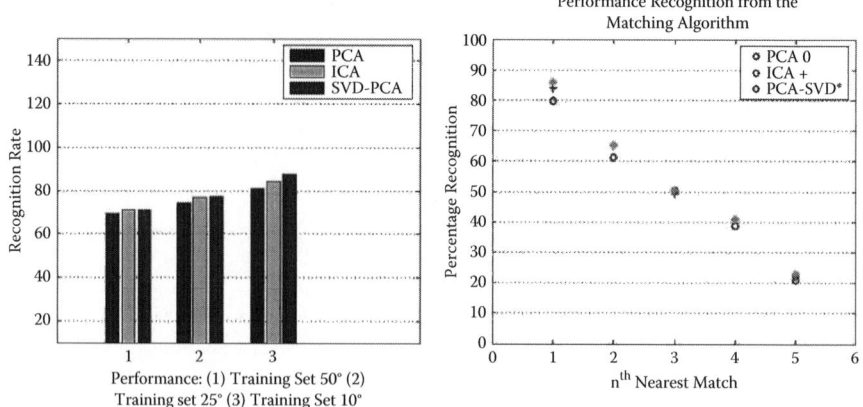

FIGURE 12.7: Comparision between PCA, ICA, SVD-PCA, and percentage recognition performance with n^{th} nearest matching

comparison between PCA, ICA, and SVD-PCA and the percentage recognition performance with n^{th} nearest matching.

12.3.4 Results with Position Angle Sampling at Every 50^0 Using ICA

The training database consists of images of all the objects. The position angle manifold is sampled at 50^0. The training is done by applying various numbers of the most dominant eigenvectors to the network. Likewise, the position angle manifold is sampled at 25^0 and the training is carried out by applying various numbers of the most dominant eigenvectors to the network. The comparative view of the eigenspace manifold for objects in the database with position angle manifold being sampled at 50^0 and 25^0 is shown in Figure 12.8.

12.3.5 Results with Position Angle Sampling at Every 10^0 Using ICA

The training database consists of images of all objects. The position angle manifold is sampled at 10^0. The training is done by applying various numbers of the most dominant eigenvectors to the network. Figure 12.9 depicts the comparison between PCA, ICA, and SVD-PCA for position angle sampling of 10^0. The number of correct recognitions by using PCA, ICA, and SVD-ICA for position angle sampling 50^0 for a total of 6400 test images is presented in Table 12.1.

SVD–PCA-Based Object Recognition Approach 335

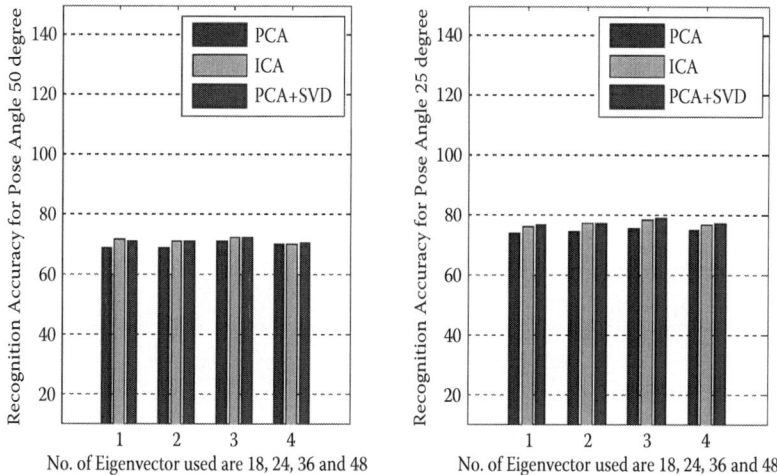

FIGURE 12.8: (a) Comparision between PCA, ICA, and PCA-SVD for position angle sampling of 50^0 (b) Comparision between PCA, ICA, and PCA-SVD for position angle sampling of 25^0

12.3.6 Comparision between PCA, ICA, and SVD-PCA

The first 30 dominant eigenvectors are used in comparision with eigenimages. The number of correct recognitions are shown in Table 12.1, Table 12.3 and Table 12.5, and the percentage (%) recognition rates are shown in Table 12.2, Table 12.4, and Table 12.6, respectively. Best performance can be observed when 24 or 36 eigenvectors were used for training. In the case when eigenvectors taken is 36, the performance is best in any technique. Recognition rate of PCA is not better than that of ICA. But SVD-PCA techniques' recognition provides better rates for object recognition than ICA. When *TrainingSet*10 is taken for compararision, the recognition rate is more than the earlier rate and SVD-PCA performs better than both PCA and ICA.

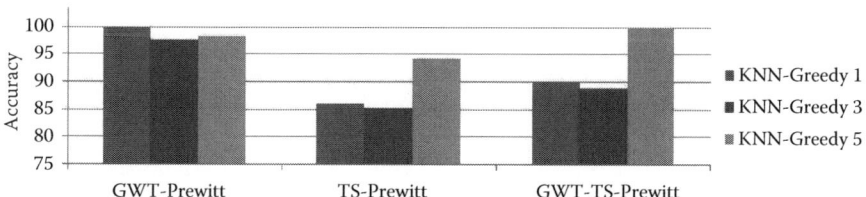

FIGURE 12.9: Comparision between PCA, ICA, and PCA-SVD for position angle sampling of 10^0

TABLE 12.1: Number of correct recognitions by using PCA, ICAs and SVD-PCA in considering position angle sampling 50^0.

Eigenvectors	18	24	36	48
PCA	4416	4403	4563	4461
ICA	4583	4531	4614	4487
SVD-PCA	4550	4563	4620	4512

TABLE 12.2: Number of (%) recognitions by using PCA, ICA, and SVD-PCA when position angle sampling is 50^0.

Eigenvectors	18	24	36	48
PCA	69%	68.8%	71.3%	69.7%
ICA	70.8%	70.8%	72.1%	70.1%
SVD-PCA	71.1%	71.3%	72.2%	70.5%

TABLE 12.3: Number of correct recognitions by using PCA, ICA, and SVD-PCA when position angle sampling is 25^0.

Eigenvectors	18	24	36	48
PCA	4203	4247	4292	4464
ICA	4338	4400	4474	4355
SVD-PCA	4383	4394	4492	4389

TABLE 12.4: Number of (%) recognitions by using PCA, ICA, and SVD-PCA when position angle sampling is 25^0.

Eigenvectors	18	24	36	48
PCA	73.3%	74.5%	75.3%	74.8%
ICA	76.1%	77.2%	78.5%	76.4%
SVD-PCA	76.9%	77.1%	78.8%	77.0%

TABLE 12.5: Number of correct recognition by using PCA, ICA, and SVD-PCA when position angle sampling is 10^0.

Eigenvectors	18	24	36	48
PCA	2884	2923	2974	2898
ICA	2974	3017	3067	3057
SVD-PCA	3129	3118	3193	3147

TABLE 12.6: Number of (%) recognitions by using PCA, ICA, and SVD-PCA when position angle sampling is 10^0.

Eigenvectors	18	24	36	48
PCA	80.1%	81.2%	82.6%	80.5%
ICA	82.6%	83.8%	85.2%	84.9%
SVD-PCA	86.9%	86.6%	88.7%	87.4%

TABLE 12.7: Average success rate for COIL-100 objects database.

Training Sets	TrainingSet 50	TrainingSet 25	TrainingSet 10
PCA	69.7%	74.6%	81.1%
ICA	71.2%	77.1%	84.1%
SVD-PCA	71.2%	77.5%	87.4%

Table 12.1 shows the number of correct recognitions by using PCA, ICA, and SVD-PCA when position angle sampling is 50^0. The recognitions are shown for a total of 6400 test images. Number of percentage (%) recognitions by using PCA, ICA, and SVD-PCA is shown in Table 12.2. Table 12.3 shows the number of correct recognitions by using PCA, ICA, and SVD-PCA when pose angle sampling is 25^0. The recognitions are shown for a total of 5700 test images. Number of percentage (%) recognitions by using PCA, ICA, and SVD-PCA is shown in Table 12.4. Table 12.5 shows the number of correct recognitions by using PCA, ICA, and SVD-PCA when position angle sampling is 10^0. The recognitions are shown for a total of 3600 test images. Number of percentage (%) recognitions by using PCA, ICA, and SVD-PCA is shown in Table 12.6. The average success rate for the $COIL-100$ objects datase set is shown in Table 12.7.

From the figures it is clear that the SVD-PCA method produced better results as compared to other results and has a significant increase as compared to PCA. The performance of the PCA, ICA, and SVD-PVD methods in the 5 nearest matches is shown in Figure 12.7(b). The COIL databases with the position angle sampled at 100, 250, and 500 are called the $TrainingSet10$, $TrainingSet25$ and $TrainingSet50$ databases. The performance of PCA, ICA, and SVD-PCA are shown in Figure 12.7(a), for all three training sets. From this figure it is clear that SVD-PCA performs better compared to other methods. Figure 12.8(a) and (b) and Figure 12.9 show the comparision between PCA, ICA, and SVD-PCA with the eigenvalues $q = 18, 24, 36, 48$, respectively, for position angle sampling of 50^0, 25^0, and 10^0 respectively.

12.3.7 Results with Position Estimation

In this experiment, modeling of the non-linear manifold formed by the object's appearance in varying positions is taken into consideration. The manifold is embedding into the low-dimensional subspace [19]. If the position angle parameter is a continuous variable, the manifold will be a smooth curve. Since it is not possible to capture images with the position angle as a continuous variable, the manifold will appear to be a line. In the experiment performed, 15 images are used for learning, one at every 25 degree of rotation. The remaining 57 images are left for evaluating the position. The line shows the location of the eigenimages in the $3D$ eigenspace with $q1$, $q2$, and $q3$ being the three most dominant eigenvectors. Finding the nearest object manifold provides an

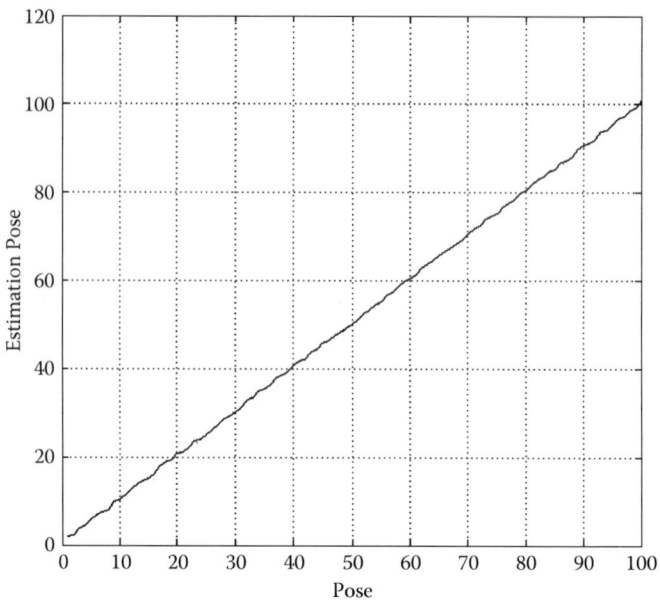

FIGURE 12.10: Results with position estimation

estimation of the object pose's. The results obtained with position estimation is shown in Figure 12.10.

12.4 Conclusion and Future Enhencement

Today, machines are able to recognize objects in applications such as surveillance, robot vision, biometrics, security tasks, etc. Appearance-based 3D object recognition in computer vision has come a long way in the last four decades. Visual appearance-based object recognition methods are usually based on feature extraction techniques such as PCA and ICA. Here, a new feature extraction method for object recognition tasks based on singular value decomposion-principal component analysis (SVD-PCA) has been used. Although PCA is intrinsically very close, SVD differs in important aspects that can affect the performance of the classifier used in the recognition system. The results are compared with other standard transforms like PCA and ICA. The images of the COIL database have been used for testing the appearance-based recognition system, based on the notion of parameteric eigenspace. An experiment is performed on a preprocessed Columbia Object Image Library ($COIL - 100$) database in which the object images are segmented from the background and are rescaled. The training images are taken by sampling the position angle of the objects so that the appearance of the object could be learned from all angles, thus resulting in 3D object recognition. Overall, it is observed that SVD-PCA performs significantly better than conventional and subsequent eigenvalue decompositons. Experimental results in appearance-based object recognition confirm that SVD-PCA offers better recognition rates than ICA and PCA.

In the appearance-based approach, learning as well as recognition are done using 3D preprocessed images. This is a strong contrast to traditional recognition algorithms that require the extraction of geometric features such as edges and lines. Such features are often difficult to compute with robustness and reliable algorithms. Dimensionality of the eigenspace is another issue. The number of eigenspace dimensions needed for representation depends on the appearance characteristics of the objects, as well as the number of objects of interest to the recognition system. If the number of objects increases a large number of dimensions may be needed for robust recognition. The exact number of dimesions required for any given set of objects is difficult to quantify since there are no simple relationships between an object's intrinsic properties and its eigenspace representation. Such relationships need to be explored.

Object recognition can be extended to mobile environments, and the system can be extended such that it also recognizes the object in real time and in much less constrained situations. This work can be extended to meet these needs for the next generation by embedding the system in cameras and microphones, which are very small lightweight wearable systems. These systems have the critical advantage that they use the modalities of objects in recognition. In the future we can include the multilinear ICA (MICA) algorithm, which can capture information to improve automatic object or face recognition, thereby increasing the recognition rate.

Bibliography

[1] V.K Ananthashayana and V. Asha. In *International Society for Photogrammetry and Remote Sensing'08: Proceedings of the international archives of the photogrammetry, remote sensing and spatial information sciences*, The Netherlands.

[2] P.J. Bell and R.C. Jain. 3D Object Recognition. *ACM Computer Surveys*, 17(1):75–145, 1985.

[3] P.J. Bell and R.C. Jain. An Information Maximization Approach to Blind Separation and Blind Deconvolution. *Neural Computation*, 7(6):1129–1159, 1995.

[4] C.L. Chowdhary. Linear Feature Extraction Techniques for Object Recognition: Study of PCA and ICA. *Journal of the Serbian Society for Computational Mechanics*, 5(1):19–26, 1991.

[5] C.L. Chowdhary. *Appearance-Based 3-D Object Recognition and Pose Estimation*. LAP Lambert Academic Publishing, first edition, 2011.

[6] C.L. Chowdhary, K. Muatjitjeja, and D.S. Jat. In *Emerging Trends in Networks and Computer Communications'15: Proceedings of the Emerging Trends in Networks and Computer Communications*, Namibia.

[7] P. Common. Independent Component Analysis, A New Concept? *Neural Networks*, 36(3):287–315, 1994.

[8] I. Dadher and R. Nachar. Face Recognition Using IPCA-ICA Algorithm. *IEEE Transactions on Pattern Analysis and Machine Intelligence*, 28(6):996–1000, 2006.

[9] K. Fukunaga. *Introduction to Statistical Pattern Recognition*. John Wiley & Sons, Inc., second edition, 1996.

[10] G.H. Golub and C. Reinsch. Singular Value Decomposition and Least Squares Solutions. *Numerische Mathematik*, 14:403–420, 1970.

[11] M. Hafed and M.D. Levine. Face Recognition Using the Discrete Cosine Transform. *International Journal of Computer Vision*, 43(3):167–188, 2001.

[12] A. Hyvrinen, J. Karhunen, and E. Oja. *Independent Component Analysis*. John Wiley & Sons, Inc., first edition, 2001.

[13] C. Jutten and J. Herault. Blind Separation of Sources, Part 1: an Adaptive Algorithm based on Neuromimetic Architecture. *Signal Processing*, 24(1):1–10, 1991.

[14] J. Karhunen and J. Joutsensalo. Representation and Separation of Signals Using Non Linear PCA Type Learning. *Neural Networks*, 7(1):113–127, 1994.

[15] D.K. Kumar and C.L. Chowdhary. *ShapeIndex based Applications of Local Features for Object Recognition*. LAP Lambert Academic Publishing, first edition, 2011.

[16] H. Murase and S. Nayar. Visual Learning and Recognition of 3-D Objects from Appearance. *International Journal of Computer Vision*, 14(1):5–24, 1995.

[17] S.A. Nene, S.K. Nayar, and H. Murase. Columbia University Web site: http://www.cs.columbia.edu/CAVE/.

[18] H.S. Sahambi and K. Khorasani. A Neural-Network Appearance-Based 3-D Object Recognition Using Independent Component Analysis. *Neural Networks*, 14(1):9–18, 2003.

[19] L. Sirovich and M. Kirby. Low-dimesional Procedure for the Characterization of Human Faces. *Journal of the Optical Society of America A*, 4(3):519–524, 1987.

[20] M. Turk and A. Pentland. Eigenfaces for Recognition. *Journal of Cognitive Neuroscience*, 3(1):71–86, 1991.

[21] M.A. Vicente, P.O. Hoyer, and K. Hyvärineni. Equivalence of Some Common Linear Feature Extraction Techniques for Appearance-based Object Recognition Tasks. *IEEE Transaction on Pattern Analysis and Machine Intelligence*, 29(5):896–900, 2007.

[22] L.W. Zhao, S.W. Luo, and L.Z. Liao. In *ITNepal'03: Proceedings of the International Conference on IT Nepal*, Nepal.

Chapter 13

The KD-ORS Tree: An Efficient Indexing Technique for Content-Based Image Retrieval

N. Puviarasan

Annamalai University, Chidamvaram, India

R. Bhavani

Annamalai University, Chidamvaram, India

13.1	Introduction	344
13.2	Related Work	344
13.3	Proposed Work	345
13.4	Hill Climbing Segmentation	347
	13.4.1 Saliency Calculation	347
13.5	Color Features Extraction	348
	13.5.1 LAB Histogram	351
13.6	Texture Feature Extraction	351
	13.6.1 Gabor Filter	352
13.7	Shape Features Extraction	353
	13.7.1 Histogram of Oriented Gradients	353
13.8	Dimensionality Reduction Using KPCA	354
13.9	Indexing	356
	13.9.1 KD-Tree Insertion Algorithm	357
13.10	Optimal Range Search Algorithm	359
13.11	Performance Evaluation	361
13.12	Results and Discussion	361
13.13	Conclusion	363
	Bibliography	365

In content-based image retrieval (CBIR) applications, the idea of indexing is mapping the extracted descriptors from images into a high-dimensional space. In this chapter, visual features like color, texture, and shape are considered. The novel idea of extracting the color features from a LAB histogram is used. Texture features are extracted by applying a Gabor filter on the image. A histogram of oriented gradients (HOG) is used to extract the shape features. After combining color, texture, and shape features, the feature vector is reduced using kernel principle component analysis (KPCA). Then, the KD-tree is used for indexing the images. A new proposed optimized range search (ORS) algorithm is used to obtain the optimal range for retrieving the relevant images from the database. The proposed KD-ORS tree is compared with the other existing trees. It is experimentally found that the proposed KD-ORS tree gives better performance than the other existing trees, with an accuracy of 82.1% on the COIL-100 database and of 82.1% on the Oxford flower-102 database.

13.1 Introduction

The amount of multimedia data has strongly increased in recent years. Therefore, new efficient and powerful applications are needed to handle the data. Particularly, most applications require efficient methods to retrieve the relevant data. In such situations, indexing of high dimensions is important. The idea of indexing is mapping the extracted descriptor from images into a high-dimensional space. Each image in the image databases is described using visual features such as color, texture, and shape. So, image signatures have high dimensions to represent the images. To find the similar images from the image database, an efficient similarity search plays a significant role. In order to achieve an efficient similarity search in an image database, a robust method to index the high-dimensional feature space is to be developed. Similarity search corresponds to range search on the indexed structure. The distance between two vectors is frequently used to estimate the similarity between the related images. Therefore, the problem of finding the most similar images to a given query image can be seen as a problem of k-nearest neighbour (k-NN) search in high-dimensional vector space. The methods that have been proposed for searching are known as high-dimensional indexing methods.

13.2 Related Work

Many approaches have been devised to index large databases. In the spatial access indexing method, an image is represented using image features, and different distances are used to find the similarity between the query image and database images [25]. KD-tree [4] and R-tree [14] are examples of the spatial access method. The R* tree [3] gives better performance than the R-tree. Rb-tree [30] introduced a method called "forced reinstart." The nearest neighbor search algorithms based on KD-trees havebeen applied to large-scale indexing and searching [2, 22]. In [24], the Bkd-tree which is the extension of the KD-tree to make the static structure of the tree dynamic, is proposed. A complete survey of multi-dimensional indexing is found in [13, 1]. Nearest neighbor (NN) search is a fundamental problem in the research communities of computational geometry [11] and machine learning [31]. The k-D-B tree is proposed in [26]. A TV-tree using telescope vectors is devised in [21]. An x-tree for indexing high-dimensional data was introduced in [5]. This X-tree uses a new organization of the directory to avoid the overlapping of regions, and performs better than the TV-tree and R* tree.

In [37], an RP-KD tree is proposed in which multiple KD trees are used to present the data points in a CBIR system into a lower-dimensional space. Space partitioning structures like quadtree [28] and LSD tree [15] are used for high-dimensional feature space. A variant optimized k-d tree called a VAM k-d tree is also presented [36]. A new dynamic index structure called the GC-tree is proposed in [7]. It is based on a subspace partitioning method optimized for a high-dimensional image dataset. In [18], an SR-tree, which is the integration of bounding spheres and bounding rectangles, is proposed. The SS-tree proposed in [35] uses minimum bounding spheres instead of rectangles. In [33], a new type of index structure called an SR tree, based on the R-Tree and inverted table is proposed. There are several other tree approaches, such as the S2-tree [34], Hybrid tree [8], A-tree [27], etc.

The proposed KD-ORS tree is the integration of the kd tree and optimal range search algorithm. This KD-ORS is a novel idea for indexing and searching the relevant images from the image database, and is not addressed by any authors in any of the literature mentioned above.

13.3 Proposed Work

Content-based image retrieval systems are used to search digital images in large databases and retrieve relevant items based on the actual content of the given query image. Content can be in the form of low-level features or any other information from the image. Figure 13.1 shows a block diagram of the proposed CBIR system. The proposed KD-ORS tree indexing method for image retrieval has the following steps:

1. The given input query image is segmented using hill climbing segmentation to extract the salient features of the image.

2. Color features of an image are obtained using the LAB color histogram.

3. Texture features are extracted using a Gabor filter.

4. Shape features are retrieved using a histogram of oriented gradients.

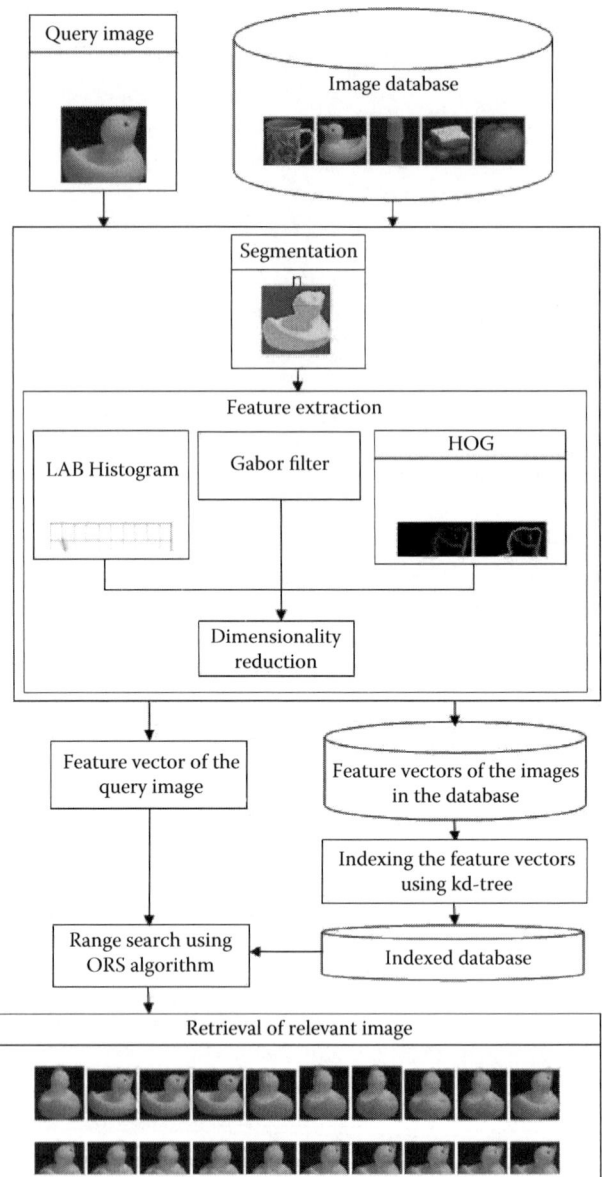

FIGURE 13.1: Block diagram of the proposed CBIR system

5. All features are fused to get the single feature vector. The dimensions of the features are reduced using the KPCA algorithm.

6. The kd tree algorithm is used to index the image features in the database after applying step 1 to step 5.

7. The proposed KD-ORS algorithm is used to search relevant images in the database.

13.4 Hill Climbing Segmentation

Image segmentation is a fundamental task in image processing. Salient regions are useful in applications such as object-based image retrieval, adaptive content delivery, adaptive region-of-interest-based image compression, and smart image resizing, which help us to extract the most features from the images. Hill climbing segmentation begins with identifying the salient regions. Saliency is determined as the local contrast of an image region with respect to its neighborhood at various scales.

13.4.1 Saliency Calculation

Saliency is calculated using the distance between the average feature vectors of the sub-region of images and its neighbors. At a given scale, the contrast-based saliency value $c_{i,j}$ for a pixel at position (i, j) in the image is determined as the distance D between the average vectors of pixel features of the inner region R_1 and those of the outer region R_2 as shown below, where N_1 and N_2 are the number of pixels in R_1 and R_2, respectively, and v is the vector of feature elements corresponding to a pixel.

$$c_{i,j} = D\left[\frac{1}{N_1}\sum_{p=1}^{N_1} v_p, \frac{1}{N_2}\sum_{q=1}^{N_2} v_q\right] \quad (13.1)$$

Figure 13.2 shows saliency feature selection using hill climbing segmentation. The distance D is a Euclidean distance if v is a vector of uncorrelated feature elements, and it is a Mahalanobis distance if the elements of the vector are correlated. In this work, a CIELab color space histogram is used to get the feature vector for color and luminance. Since perceptual differences in CIELab color space are approximately Euclidean, then D in Equation 14.1 is given below, where $v_1 = [L_1; a_1; b_1]^T$ and $v_2 = [L_2; a_2; b_2]^T$ are the average vectors for regions R_1 and R_2, respectively.

$$c_{i,j} = \|v_1 - v_2\| \quad (13.2)$$

If the image is noisy, then R_1 can be a small region of $N \times N$ pixels. If the width of image is w pixels and the height of image is h pixels, then the width of region R_2, namely w_{R_2}, is varied as:

$$\frac{w}{2} \geq (w_{R_2}) \geq \frac{w}{8} \quad (13.3)$$

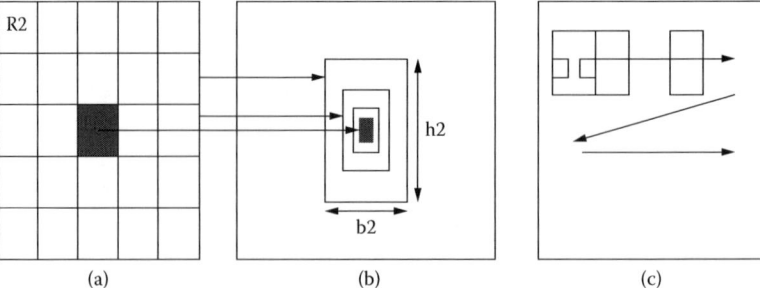

FIGURE 13.2: (a) Contrast detection filter showing inner square region R_1 and outer square region R_2. (b) The width of R_1 remains constant while that of R_2 ranges according to Equation 14.3 by halving itself for each new scale. (c) Filtering the image at one of the scales in a raster scan fashion.

assuming w is smaller than h. Otherwise, choose h to decide the dimensions of R_2. Depending on the size of the regions, only the salient features are extracted correctly. Otherwise, the non-salient features are detected as salient features. So, for each image, filtering is performed at three different scales, and the final saliency map is determined as a sum of saliency values across the scales S, where $m_{i,j}$ is an element of the combined saliency map M obtained by point-wise summation of saliency values across the scales [20, 6].

$$m_{i,j} = \Sigma_s c_{i,j}; \quad i \in [1, w], j \in [1, h] \qquad (13.4)$$

In this chapter, the CIELab histogram of the input image is computed. Starting from the non-zero bin of the color histogram, the uphill is moved until the peak is reached. The process of finding the peaks is repeated. The neighboring pixels that lead to the same pixels are grouped together as one segment. Figure 13.3 shows the sample input images from the COIL-100 database. It has objects rotated at different angles, like 0°, 50°, 100°, 150°, 200°, 250°, and 300°. Figure 13.4 shows the sample input images from the Oxford flower-102 database with different orientations. Figure 13.5 and Figure 13.6 show their corresponding segmented images using hill climbing segmentation. The components of the proposed approach are shown in Table 13.1.

13.5 Color Features Extraction

Typically, there are two types of visual features in CBIR: primitive features that include color, shape, and texture, and domain specific features. Feature extraction is concerned with capturing visual content of images for search and retrieval. In this research work, for color features, LAB histogram values are considered.

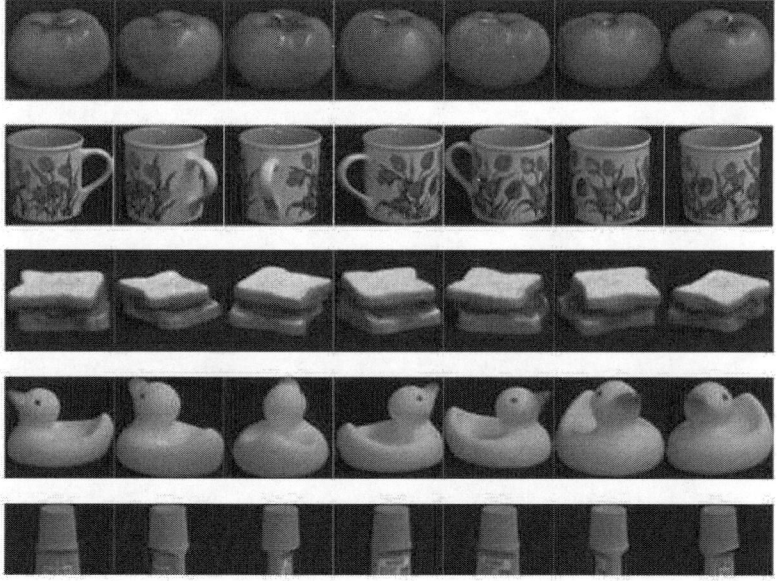

FIGURE 13.3: Sample input images from the COIL-100 database

FIGURE 13.4: Sample input images from the Oxford flower-102 database

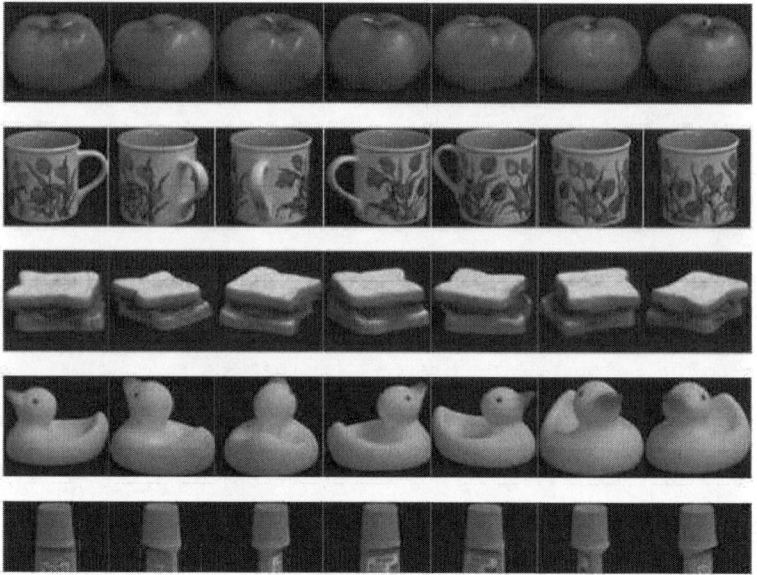

FIGURE 13.5: Sample segmented images of the COIL-100 DB shown in Figure 13.3

FIGURE 13.6: Sample segmented images of the Oxford flower-102 database shown in Figure 13.4

TABLE 13.1: Components of the proposed approach.

Sl. No.	Type	Method	Description
1	Color	LAB Histogram	Used to extract color features
2	Texture	Gabor filter	Used to extract texture features
3	Shape	Histogram	Used to extract shape features
4	Feature	KPCA algorithm	Used to reduce the dimension of feature
5	Indexing	Proposed KD-ORS tree	Used to index the image features

13.5.1   LAB Histogram

CIE L*a*b* (CIELab) describes all the colors visible to the human eye and serves as a device-independent model to be used as a reference. The three coordinates L*, a*, and b* of CIELab represent the lightness of the color. When the value of L* is 0, it becomes black, and when L* is 100, it indicates diffuse white. The position of a* is between red/magenta and green. When a* has negative values, it indicates green, and when a* has positive values, it indicates magenta. The position of b* is between yellow and blue. While b* has negative values, it indicates blue, and while b* has positive values, it indicates yellow. The LAB histogram uses a 3D binned space where each bin is the proportion of pixels in the bin's color range. The color descriptors are the dominant colors in the given image, their variance and percentage. In this research work, 32 bins are used in the LAB histogram. It is a novel idea to use LAB histogram values as the feature vector for describing the color features of an image. The 32 × 32 size feature vector is obtained from this histogram.

13.6   Texture Feature Extraction

Texture means visual pattern, which is one of the main features utilized in image processing. It describes the distinctive physical composition of a surface and contains information about structural arrangements of the surface, like clouds, leaves, bricks, etc. Since an image is composed of pixels, texture can be defined as the entity consisting of mutually related pixels and group of pixels. This group of pixels is called texture primitives or texture elements, referred to as texels. The methods of characterizing texture fall into two major categories: Statistical and Structural. The former qualifies texture by the statistical distribution of the image density, and the latter describes texture by identifying structural primitives and their placement rules. The Gabor filter belongs to the structural category, and is used to extract the texture features in the proposed system.

13.6.1 Gabor Filter

The Gabor filter provides a useful means for analyzing the texture information of an image. Gabor wavelets are widely adopted to extract texture from the images for retrieval. Basically, Gabor filters are a group of wavelets, with each wavelet capturing energy at a specific frequency and specific orientation. The scale and orientation tuneable property of a Gabor filter make it especially useful for texture analysis. To extract the Gabor texture feature of an image I, I is first filtered with a bank of scale and orientation, then computes the mean and standard deviation. When an image is processed by a Gabor filter, the output is the convolution of the image $I(x, y)$ having a size of $P \times Q$ with the Gabor function $g(x, y)$ defined in Equation 14.5, where $*$ represents the $2D$ convolution.

$$G_{mn}(x, y) = I(x, y) * g(x, y) \tag{13.5}$$

After applying Gabor filters on the image by scale and orientation, we are able to obtain an array of magnitudes $E(m, n)$, where m represents the scale and n represents the orientation of the wavelet, respectively.

$$E(m, n) = \Sigma_x \Sigma_y |G_{mn}(x, y)| \tag{13.6}$$

where $m = 0, 1, \cdots, M-1$ and $n = 0, 1, \cdots, N-1$. The M and N are number of scales and orientations. The magnitudes represent the energy content at different scale and orientation of image. The filter is applied with 4 scales and 3 orientations. In the proposed work, 4 scales and 3 orientations at $0°$, $45°$, and $90°$ are implemented to extract texture features. The following mean μ_{mn} (14.7) and standard deviation σ_{mn} (14.8) of the magnitude of the transformed coefficients are used to represent feature components of the texture features [23].

$$\mu_{mn} = \frac{E(m, n)}{P \times Q} \tag{13.7}$$

$$\sigma_{mn} = \sqrt{\Sigma_x \Sigma_y (|G_{mn}(X, Y)| - \mu_{mn})^2 / P \times Q} \tag{13.8}$$

These feature components are saved into two feature vectors having 12 values each, and then these two vectors are combined in order to make the single feature vector that will be treated as an image texture descriptor. In this proposed system, 24 feature values comprising mean and standard deviation of 12 values each are used.

13.7 Shape Features Extraction

The shape plays a critical role in searching for similar images. Shape features are more complicated to define because of their inherent complexities. For shape features, a histogram of oriented gradients is extracted in this research work.

13.7.1 Histogram of Oriented Gradients

The concept of dense and local histograms of oriented gradients (HOG) is introduced by Dalal and Triggs [9]. The histogram is used to count the occurrences of gradient orientation in a local part of the image. In this chapter, a local histogram of oriented gradients is used to obtain the complete description of an image.

Gradient Computation: The gradient of an image has been simply obtained by filtering it with two one-dimensional filters, such as horizontal: (-101) and vertical: $(-101)^T$. Gradient could be signed or unsigned. The direction of the contrast is of no importance for the gradient. In this work, the unsigned gradient is varied from $0°$ to $180°$.

Cell and Block Descriptors: Each cell consists of a fixed number of pixels. For each cell, HOG is computed for orientation and the values are stored into a number of bins. A greater number of bins gives more details of an image. It is necessary to normalize the cell histograms because of the illumination variations and other noises in images. Cell histograms are normalized according to the values of the neighborhood cell histogram. Then, normalization is done for a block, which is a group of cells. After the normalization process, all histogram values are combined as a single feature vector. Different normalization schemes are possible for a vector V containing all histograms of a given block. The normalization factor nf, could be obtained along these schemes, such as L1-norm (14.9) and L2-norm (14.10), where ϵ is a small regularization constant. It is necessary to evaluate when the gradient is empty. Then, the value of ϵ has no influence on the results [32, 19]. In addition, no normalization is applied on the cells $nf = 1$.

$$nf = \frac{V}{\|V\|_1 + \epsilon} \qquad (13.9)$$

$$nf = \frac{V}{\|V\|_2^2 + \epsilon} \qquad (13.10)$$

In this chapter, the image is divided into (3×3) cells, and a 4-bin histogram is drawn for each cell. So, there are 36-bin histograms of oriented gradient values available. Therefore, the HOG features of an image contain 36 feature values. Figure 13.7 shows orientation vector visualization of the COIL-100

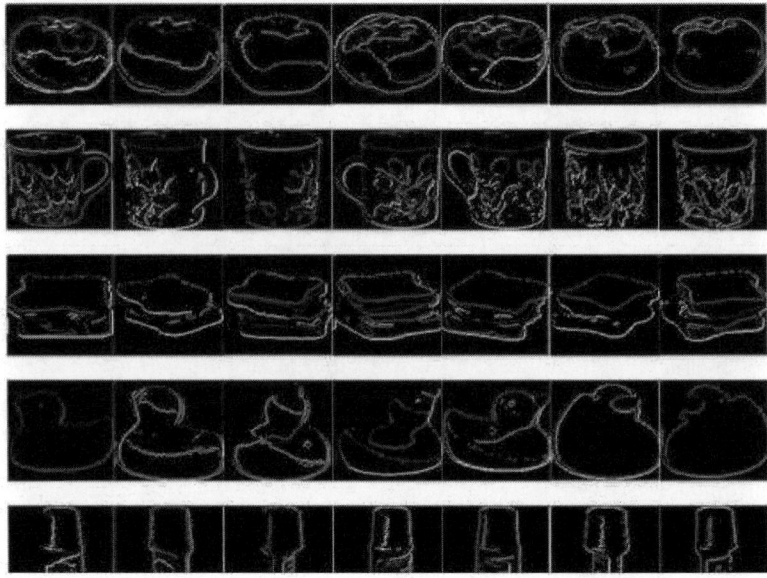

FIGURE 13.7: Orientation vector visualization of the COIL-100 database using HOG

database using HOG. Figure 13.8 shows orientation vector visualization of the Oxford flower-102 database using HOG. Figure 13.9 shows magnitude vector visualization of the COIL-100 database using HOG. Figure 13.10 shows orientation vector visualization of the Oxford flower-102 database using HOG.

13.8 Dimensionality Reduction Using KPCA

Given a dataset S in an input space χ, KPCA maps S into a kernel space F, also called feature space, through a possibly nonlinear mapping f, associated with a given kernel function k, where $k(p,q) = \phi(p), \phi(q), p, q, S and <,>$ denotes the dot product. KPCA finds the set of eigenvectors e_i, and their corresponding eigenvalues λ that satisfy Equation 13.11.

$$Me_i = \lambda_i e_i \qquad (13.11)$$

The covariance matrix M is defined as in Equation 13.12 with N as the cardinal of S.

$$M = \frac{1}{N} \sum_{p \in S} \phi(p)\phi(p)^t \qquad (13.12)$$

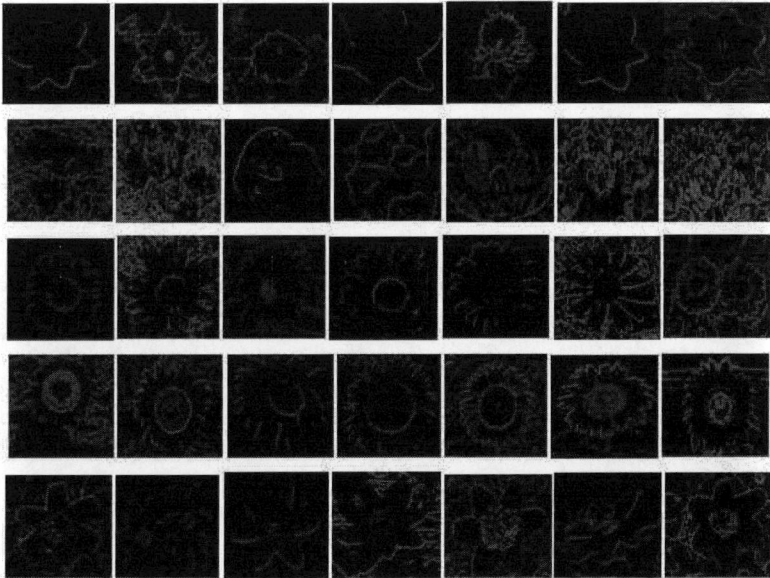

FIGURE 13.8: Orientation vector visualization of the Oxford flower-102 database using HOG

The problem in Equation 14.9 can be reformulated as below, where $\omega_i = (\omega_{i,1}, \omega_{i,2}, \cdots, \omega_{i,N})$ is a vector such that e_i is defined as in Equation 13.14. The symmetric matrix K is defined as in Equation 13.15:

$$k\omega_i = \lambda_i \omega_i \qquad (13.13)$$

$$e_i = \sum_{p \in S} \omega_{i, rank(p)} \phi(p) \qquad (13.14)$$

$$K_{p,q \in s} = k(p, q) \qquad (13.15)$$

The solution to Equation 7.12, ω_i is obtained, and the projection α_i^p of vector p along the ith kernel principal component is given by

$$\alpha_i^p = <e_i, \phi(p)> = \Sigma_{q \in S} W_{i, rank(q)} k(p, q) \qquad (13.16)$$

In this paper, the Gaussian radial basis function (GRBF) kernel is used as a kernel function. The GRBF kernel is defined as $(p, q) = e^{-\|p-q\|^2 / 2\delta^2}$, where p and q are two vectors in the input space. In CBIR context, p and q are image descriptors containing heterogeneous attributes such as color, texture, shape, etc.; δ is known as the width of the GRBF, often set by users beforehand. This kernel parameter determines the nature of the nonlinear mapping $\phi(.)$ to the kernel space. The feature space dimensionality, denoted d, corresponds to the

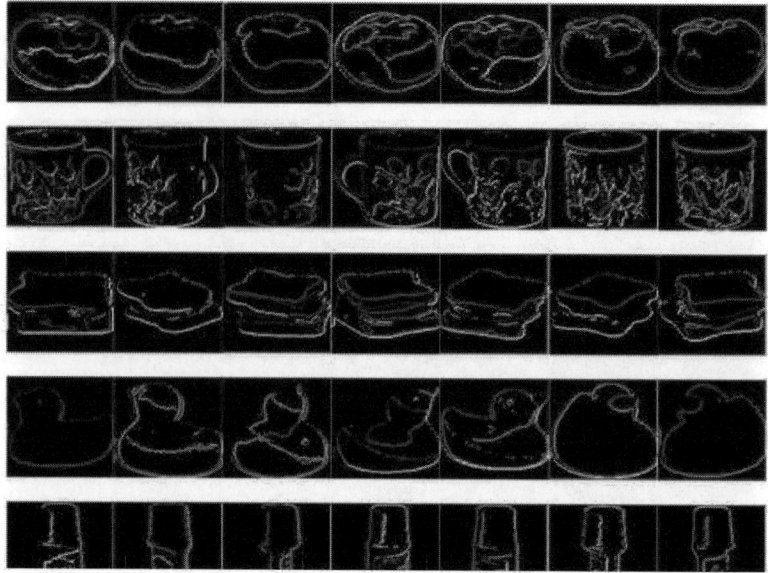

FIGURE 13.9: Magnitude vector visualization of the COIL-100 database using HOG

number of eigenvectors that are kept in the reduced space, also has an influence on kernel data mapping. ϕ and d are two parameters to set while dealing with the GRBF kernel [10]. Thus, different values of these parameters will induce a series of different kernel spaces, and therefore different representation of the vectors [29, 12]. In this research work, 32×32 feature values are obtained from the LAB histogram to represent the color features. The texture features are retrieved using a Gabor filter with 24 values. The histogram of oriented gradient method is used to extract the shape features of an image with 36 feature values. The KPCA algorithm is used to reduce the dimensions of the color feature vector. After applying KPCA, the size of the color feature is 32×1 and hence the size of the feature vector of an image is 92×1.

13.9 Indexing

As the size of the feature vector is very high, the mean and median values play an important role, and describe the features of an image in an abstract manner. Only two values, i.e., mean and median values of the features, are considered for indexing. As the "k" value is 2, it becomes a $2d$-tree. It can be extended to any value of k. The KD-tree insertion algorithm is used to insert

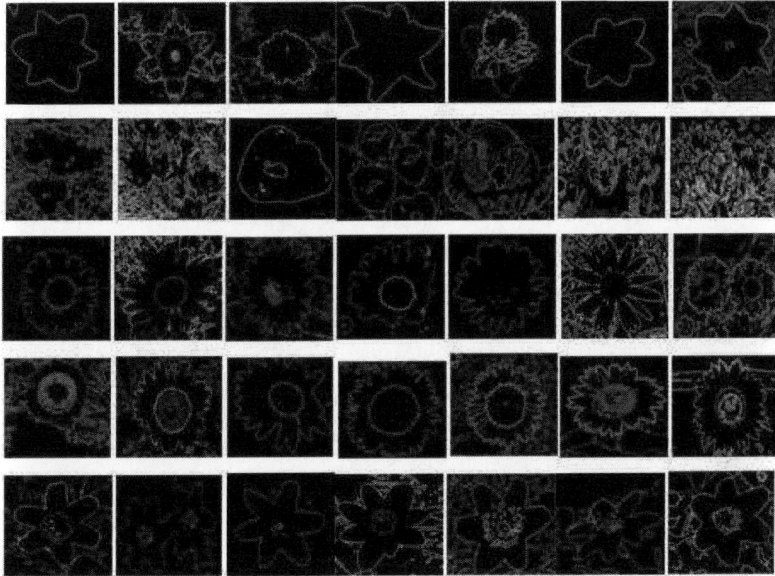

FIGURE 13.10: Magnitude vector visualization of the Oxford flower-102 database using HOG

the node in indexing. A newly proposed search algorithm called optimal range search (ORS) is used for efficient searching of the relevant images.

13.9.1 KD-Tree Insertion Algorithm

The proposed indexing scheme is based on the KD-tree indexing method. Here, a 2-dimensional tree structure is built. Each node of the tree contains the left pointer, the data, and the right pointer. The data node of the tree contains two values: The first value is the mean of the feature vector, which is obtained as described in the previous section, and the second value is the median of the same feature vector. Consider a 2-dimensional rectangular range-searching problem. Let P be the set of n-points in the plane. A 2-dimensional rectangular range query on P asks for the points from P lying inside the query rectangle $[x, x'] * [y, y']$. A point $t := (t_x, t_y)$ lies inside this rectangle if and only if $t_x \in [x, x']$ and $t_y \in [y, y']$.

Let us consider the following recursive definition of the binary search tree: The set of points is split into two subsets of roughly equal size, one subset contains the points smaller than or equal to root, the other contains the points greater than the root. The two subsets are stored recursively in two subtrees. Figure 13.11 shows the example for the 2d tree. Figure 13.12 shows the insertion of a new node in the 2d tree.

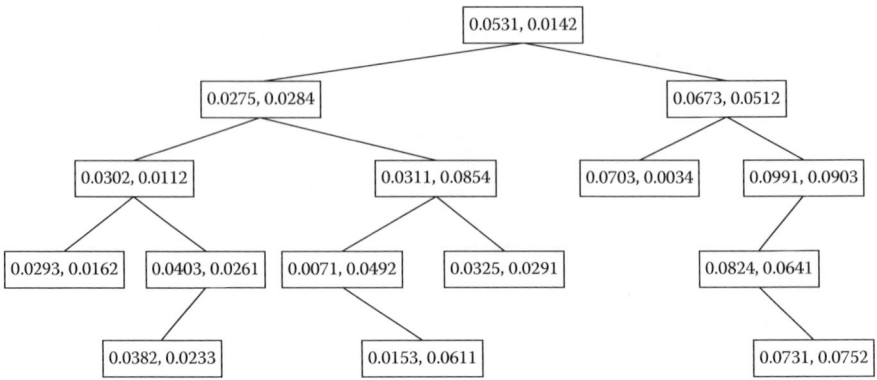

FIGURE 13.11: Example for 2d-tree

The KD-tree used in this work is the 2-dimensional tree. Each data node has mean and median values of the feature vector of the image. The mean value is the x-coordinate and the median value is the y-coordinate. The tree is split on the x-coordinate and then on the y-coordinate, and again on the x-coordinate, and so on. At the root, the set "t" is split with a vertical line into two subsets of roughly equal size. The mean values less than or equal to the root are stored in t_{left} subtree, and those greater than the root are stored in the t_{right} subtree. Similarly, t_{right} is split with a horizontal line, and then the values are stored in the left and right subtrees of the right child [17, 16].

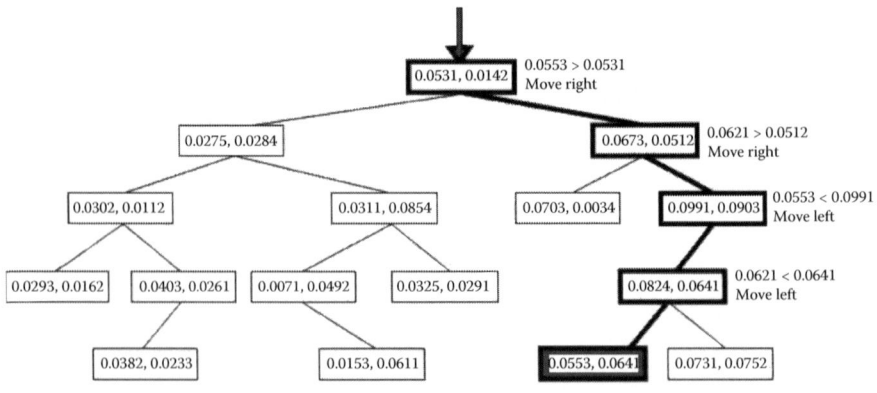

FIGURE 13.12: Insertion of a new node $(0.0553, 0.0621)$ in the 2d-tree

Algorithm [Insertion of new node into KD-ORS Tree]

Input: Mean and median feature vectors fm_i and fmd_i in feature database (Mean and median feature vector $f_d(f_{(d_i)}) \rightarrow [fm_i; fmd_i])$
Output: Created tree structure
1. For each feature vector in the feature database f_d:
2. Read $[fm_i; fmd_i]$ and
3. Check for the empty tree ($t = null$)
4. If the tree is empty, insert the $[fm_i; fmd_i]$ as the root of the tree $[fm_t; fmd_t]$.
5. Else check for level
6. If the height h of node ($h/2 \neq 0$)
7.   If $fm_i < fm_t$, then move towards left t_{left} of the tree
8.   Repeat steps from step 3
9.   Else move towards right t_{right} of the tree $fm_i > fm_t$
10.   Repeat steps from step 3
11. Else if ($h/2 = 0$)
12.   If $fmd_i < fmd_t$ then move towards left t_{left} of the tree
13.   Repeat steps from step 3
14.   Else move towards right t_{right} of the tree $fmd_i > fmd_t$
15.   Repeat steps from step 3
16. End
17. Return the tree structure with all nodes inserted.

13.10 Optimal Range Search Algorithm

In this chapter, a novel idea of searching the tree using optimum range is proposed. The new method range search looks for a range of values. Consider the KD-tree in which each data node contains mean and median values of the feature vector of the image. A range search looks for all members of the tree that have a mean value between fm_{low} and fm_{high}, and a median value between fmd_{low} and fmd_{high}. The optimal values for the search, fm_{opt}, fmd_{opt} are fixed as 0.0015. The values fm_i - fm_{opt} are fixed as fm_{low}, and $fm_i + fm_{opt}$ is fixed as fm_{high}. Similarly, fmd_{low} and fmd_{high} are found. The nodes in the tree having the mean values between fm_{low} and fm_{high} and median values between fmd_{low} and fmd_{high} are retrieved. This is called an optimal range search tree because, depending upon the mean and median values of the query image, the optimal values fm_{opt} and fmd_{opt} are fixed. Since kd trees divide the range of a domain in half at each level of the tree, they are useful for performing range searches. This algorithm takes the time complexity of $O(n \log n)$ to construct the KD-tree and $O(\log n)$ to search the tree and retrieve the relevant images.

Algorithm [The proposed KD-ORS algorithm for searching a node]

Input: Tree structure t, query feature vector fq.
Output: List of feature vectors within the range.
1. Read the optimal values of feature vectors fm_{opt}, fmd_{opt}.
2. Calculate fm_{min}, fm_{max}, fmd_{min}, and fmd_{max}.
3. While tree t is not empty
4. Check for level
5. If the height $'h'$ of the node $(h/2 \neq 0)$
6.   If $(fm_i < fm_{max})$ then
7.   If $(fm_i > fm_{min})$ then
8.   If $(fmd_i < fmd_{max})$ and $(fmd_i > fmd_{min})$
9.   Add the node into relevance array $a_{rel}[i]$
10.   Else
11.   Move towards the left of the tree
12.   Repeat steps from 2
13.   Else
14.   Move towards the left of the tree
15.   Repeat steps from 2
16.   Else
17.   Move towards the right of the tree
18.   Repeat steps from 2
19. Else if $(h/2 = 0)$
20.   If $(fmd_i < fmd_{max})$
21.   If $(fmd_i > fmd_{min})$ then
22.   If $(fm_i < fm_{max})$ and $(fm_i > fm_{min})$
23.   Add the node into relevance array $a_{rel}[i]$
24.   Else
25.   Move towards the left of the tree
26.   Repeat steps from 2
27.   Else
28.   Move towards the left of the tree
29.   Repeat steps from 2
30.   Else
31.   Move towards the right of the tree
32.   Repeat steps from 2
33. End
34. Print the elements in the array $a_{rel}[i]$.

13.11 Performance Evaluation

In information retrieval, precision, also called positive predictive value, is the fraction of retrieved instances that are relevant over all images retrieved, as defined in Equation 13.17. Recall, also known as sensitivity, is the fraction of relevant instances that are retrieved over all relevant images, as defined in Equation 13.18. Both precision and recall are therefore based on an understanding and measure of relevance.

$$Precision(P) = \frac{\text{Number of relevant images retrieved}}{\text{Number of all images retrieved}} \quad (13.17)$$

$$Recall(R) = \frac{\text{Number of relevant images retrieved}}{\text{Number of all relevant images}} \quad (13.18)$$

In simple terms, high recall means that an algorithm returned most of the relevant results, while high precision means that an algorithm returned substantially more relevant than irrelevant results. To evaluate the most efficient image retrieval, precision and recall scores are combined into a single measure of performance known as the F-score. Higher values of the F-score are obtained when both precision and recall are higher. Equation 13.19 is used to calculate the F-score. The accuracy is calculated as defined in Equation 13.20.

$$F - Score = 2 * \frac{P * R}{P + R} \quad (13.19)$$

$$Accuracy = \frac{(P + R)}{2} \quad (13.20)$$

13.12 Results and Discussion

The experiment is performed on two databases. In order to evaluate the performance of the proposed KD-ORS tree, two experiments have been conducted.

Experiment 1: The first experiment used the Columbia object image library (COIL-100), which is a database of color images of 100 objects. The object is placed on a motorized turntable against a black background. The turntable was rotated through 360 degrees to vary the position of the object. Images of the object are taken at position intervals of 5. This corresponds to 72 position per object. COIL-100 is a database of 7,200 color images of 100 objects. The image size is 128 × 128.

TABLE 13.2: Average precision and average recall for all indexing methods.

Database	kd with k-NN		R Tree		SS Tree		KD-ORS Tree	
	Prec	Rec	Prec	Rec	Prec	Rec	Prec	Rec
COIL 100	0.741	0.691	0.719	0.653	0.703	0.631	0.827	0.753
Oxford Flower-102	0.763	0.712	0.742	0.683	0.726	0.693	0.843	0.762

Experiment 2: The database used for this experiment is the Oxford flower-102 database. The flowers in the database are the flowers commonly occurring in the United Kingdom. It has 102 categories. Each class consists of between 40 and 258 images. The total number of images in the database is 8189. The size of the image ranges from 300 × 500 to 500 × 700.

The color feature used for these experiments is the LAB histogram. The texture features are extracted using a Gabor filter. The shape features considered in these experiments is a histogram of oriented gradients. These shape features are rotation invariant. The performance of the proposed KD-ORS tree is compared with other existing trees, like the kd tree with k-NN search, R tree, and SS tree. Figure 13.13 (a) depicts the given query image of the COIL-100 database, whereas (b) to (e) give the retrieved results of the kd with the k-NN search algorithm, R tree, SS tree, and the proposed KD-ORS tree for the COIL-100 database, respectively. Figure 13.14 (a) depicts the given query image of the Oxford flower-102 database, whereas (b) to (e) give the retrieved results of the kd with k-NN search algorithm, R tree, SS tree, and the proposed KD-ORS tree for Oxford flower-102 database, respectively. Table 13.2 shows the average precision and average recall values of different indexing methods for both the COIL-100 and Oxford flower-102 databases. It is found that the average precision values of the proposed KD-ORS tree are 0.82 on the COIL-100 database and 0.84 on the Oxford flower-102 database. Similarly, the average recall values of the proposed KD-ORS tree are 0.75 on the COIL-100 database and 0.76 on the Oxford flower-102 database. The results of the other existing indexing methods are also given.

Figure 13.15(a) and Figure 13.16(a) depict the average F-score values of different indexing methods for the images in different categories of the COIL-100 and Oxford flower-102 databases, respectively. It is found that the average F-score of KD-ORS on the COIL-100 database is 82.1%, and on the Oxford flower-102 database, it is 81.6%. It is better than the existing indexing methods, such as kd with k-NN search, the R tree, and the SS tree. Similarly, Figure 13.15(b) and Figure 13.16(b) show the average accuracy of the indexing methods for the images in different categories of the Oxford flower-102 database, respectively. After experimenting with the different indexing methods, it is found that the average retrieval accuracy of the proposed KD-ORS tree is 82.6 % for the COIL-100 database and 81.1% for the Oxford flower-102 database. It is better than the other indexing methods, such as the kd with

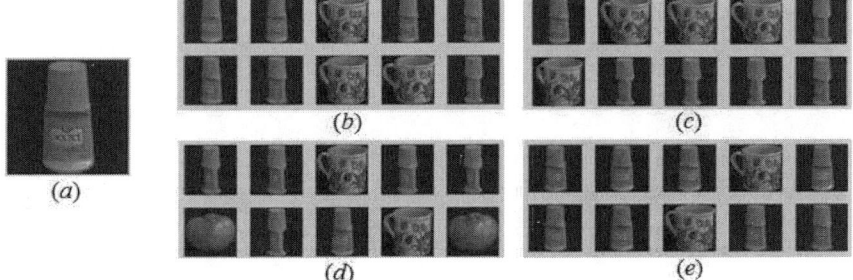

FIGURE 13.13: (a) Query images, (b) retrieval of images by KD-tree with k-NN search algorithm, (c) retrieval of images by R-tree algorithm, (d) retrieval of images by SS-tree algorithm, (e) retrieval of images by KD-ORS tree algorithm of the COIL-100 database

k-NN search, R tree, and SS tree. Table 13.3 shows the computation time of various indexing methods experimented with on the COIL-100 and Oxford flower-102 databases. It is found that the KD-ORS tree takes 2.81s for the COIL-100 database and 2.42s for the Oxford flower-102 database.

13.13 Conclusion

In this chapter, hill-climbing segmentation is done to find the salient features of the query image. The color features are extracted by the LAB histogram. The texture features are obtained by applying a Gabor filter. The shape features are extracted by a histogram of the oriented gradients. The extracted features

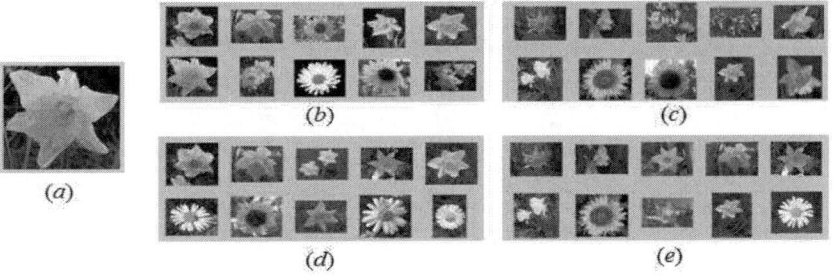

FIGURE 13.14: (a) Query images, (b) retrieval of images by KD-tree with k-NN search algorithm, (c) retrieval of images by R-tree algorithm, (d) retrieval of images by SS-tree algorithm, (e) retrieval of images by KD-ORS tree algorithm of the Oxford flower-102 database

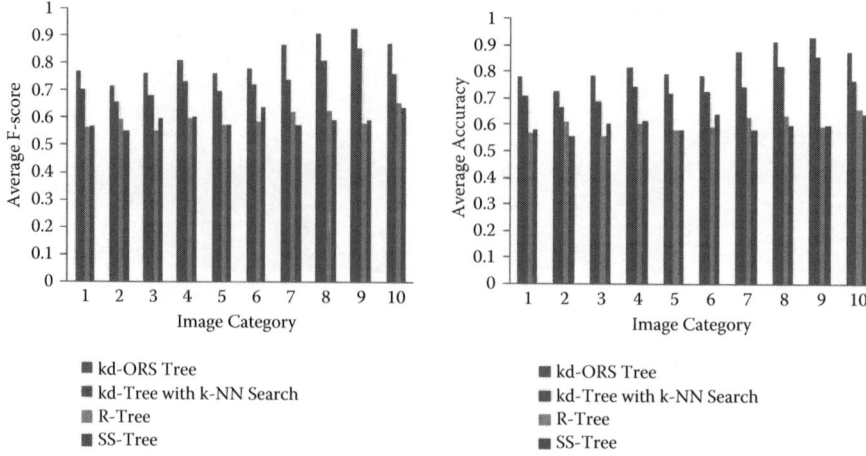

FIGURE 13.15: (a) Average F-score values for image categories in the COIL-100 database, (b) average accuracy values for image categories in the COIL-100 database

are passed to a nonlinear dimensionality reduction technique called kernel principle component analysis (KPCA). Then, the above processes are done for all the images in the database, and the obtained feature vectors are stored into a database. Next, the proposed KD-ORS tree indexing with optimal range search is applied to the feature vectors in the database. In order to compare

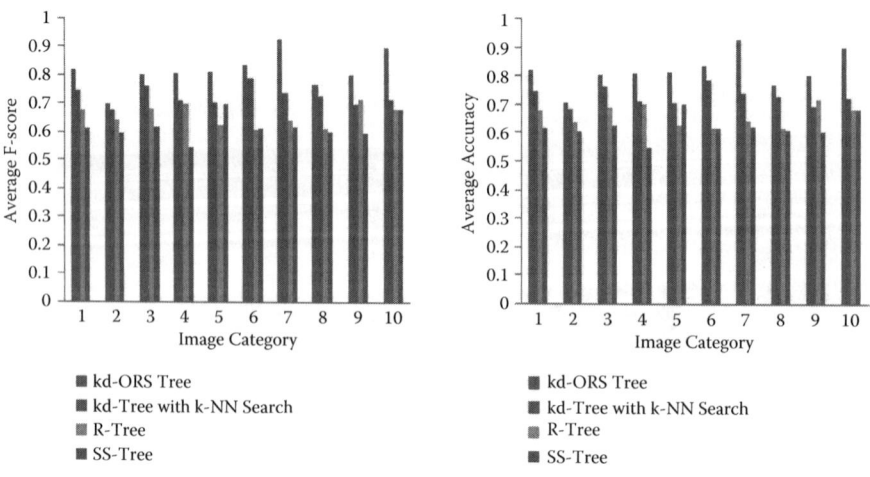

FIGURE 13.16: (a) Average F-score values for image categories in the Oxford flower-102 database, (b) average accuracy values for image categories in the Oxford flower-102 database

TABLE 13.3: Computation time of different indexing methods in seconds.

Indexing Methods	COIL-100	Oxford flower-102
Kd with k-NN search	3.54	3.48
R tree	3.65	3.59
SS tree	3.78	3.66
Proposed KD-ORS tree	2.81	2.42

the performance, experiments have been done in the COIL-100 database and Oxford flower-102 database for the proposed KD-ORS tree and other existing trees, such as kd with k-NN search, R tree, and SS tree. The results of the experiments are evaluated based on the different performance measures. It is observed that the average accuracy of the proposed KD-ORS tree is 82.6% on the COIL-100 database and 82.1% on the Oxford flower-102 database. Also, the proposed KD-ORS tree outperforms all other existing trees for retrieving more relevant images.

Bibliography

[1] Lars Arge. External memory data structures. In *Handbook of Massive Data Sets*, pages 313–357. Springer, 2002.

[2] Sunil Arya, David M. Mount, Nathan S. Netanyahu, Ruth Silverman, and Angela Y. Wu. An optimal algorithm for approximate nearest neighbor searching fixed dimensions. *Journal of the Association for Computing Machinery (JACM)*, 45(6):891–923, 1998.

[3] Norbert Beckmann, Hans-Peter Kriegel, Ralf Schneider, and Bernhard Seeger. The r*-tree: an efficient and robust access method for points and rectangles. In *Association for Computing Machinery Sigmod Record*, volume 19, pages 322–331. ACM, 1990.

[4] Jon Louis Bentley. Multidimensional binary search trees used for associative searching. *Communications of the Association for Computing Machinery*, 18(9):509–517, 1975.

[5] Stefan Berchtold, A. Keim Daniel, and Hans-Peter Kriegel. The x-tree: An index structure for high-dimensional data. In *Proceedings of the International Conference on Very Large Databases*, pages 28–39. VLDBr, 1996.

[6] Sukhwinder Bir and Amanjot Kaur. Color image segmentation in cielab space using hill climbing algorithm. *International Journal of Computer Applications*, 7(3):48–53, 2010.

[7] Guang-Ho Cha and Chin-Wan Chung. The gc-tree: a high-dimensional index structure for similarity search in image databases. *IEEE Transactions on Multimedia*, 4(2):235–247, 2002.

[8] Kaushik Chakrabarti and Sharad Mehrotra. The hybrid tree: An index structure for high dimensional feature spaces. In *15th International Conference on Data Engineering, 1999. Proceedings*, pages 440–447. IEEE, 1999.

[9] Navneet Dalal and Bill Triggs. Histograms of oriented gradients for human detection. In *IEEE Computer Society Conference on Computer Vision and Pattern Recognition, 2005. CVPR 2005*, volume 1, pages 886–893. IEEE, 2005.

[10] Imane Daoudi, Khalid Idrissi, S.E. Ouatik, Atilla Baskurt, and Driss Aboutajdine. An efficient high-dimensional indexing method for content-based retrieval in large image databases. *Signal Processing: Image Communication*, 24(10):775–790, 2009.

[11] Mark De Berg, Marc Van Kreveld, Mark Overmars, and Otfried Cheong Schwarzkopf. Computational geometry. In *Computational geometry*, pages 1–17. Springer, 2000.

[12] Murat Ekinci and Murat Aykut. Palmprint recognition by applying wavelet-based kernel pca. *Journal of Computer Science and Technology*, 23(5):851–861, 2008.

[13] Volker Gaede and Oliver Günther. Multidimensional access methods. *ACM Computing Surveys (CSUR)*, 30(2):170–231, 1998.

[14] Antonin Guttman. *R-trees: A Dynamic Index Structure for Spatial Searching*, volume 14. Association for Computing Machinery, 1984.

[15] Andreas Henrich, Hans-Werner Six, Hagen, and Peter Widmayer. The lsd tree: spatial access to multidimensional point and non-saint objects. In Proceedings of the 15th International Conference on Very Large Data Bases. San Francisco, USA: Morgan Kaufmann Publishers Inc., pages 45–53, 1989.

[16] You Jia, Jingdong Wang, Gang Zeng, Hongbin Zha, and Xian-Sheng Hua. Optimizing kd-trees for scalable visual descriptor indexing. In *2010 IEEE Conference on Computer Vision and Pattern Recognition (CVPR)*, pages 3392–3399. IEEE, 2010.

[17] Hemant M. Kakde. Range searching using kd tree, 2005.

[18] Norio Katayama and Shin'ichi Satoh. The sr-tree: An index structure for high-dimensional nearest neighbor queries. In *Association for Computing Machinery Special Interest Group on the Management of Data Record*, volume 26, pages 369–380. Association for Computing Machinery, 1997.

[19] Takumi Kobayashi. BFO meets hog: feature extraction based on histograms of oriented pdf gradients for image classification. In *Proceedings of the IEEE Conference on Computer Vision and Pattern Recognition*, pages 747–754, 2013.

[20] Shailesh Kochra and Sanjay Joshi. Study on hill climbing algorithm for image segmentation. *International Journal of Engineering Research and Applications*, 2(3):2171–2174, 2012.

[21] King Ip Lin, Hosagrahar V. Jagadish, and Christos Faloutsos. The tv-tree: An index structure for high-dimensional data. *The International Journal on Very Large Data Bases*, 3(4):517–542, 1994.

[22] David G. Lowe. Distinctive image features from scale-invariant keypoints. *International Journal of Computer Vision*, 60(2):91–110, 2004.

[23] C.B. Richard Ng, Guojun Lu, and Dengsheng Zhang. Performance study of Gabor filters and rotation invariant Gabor filters. In *Multimedia Modelling Conference, 2005. MMM 2005. Proceedings of the 11th International*, pages 158–162. IEEE, 2005.

[24] Octavian Procopiuc, Pankaj Agarwal, Lars Arge, and Jeffrey Vitter. BKD-tree: A dynamic scalable kd-tree. *Advances in Spatial and Temporal Databases*, pages 46–65, 2003.

[25] Guoping Qiu. Indexing chromatic and achromatic patterns for content-based colour image retrieval. *Pattern Recognition*, 35(8):1675–1686, 2002.

[26] John T. Robinson. The kdb-tree: a search structure for large multidimensional dynamic indexes. In *Proceedings of the 1981 Association for Computing Machinery Special Interest Group on Management of Data International Conference on Management of Data*, pages 10–18. Association for Computing Machinery, 1981.

[27] Yasushi Sakurai, Masatoshi Yoshikawa, Shunsuke Uemura, Haruhiko Kojima, et al. The a-tree: An index structure for high-dimensional spaces using relative approximation. In *Very Large Data Base*, volume 2000, pages 516–526. Citeseer, 2000.

[28] Hanan Samet. The quadtree and related hierarchical data structures. *Association for Computing Machinery Computing Surveys (CSUR)*, 16(2):187–260, 1984.

[29] Bernhard Schölkopf, Alexander Smola, and Klaus-Robert Müller. Nonlinear Component analysis as a kernel eigenvalue problem. *Neural computation*, 10(5):1299–1319, 1998.

[30] Timos Sellis, Nick Roussopoulos, and Christos Faloutsos. The r+-tree: A dynamic index for multi-dimensional objects. Technical report, 1987.

[31] G. Shakhnarovich, T. Darrell, and P. Indyk. Nearest-Neighbor Methods in Learning and Vision: Theory and Practice, chapter 3, 2006.

[32] Frédéric Suard, Alain Rakotomamonjy, Abdelaziz Bensrhair, and Alberto Broggi. Pedestrian detection using infrared images and histograms of oriented gradients. In *Intelligent Vehicles Symposium, 2006 IEEE*, pages 206–212. IEEE, 2006.

[33] Xing Tong, Yang Liu, Zhao Shi, Peng Zeng, and Hai Bin Yu. SR-tree: An index structure of sensor management system for spatial approximate query. In *Advanced Materials Research*, volume 756, pages 885–889. Trans Tech Publication, 2013.

[34] Haixun Wang and Chang-Shing Perng. The s 2-tree: An index structure for subsequence matching of spatial objects. In *Pacific-Asia Conference on Knowledge Discovery and Data Mining*, pages 312–323. Springer, 2001.

[35] David A. White and Ramesh Jain. Similarity indexing with the ss-tree. In *Proceedings of the Twelfth International Conference on Data Engineering, 1996*, pages 516–523. IEEE, 1996.

[36] David A. White and Ramesh C. Jain. Similarity indexing: Algorithms and performance. In *Electronic Imaging: Science & Technology*, pages 62–73. International Society for Optics and Photonics, 1996.

[37] Pengcheng Wu, Steven Hoi, Duc Nguyen, and Ying He. Randomly projected kd-trees with distance metric learning for image retrieval. *Advances in Multimedia Modeling*, pages 371–382, 2011.

Chapter 14

An Efficient Image Compression Algorithm Based on the Integration of a Histogram Indexed Dictionary and the Huffman Encoding for Medical Images

D.J. Ashpin Pabi
Annamalai University, Chidamvaram, India

P. Aruna
Annamalai University, Chidamvaram, India

N. Puviarasan
Annamalai University, Chidamvaram, India

14.1	Introduction	370
14.2	Background and Related Works	372
14.3	The Proposed Histogram Indexed Dictionary (HID) Compression Technique	374
	14.3.1 The Discrete Cosine Transform	376
	14.3.2 Sliding Neighborhood Filtering	377
	14.3.3 Quantization	378
	14.3.4 The Proposed Histogram Indexed Dictionary	379
	14.3.5 Huffman Encoding	380
14.4	Results and Discussion	381
	14.4.1 Peak Signal-to-Noise Ratio	381
	14.4.2 Compression Ratio and Bits Per Pixel	382
14.5	Conclusion	391
	Bibliography	391

Compression plays a vital role in the medical field for diagnosis in terms of efficient storage and data transfer. In this chapter, an efficient image compression technique constructed by integrating the proposed histogram indexed dictionary (HID) into the standard Huffman encoding has been proposed for magnetic resonance imaging (MRI) of brain images. The proposed algorithm consists of three stages: initialization, quantization, and encoding. At the initialization stage, the given input medical image is decomposed into 8×8, or 16×16, or 32×32 blocks and then the discrete cosine transform (DCT) is applied on each of these blocks with sliding neighborhood filtering. The second stage quantization truncates the high frequency content of the filtered DCT coefficients by applying the thresholding. The final stage is an encoding scheme. A novel HID is created and integrated into the standard Huffman encoding. The HID improves the performance of the Huffman encoding. The proposed algorithm assists the radiologists in deciding and planning the treatment. It also helps to reduce the size of an image as well as transmission time, and increases the transmission speed. The level of compression suits the needs of the examination, and this does not affect the examination's diagnostic value. The experimental results show that the proposed method provides greater improvement in PSNR values than the other existing techniques.

14.1 Introduction

Image compression is a process to reduce the redundancy of the image and to store or transmit data in an efficient form. Digital images are large in size and hence require larger storage space. Due to their larger size, they might take larger bandwidth and more time to upload or download through the Internet. This will be inconvenient for storage as well as file sharing. To avoid this problem, the images are compressed in lower byte size with special techniques [5, 7]. The challenge in compression is to reduce the storage quantity, and the decoded image displayed on the monitor can be as similar to the original image as possible. The compression algorithm benefits someone who owns a digital camera, surfs the web, or watches Hollywood movies on digital video disks (DVDs). The high-resolution images and the Web page images are compressed to save storage space and reduce transmission time. Medical images including computed tomography (CT) and MRI images are also compressed for long-term storage, which helps the radiologist to treat diseases by storing large images with a minimum storage capacity [13, 17, 10]. Radiologists are medical doctors who specialize in diagnosing and treating diseases and injuries using medical imaging techniques, such as X-rays, MRI, CT, positron emission tomography (PET), ultrasound, and nuclear medicine. Imaging devices continue to generate more data per patient, often 1000 images or $500MB$. These data need long-term storage and efficient transmission. As there is increasing demand for transmitting diagnostic medical imagery in tele-radiology-telemedicine, the compression has to be performed. The medical images are usually rich in radiological content especially in slicing modalities, and the associated large file sizes must be compressed into minimal file size to minimize transmission time, and robustly coded to withstand the required network medium. Image compression can be lossy, lossless, and near-lossless. Lossy compression techniques are used to reduce data size for storing, handling, and transmitting content, which uses inexact approximations to represent the content. Lossy compression reduces file sizes significantly before degradation is viewed by the end-user. These methods are suitable for natural images such as photographs in applications where minor loss of fidelity is acceptable to achieve a substantial reduction in bit rate. Medical images are usually compressed using lossless compression methods. Nowadays these compression methods are also suited to reducing the size of the medical images as long as the image quality is not degraded [9]. Some of the methods for lossy compression are chroma subsampling, transform coding, and fractal compression.

A lossless compression technique encodes and decodes the data perfectly, and the resulting image matches exactly to the original image. Therefore, there is no degradation or distortion. The lossless coding techniques are often used for medical imaging [1], technical drawings, clip art, or comics, and are also preferred for archival purposes. Techniques involved in lossless image compression are run-length encoding, area image compression, differential pulse code modulation, entropy encoding, adaptive dictionary algorithms, deflation, and chain codes.

In near-lossless compression, the maximum absolute difference between the reconstructed image and the original image does not exceed a defined peak value [3]. The bit rate is reduced in near-lossless compression by introducing a limit peak error to preserve the quality of images. The near-lossless coding techniques are highly preferable for onboard processing. Methods of near-lossless compression include JPEG 2000, LOCO-I, and CALIC.

This chapter focuses on a novel algorithm for medical image compression, especially for MRI images. The algorithm that was developed can also be implemented on natural images. In the case of medical images, the quality is very important. The medical image compression algorithm must yield highest-fidelity reconstructed images with lower bits. Figure 14.1 depicts the flow

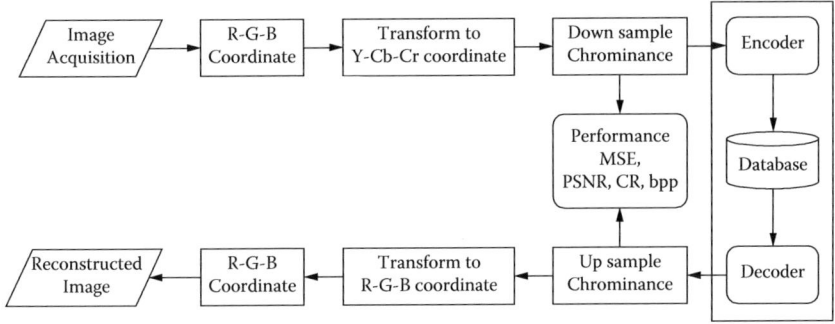

FIGURE 14.1: Block diagram of general image storage system

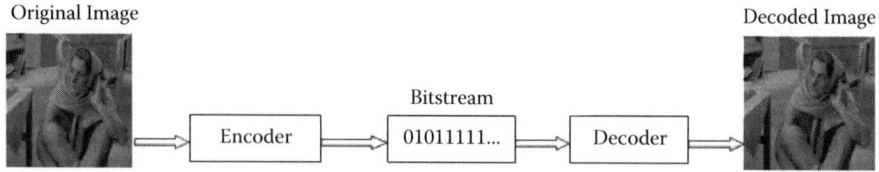

FIGURE 14.2: The basic flow of image compression coding

diagram of the general image storage system [14]. Figure 14.2 shows the basic flow of image compression coding.

14.2 Background and Related Works

Medical images, specifically CT and MRI sequence are data sets representing $3D$ sampling of an organ, and contain both the intraframe and interframe redundancies. Therefore, better performance could be expected from the techniques that exploit both redundancies. Lots of work has been carried out by different authors to compress medical images. As medical imaging relates to saving lives, they mainly focus on the quality of the reconstructed image. Many advanced image compression techniques have been developed in accordance with increasing demands for medical images. Sujitha Juliet et al. developed a novel image compression algorithm for medical images using geometrical regularity of image structure. They used the sparse representation approach, which exploits the geometrical regularity of image structure. The geometrical flows represented the direction in which the gray levels have regular variations. Thus, the wavelet decomposition of these data regularities results in less significant coefficients. Their experimental results demonstrated higher compression performance over other methods [4].

K. M. M. Prabhu et al., suggested $3D$ warped discrete cosine transform for MRI medical images. They proposed a complete image coding scheme for volumetric data sets based on the $3D$ WDCT scheme. The warped discrete Fourier transform computes the frequency samples at unequally spaced frequency samples. They extended these concepts to an adaptive discrete cosine transform to WDCT, which computes the DCT samples by warping the frequency axis by means of an all-pass transform. The complexity in their work is the all pass filters that they have used. The work we carried out in adapting the filtering reduces this complexity as the filter is passed as a mask [9].

Jin Wang et al., proposed an adaptive spatial post processing algorithm for use in block-based discrete cosine transform (BDCT) coded images. Their method is comprised of three procedural steps: a thresholding step, a model classification step, and a deblocking filtering step. First, they apply adaptive thresholding to extract the pixel vector containing the blocking artifacts. Next, they update block types using a simple rule, thus influencing deblocking performance. Finally, they were able to apply a suitable filter to each model with different local properties. Their method exhibited both significant visual quality improvement and PSNR gain, with fairly low computational complexity [15]. Jing-Ya Zhu et al., proposed a surveillance image compression technique via dictionary learning to fully exploit the constant characteristics of a target scene. The method transformed images over dictionaries directly from image samples, which approximated an image with fewer coefficients. These dictionaries are applied for sparse representation. An adaptive image blocking method was developed in [19]. The authors significantly encoded an image in a texture-aware way and found that their method outperforms PEG and JPEG 2000 in terms of both quality of reconstructed images and compression ratio.

Tim Bruylants et al., investigated techniques for improving the performance of JPEG 2000 for volumetric medical image compression. They made a generic codec framework that supports JPEG 2000 with its volumetric extension (JP3D), plus various directional wavelet transforms as well as a generic intra-band prediction mode. Their work optimally compresses medical volumetric images at an acceptable complexity level [2]. Selvi and Nadarajan proposed a compression method and experimented with real-time CT and MRI images. They constructed a $2D$ lossy compression technique using wavelet-based contourlet transform (WBCT) and a binary array technique (BAT). The high-frequency sub-band obtained from the wavelet-based contourlet transform is decomposed into multiple directional sub-bands by the directional filter bank. The results obtained by this method indicated that the proposed method reproduced the diagnostic features of CT and MRI images precisely [12].

Hualei Shen et al., developed a Web-based system to access and present medical images remotely. The system they developed is MIAPS (medical image access and presentation system). MIAPS provided four features including Digital Imaging and Communications in Medicine image retrieval, maintenance, presentation, and output. They mainly provided a Web-based solution for teleradiology and medical image sharing [13]. A reversible integer-to-integer time domain lapped transform (RTDLT) was also introduced. RTDLTs are realized from integer to integer by multi-lifting implementations after factorizing the filtering and transforming matrices into triangular elementary reversible matrices. The results the authors obtained have given higher compression ratios than that of JPEG2000 and JPEG-LS [16].

Xin Zhan et al., focused on a new compression scheme for synthetic aperture radar (SAR) amplitude images. They have shown interest in the study of dictionary learning and sparse representation, and proposed a novel SAR image compression using multiscale dictionaries [18]. Nibedita Pati et al., made an investigation using the sparse coding method by the orthogonal matching pursuit (OMP) algorithm. In their proposed work, conventional DCT can be replaced by a set of trained dictionaries. For dictionary construction, they

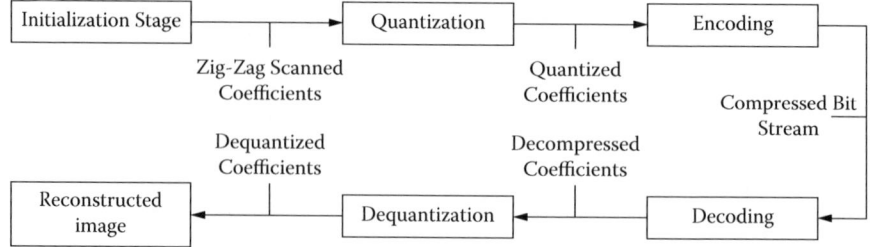

FIGURE 14.3: The stages of the proposed HID system

used a combination of DCT and Gabor basis to encode the trained dictionary elements. Their experimental results demonstrated that the OMP algorithm gains in rate distortion (RD) performance and improvements in perceptual quality [8].

14.3 The Proposed Histogram Indexed Dictionary (HID) Compression Technique

In this section, an efficient image compression algorithm with high-quality reconstruction based on histogram and Huffman encoding is presented. The algorithm is tested for MRI brain images to help the radiologists diagnose and make surgical planning about diseases. Later this is also implemented to natural standard color images. Thus, the proposed work focuses on different aspects of application, but importance is given only to reconstructing high-quality compressed medical images. The performance of the algorithm has been evaluated by comparing their experimental results with the other compression techniques. The proposed algorithm consists of three stages, as depicted in Figure 14.3. At the initialization stage the medical image $f(x,y)$ is given as an input. The quantization process is carried out to reduce the redundancy of an image. Finally the Huffman encoding encodes the bits of an image to a lower bit rate. Figure 14.4 and Figure 14.5 shows the schema for the encoder and the decoder respectively to the proposed compression technique using a novel histogram indexed dictionary (HID).

As explained earlier, the three stages of the compression algorithm hold several processes. First, at the initialization stage, 256×256 medical images are given as an input to the encoder. Afterwards, those images are decomposed into 8×8, 16×16, 32×32 blocks. DCT is performed over each of those blocks. The filtering technique called sliding neighborhood filtering adjusts the frequency response according to the frequency content of the inputs. Next, zig-zag scanning is performed over the filtered DCT coefficients. The high

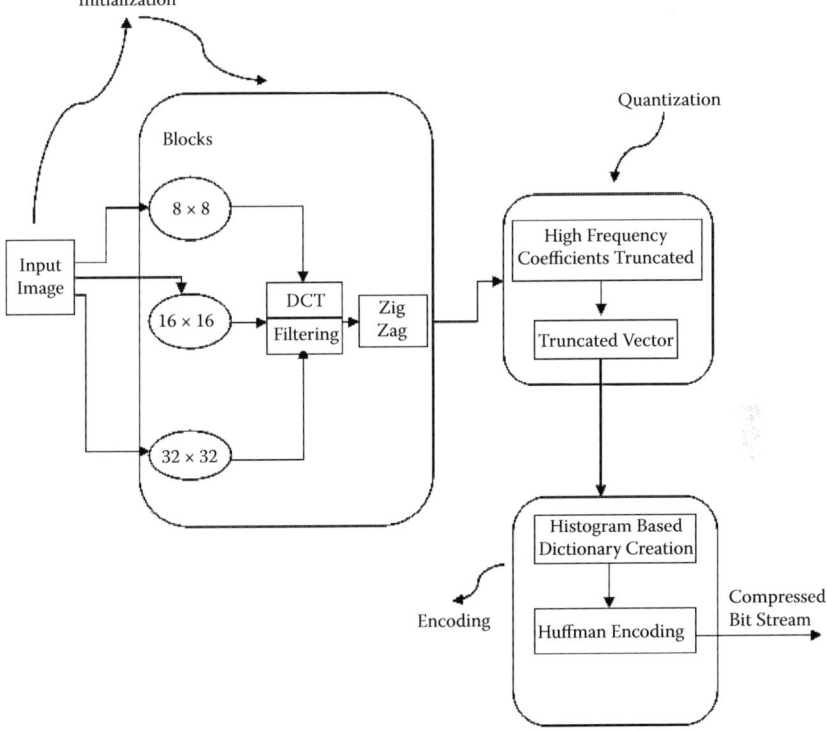

FIGURE 14.4: The block diagram for the encoder of the proposed HID image compression

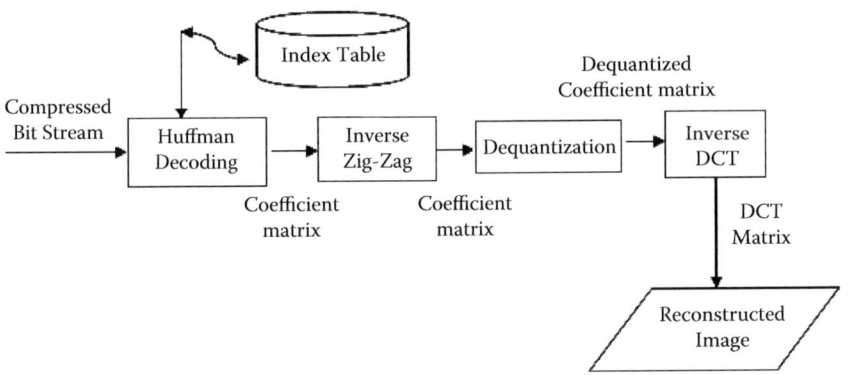

FIGURE 14.5: The block diagram for the decoder of the proposed HID image compression

frequency coefficients are truncated using thresholding in the quantization stage. At the encoding stage, the proposed histogram indexed dictionary is generated to increase the performance of the Huffman encoding. It also compresses medical images with lower bit rate and higher quality. The fidelity of the reconstructed images is evaluated through objective fidelity criteria. The objective fidelity criterion is one whose information loss can be expressed as a mathematical function of the input and output of a compression process.

14.3.1 The Discrete Cosine Transform

The trivial steps carried out to perform transform-based encoding and decoding include transformation, quantization, and entropy coding at the encoder, and the corresponding inverse operations are performed at the decoder. The discrete cosine transform is a technique that converts a signal into elementary frequency components [15]. It is widely used in image compression due to its energy compaction property. The discrete cosine transform is a linear, invertible function $f : R^N \to R^N$, or equivalently, an invertible $N \times N$ square matrix. The type II DCT is defined as follows, where u, v are discrete frequency variables that vary from $(0, 1, 2, \cdots, N-1)$; $f(m, n)$ is an $N \times N$ image pixel; and $F[u, v]$ is the DCT output.

$$F[u,v] = \frac{1}{N^2} \sum_{m=0}^{N-1} \sum_{n=0}^{N-1} f(m,n) \cos\left[\frac{u\Pi}{2N}(2m+1)\right] \cos\left[\frac{v\Pi}{2N}(2n+1)\right] \quad (14.1)$$

The inverse $2D$ DCT is formulated as below, where m, n are the image result pixel indices $(0, 1, 2, \cdots, N-1)$; $F(u, v)$ is the $N \times N$ DCT output; $f[m, n]$ is the $N \times N$ IDCT result; and $c(\lambda)$ is defined as in Equation 14.3.

$$f[m,n] = \sum_{m=0}^{N-1} \sum_{n=0}^{N-1} c(u)c(v)F(u,v) \cos\left[\frac{u\Pi(2m+1)}{2N}\right] \cos\left[\frac{v\Pi(2n+1)}{2N}\right]$$

$$(14.2)$$

$$c(\lambda) = \begin{cases} 1 & \text{for } \lambda = 0 \\ 0 & \text{for } \lambda = 1, \cdots, N-1 \end{cases} \quad (14.3)$$

DCT decorrelates the high correlated pixels with the adjacent pixels and results in the number of coefficients becoming zero. After quantization those coefficients are coded using entropy coding. The proposed algorithm is initialized by decomposing an input image into (8×8), (16×16), and (32×32) blocks individually. Thus, the input image $f(x, y)$ is blocked into (8×8) and then Equation 14.1 is applied to each pixel $(i, j)^{th} \in f(x, y)$. Figure 14.6 shows how $2D$ DCT has performed. The 64 (8×8) DCT basis functions are illustrated in Figure 14.7. The DCT basis functions can be represented as a matrix of basis functions. Next, the sliding neighborhood filtering is applied to the coefficents that adjust the frequency response, according to the frequency content

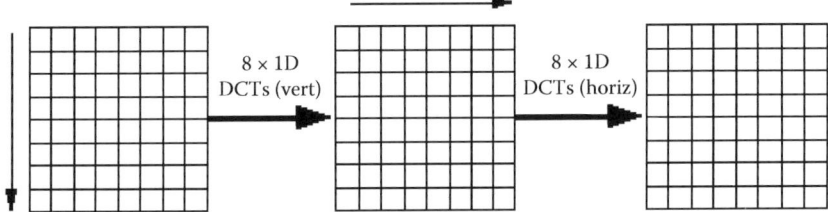

FIGURE 14.6: 2D-DCT

of the inputs. The filtering incorporated into the DCT results in a high quality reconstructed image. Thus, this filtering reduces the distortion of the images.

14.3.2 Sliding Neighborhood Filtering

A sliding neighborhood filtering is an operation that is performed on one coefficient at a time, with the value of any given coefficient in the output matrix being determined by the application of an algorithm to the values of the corresponding input coefficient's neighborhood. A coefficient's neighborhood is some set of values, defined by their locations relative to that value, which is called the center value. The neighborhood is a rectangular block that can move from one element to the next in an image matrix; the neighborhood block slides in the same direction. To operate on an image a block at a time, rather than a coefficient value at a time, the distinct block processing function is used.

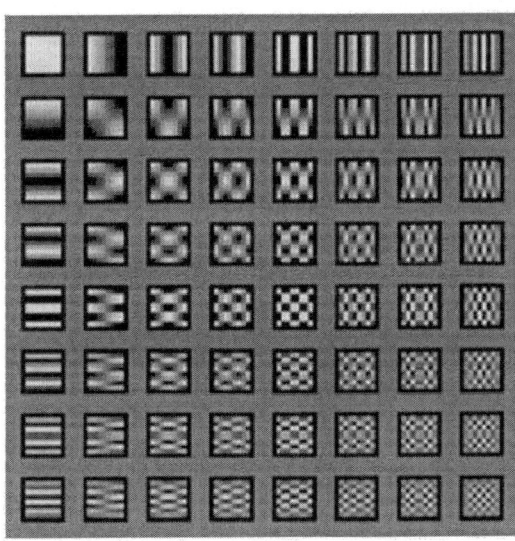

FIGURE 14.7: DCT basis functions

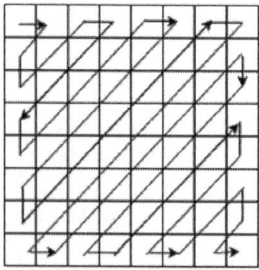

FIGURE 14.8: Zig-zag scan

In distinct block processing, a matrix is divided into $m \times n$ sections. These sections, or distinct blocks, overlay the image matrix starting in the upper left corner, with no overlap. If the blocks do not fit exactly over the image, it is necessary to add padding to the image or work with partial blocks on the right or bottom edges of the image [9]. Thus, the $1D$ DCT filter coefficients can be represented as follows, where $U(k)$ is defined as in Equation 14.5 and k varies from $0, 1, \cdots, 7$.

$$F_k(Z^{-1}) = U(k)\left(\cos\frac{k\pi}{16} + \cos\frac{3k\pi}{16}Z^{-1} + \cdots + \cos\frac{15k\pi}{16}Z^{-7}\right) \quad (14.4)$$

$$U(k) = \begin{cases} \frac{1}{2\sqrt{2}}, & k = 0 \\ \frac{1}{2}, & \text{otherwise} \end{cases} \quad (14.5)$$

Zig-zag scanning is not a property of DCT itself, but it is a part of the transform-based coding. Lower, frequency coefficients contain most of the information, and the high-frequency coefficients contain less information. In a DCT matrix, the frequency is increased in the first dimension along a row, and the frequency increases in the second dimension along the column. A high frequency in any dimension probably contains less energy. Hence, these high-frequency coefficients can be quantized based on any of the quantization methods. Before applying quantization, the zig-zag scanning must be performed to order the DCT matrix. Figure 14.8 illustrates the zig-zag scanning process.

14.3.3 Quantization

Quantization reduces the number of bits per samples [17]. It is a lossy compression technique that gives single quantum value by compressing a range of values. Here the high-frequency coefficients are truncated. The quantization round offs the values. The process used here is thresholding. The mathematical expression for the thresholding is given in Equation 14.6, where $Xdct$ is the $DCTMatrix$; $DCTmax$ is $max(max(Xdct))$; and $DCTmin$ is

$min(min(Xdct))$.

$$XdctQ = round\left((-1+2^Q) * \frac{Xdct - DCTmin}{DCTmax - DCTmin}\right) \quad (14.6)$$

The decoder Equation 14.6 becomes

$$IXdctQ = round\left((DCTmax - DCTmin) * \frac{Xdct - DCTmin}{-1+2^Q}\right) \quad (14.7)$$

14.3.4 The Proposed Histogram Indexed Dictionary

The histogram of an image refers to a histogram of the pixel intensity values [6]. This histogram is a graph showing the number of pixels in an image at each different intensity value found in that image. The histogram of a digital image with the intensity levels ranges [0, 255] for medical images is a discrete function defined as in Equation 14.8, where x_k is the kth intensity level and n_k is the number of pixels in the image with intensity x_k

$$hist(x_k) = n_k \quad (14.8)$$

The normalized histogram is given as in Equation 14.9, where M is the row dimension of the image; N is the column dimension of the image, and $p(x_k)$ is the probability of occurrence of intensity level x_k.

$$P(x_k) = \frac{n_k}{MN} \quad (14.9)$$

Histogram manipulation is the basis for numerous spacial domain processing technologies that can be used for image enhancement. This provides useful image statistics that are simple to calculate in software. The histogram of the DCT matrix is the probability of occurrence of particular coefficient values. A novel indexed dictionary is generated based on these histogram statistics, in order to increase the performance of the Huffman encoding.

The proposed histogram-based indexed dictionary stores the position of coefficient values that occurred at the same number of times. Then, the same number of bits are assigned to those coefficients that are similar in number. Thus, this reduces the time taken by the Huffman encoding to process the entire Huffman tree. Table 14.1 shows the proposed indexed dictionary and the bit allocation based on the histogram of the coefficient values. While processing the Huffman tree, the condition is imposed to assign the code words to the symbols. The symbols are the coefficient values in this work. The assigned condition is that code words are not allotted to symbols that are already coded by Huffman process. Code words are bits of the pixel. The complexity occurs at the decoder while creating indexed dictionaries. Thus, the indexed dictionary has to be scanned at the decoder while doing the Huffman decoding.

TABLE 14.1: Indexed dictionary.

Index	Location	Coeff. Value	No. of Occurences	Bit Allocation
0	(10,12)	42	31	0
1	(45,12)	43	31	0
2	(34,25)	67	78	10
3	(24,76)	34	34	1001
⋮	⋮	⋮	⋮	⋮
255	(255,255)	...	...	1011101

14.3.5 Huffman Encoding

The Huffman encoding algorithm starts by constructing a list of all the alphabet symbols in descending order of their probabilities. It then constructs, from the bottom up, a binary tree with a symbol at every leaf. This is done in steps, where at each step, two symbols with the smallest probabilities are selected, added to the top of the partial tree, deleted from the list, and replaced with an auxiliary symbol representing the two original symbols. When the list is reduced to just one auxiliary symbol (representing the entire alphabet), the tree is complete. The novelty in the proposed work is that the symbols that are already assigned by code words are not passed for Huffman encoding, which is usually done in other compression techniques. The tree is then traversed to determine the code words of the symbols. This process is best illustrated by an example. Given five symbols with probabilities as shown in Figure 14.9, they are paired in the following order:

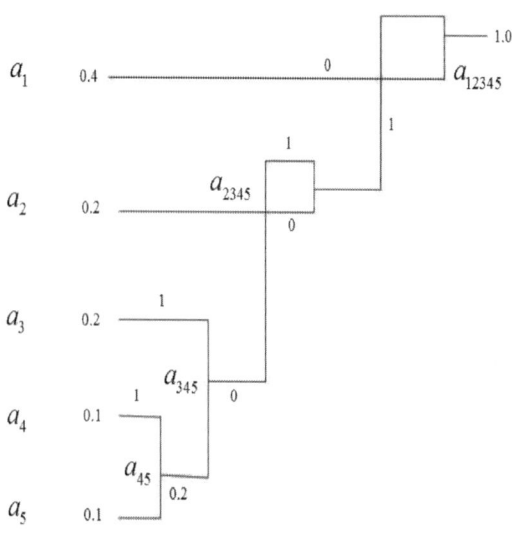

FIGURE 14.9: Huffman encoding

1. a_4 is combined with a_5 and both are replaced by the combined symbol a_{45}, whose probability is 0.2.

2. There are now four symbols left: a_1, with probability 0.4, and a_2, a_3, and a_{45}, with probabilities 0.2 each. We arbitrarily select a_3 and a_45 as the two symbols with smallest probabilities, combine them, and replace them with the auxiliary symbol a_{345} whose probability is 0.4.

3. Three symbols are now left, a_1, a_2, and a_{345}, with probabilities 0.4, 0.2, and 0.4, respectively. Arbitrarily select a_2 and a_{345}, combine them, and replace them with the auxiliary symbol a_{2345} whose probability is 0.6.

4. Finally, combine the two remaining symbols, a_1 and a_{2345}, and replace them with a_{12345}, with probability 1. The tree is now complete. It is shown in Figure 14.9 "lying on its side" with its root on the right and its five leaves on the left. To assign the code words, arbitrarily assign a bit of 1 to the top edge, and a bit of 0 to the bottom edge, of every pair of edges. This results in the code words 0, 10, 111, 1101, and 1100. The assignments of bits of the edges are arbitrary. The average size of this code is $0.4 \times 1 + 0.2 \times 2 + 0.2 \times 3 + 0.1 \times 4 + 0.1 \times 4 = 2.2$ bits/symbol. But even more importantly, the Huffman code is not unique. Some of the steps above were chosen arbitrarily, because there were more than two symbols with smallest probabilities.

The encoding and decoding algorithm using the proposed HID compression technique is presented below.

14.4 Results and Discussion

The proposed compression algorithm for medical images based on the integration of a novel histogram indexed dictionary and the Huffman encoding has been implemented. The histogram indexed dictionary is created to yield lower-bit-rate medical images, and also increases the performance of the Huffman encoding scheme, thus increasing the performance of the compression technique. The quality of the reconstructed images is higher compared to other compression techniques. The proposed compression algorithm has been tested for MRI brain images. Later, it was also tested over standard color images. The size of the images used for testing is 256×256. Figure 14.10 shows the tested images.

14.4.1 Peak Signal-to-Noise Ratio

The peak signal-to-noise ratio (PSNR) measures the difference in pixel value between two images, and is widely used to measure the quality of compressed

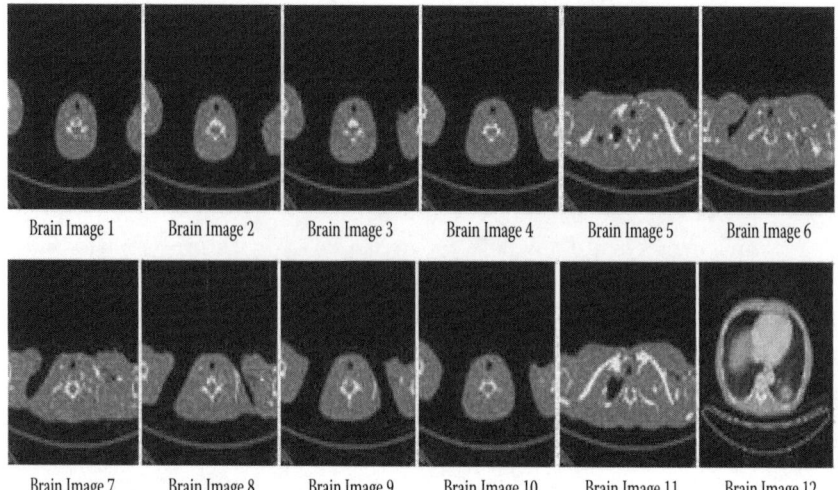

FIGURE 14.10: Tested medical images

or reconstructed images [17, 11]. The definition of PSNR is given below in Equation 14.10. The mean square error (MSE) is the cumulative squared error between the original and the compressed image, and is defined in Equation 14.11.

$$PSNR = 10 * log\left(\frac{255^2}{MSE}\right) \qquad (14.10)$$

$$MSE = \frac{1}{MN}\sum_{i=0}^{M}\sum_{j=0}^{N}(I(i,j) - (i,j))^2 \qquad (14.11)$$

14.4.2 Compression Ratio and Bits Per Pixel

This is the ratio between uncompressed image size and the compressed image size. Compression is defined as in Equation 14.12. Number of bits used to represent single pixels, termed as bits per pixel. It is defined as in Equation 14.13.

$$\text{Compression Ratio} = \frac{\text{Uncompressed Size}}{\text{Compressed Size}} \qquad (14.12)$$

$$\text{bpp} = \frac{\text{Size of the compressed image in bits}}{\text{Total No.of pixels}} \qquad (14.13)$$

The loss of quality caused by the influence of the control parameters needs to be tested. Table 14.2 and Table 14.3 depict the quality metrics for a medical image, Brain image 1, and a standard Cameraman image. It is found that the PSNR value increases on increasing the threshold. The average PSNR value is

Algorithm [Encoding algorithm using the proposed HID compression technique]

Input: MRI brain image $f(x_i, y_j)$, $(i,j) \to ij^{th}$ position of intensity level
Output: Compressed bit stream

1. Get the input image $f(x_i, y_j)$.

2. For $i = 1$ to N and $j = 1$ to M decompose $f(x,y)$ into 8×8 blocks. Hence an input image with size 256×256 is divided into $32(8 \times 8)$, $16(16 \times 16)$, $8(32 \times 32)$ sublocks, respectively.

3. Apply Equation 14.1 to each block to compute the discrete cosine transform $DCT(f(x,y))$.

4. Compute filtered coefficient matrix $fDCT(f(x,y))$.

5. Compute quantization matrix $Qfm(f(x,y))$ using Equation 14.3.

6. Perform HID-based assignment.

7. Perform Huffman encoding to the symbols that are not assigned by HID, such that if $a_n = c$, $c \to$ codeword assigned using HID where a_n is not performed using Huffman.

8. Update all the codewords of all the symbols to HID after doing Huffman encoding.

9. If $a_n(c) = a_m(c)$ where m is another symbol word, then again do Huffman encoding by exchanging the assignment of the bits on each side of the Huffman tree and go to 8.

10. Remove the symbols from HID that were assigned by Huffman.

11. Write the compressed bits to the database.

12. Repeat the steps from 2 to decompose $f(x,y)$ into 16×16 and 32×32 blocks, respectively.

13. Save the compressed bit stream in the database.

59.22175 for Brain image 1 and 56.5195 for the standard Cameraman image. The compression ratio increases for minimum threshold value.

Table 14.4 shows the quality of the compressed image for various image blocks when different filtering methods are applied. The filtering method used

Algorithm [Decoding algorithm using the proposed HID compression technique]

> Input: Compressed bit streams
> Output: Reconstructed image
>
> 1. Input compressed bit streams to the decoder.
> 2. For all codewords C_n, check the index table
> 3. If C_n exists in I, assign the coefficient value that is based on the coefficient location (x_i, y_j) and then index value I_m; else go to step 3.
> 4. Decode the compressed bit streams using Huffman decoding.
> 5. Dequantize the coefficient matrix $Q^{-1} fm(f(x,y))$ using Equation 14.7.
> 6. Generate DCT filtered coefficient matrix.
> 7. Apply inverse DCT ($IDCT$) to the DCT filtered coefficient matrix using Equation 14.2.
> 8. Get the reconstructed medical image

for the proposed work is the sliding neighborhood filtering. The results using the sliding neighborhood filtering is compared with the other filtering like the mean and the median filtering. From the table it is found that the filtering using sliding neighborhood gives the highest quality image with block size 8×8. The average PSNR value for the Brainimg1 image is $55.4139dB$ for an image block with size 8×8 and it is $54.707175dB$ for 16×16 block size. It is also observed that the average PSNR value is 55.8712, when the sliding neighborhood filtering is applied and $27.22443dB$ for the mean filtering. Thus, the DCT with sliding neighborhood filtering gives highest quality image.

Figure 14.11 shows the quality of the proposed compression technique for various filtering models. The PSNR value is high for the proposed method

TABLE 14.2: Influence of threshold over compression, bits per pixel, and PSNR for a medical image, Brain image 1.

Threshold	CR	bpp	PSNR
10	0.20000	0.1053	56.674
30	0.18372	0.0967	57.734
50	0.16500	0.0870	59.023
80	0.13413	0.0706	63.456

TABLE 14.3: Influence of threshold over compression, bits per pixel, and PSNR for a standard Cameraman image.

Threshold	CR	bpp	PSNR
10	0.224	0.112	55.723
30	0.195	0.094	56.543
50	0.187	0.090	56.567
80	0.124	0.067	57.245

when the sliding neighborhood filtering is imposed. The proposed method with the other two filters yields a poor-quality reconstructed image.

Figure 14.12 (a) and (b) show the comparison of the PSNR over different blocks of images at different threshold values. The PSNR value is increasing over the increase in threshold values. Block size with 8×8 gives higher PSNR value for both the images.

Figure 14.13 and Figure 14.14 show the reconstructed output images of various threshold values when the input images are decomposed into 8×8, 16×16, and 32×32, respectively.

Table 14.5 shows the quality metrics of various medical images for different threshold values and the block sizes. Images divided into (8×8) blocks give better result than the other blocks. For example, Brain image 2 gives the PSNR value of $54.5864 dB$ when it is divided into 8×8 blocks, whereas it gives $51.9713 dB$ when it is divided into (16×16) blocks. Thus, the medical images yield better PSNR value for (8×8) blocks.

TABLE 14.4: The quality of the compressed image for various image blocks when applying different filtering models.

Image Name	Filtering Types	PSNR(dB)		
		8×8	16×16	32×32
Brain image 1	Sliding Neighborhood Filtering	57.5363	55.5259	54.5516
	Mean Filtering	27.1143	27.3345	27.2245
	Median Filtering	27.6996	26.8765	26.6554
Brain image 2	Sliding Neighborhood Filtering	56.6025	52.6058	55.9339
	Mean Filtering	26.7654	26.8765	26.5647
	Median Filtering	26.8765	26.9831	26.8435
Brain image 3	Sliding Neighborhood Filtering	51.4193	55.4888	56.0745
	Mean Filtering	25.4536	25.7643	25.2887
	Median Filtering	25.8793	25.9987	25.7652
Brain image 4	Sliding Neighborhood Filtering	56.0975	55.2082	49.3274
	Mean Filtering	22.4567	22.7654	22.5463
	Median Filtering	22.6574	22.9876	22.6875

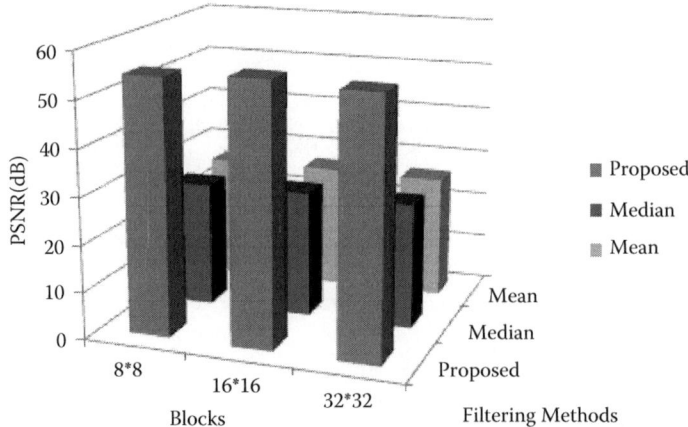

FIGURE 14.11: Quality metrics of the proposed compression technique over some other filtering methods

Table 14.6 shows the quality metrics of various medical images for threshold TH = 30. Images divided into 8×8 blocks give better results than the other blocks. For example, Brain image 2 gives the PSNR value of $59.9dB$ when it is divided into 8×8, whereas it is $58.7dB$ when it is divided into 16×16 blocks. Thus, the medical images yield better PSNR value for 8×8 blocks with the threshold TH = 30.

Table 14.7 shows the quality metrics of various medical images for threshold TH = 80. Images divided into 8×8 blocks give better results than the other blocks. The algorithm reaches the best PSNR value at threshold 80. Images divided into 8×8 blocks give better results than the other blocks. For

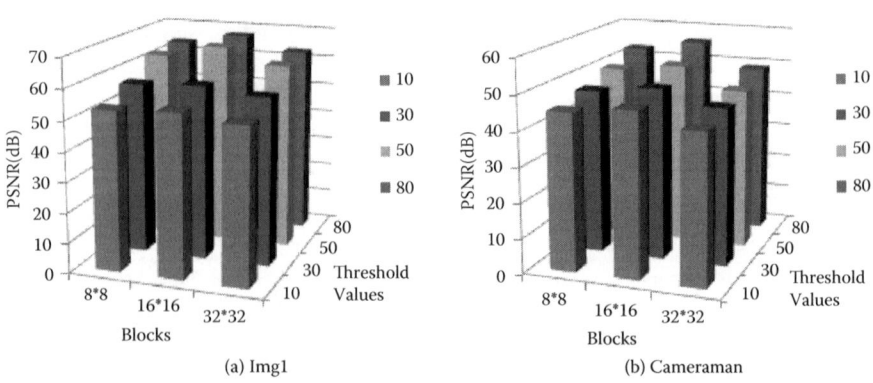

FIGURE 14.12: Comparison of PSNR over different blocks of images at different threshold values for (a) Brain image 1, (b) Cameraman

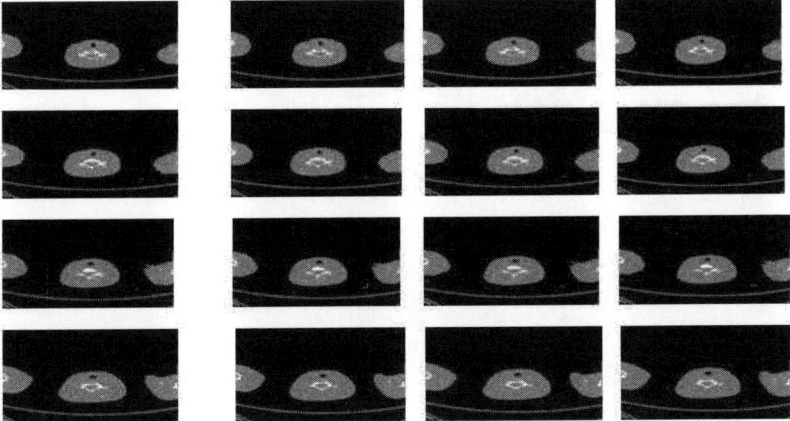

FIGURE 14.13: Reconstructed images when the threshold TH = 10. Columns from left to right: first column: input images, second and succeeding: reconstructed output images when the images are divided into (8×8), (16×16), and (32×32), respectively.

example, Brain image 1 gives the PSNR value of $63.1783 dB$ when it is divided into 8×8 blocks, whereas it gives $62.643 dB$ when it is divided into 16×16 blocks. Thus, the medical images yield better PSNR values for 8×8 blocks when the threshold TH = 80.

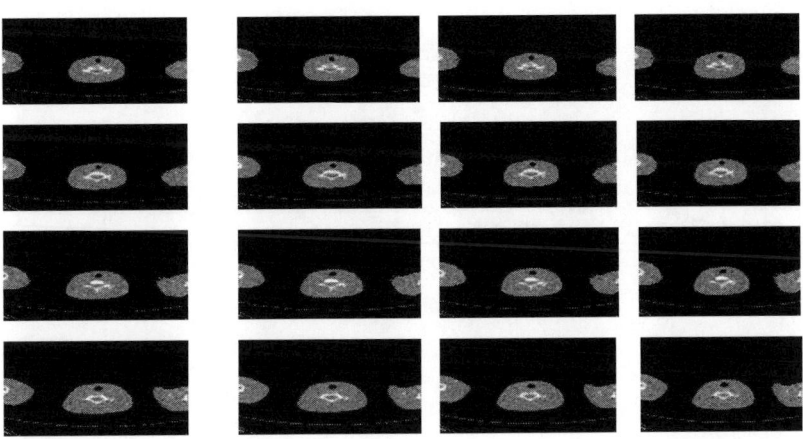

FIGURE 14.14: Reconstructed images when the threshold TH = 80. Columns from left to right: first column: input images, second and succeeding: reconstructed output images when the images are divided into (8×8), (16×16), and (32×32), respectively.

TABLE 14.5: Objective quality metrics of the compressed image for various image blocks when the threshold TH = 10.

Image Name	Block Size	PSNR(dB)	CR	bpp
Brain image 1	8 × 8	54.5835	0.1915	0.1008
	16 × 16	55.5259	0.2000	0.1053
	32 × 32	54.5516	0.2182	0.1149
Brain image 2	8 × 8	54.5864	0.1993	0.1049
	16 × 16	51.9713	0.2771	0.1459
	32 × 32	54.2665	0.2326	0.1224
Brain image 3	8 × 8	53.9589	0.2154	0.1134
	16 × 16	54.0786	0.2420	0.1274
	32 × 32	52.8908	0.2605	0.1371
Brain image 4	8 × 8	50.4735	0.2576	0.1356
	16 × 16	54.5046	0.2544	0.1339
	32 × 32	53.6732	0.2699	0.1421

Figure 14.15 shows the evolution of the PSNR over bpp with different block sizes. The quality of the image is influenced by the compression ratio and the bits per pixel. The quality of the image becomes higher for higher CR and lower bpp. From the figure it is found that the average PSNR value is $61.723dB$ for the images with block size 32 × 32, with the bpp 0.03 when the threshold is 80.

Figure 14.16 illustrates the results of the various tested images for the proposed (HID) method over the other existing methods SVD, WDR, and the JPEG. The simulation process aims to evaluate the performance of the proposed compression method. To that end, a set of tests were scheduled to measure the compressed image quality, compression ratio, and bits per pixel. The tests have been performed over the entire test image set and the results shown in the following graphs are the average of the obtained results.

TABLE 14.6: Objective quality metrics of the compressed image for various image blocks when the threshold TH = 30.

Image Name	Block Size	PSNR(dB)	CR	bpp
Brain image 1	8 × 8	59.9	0.128	0.067
	16 × 16	58.7	0.165	0.087
	32 × 32	59.3	0.170	0.089
Brain image 2	8 × 8	59.9798	0.1342	0.0706
	16 × 16	53.2271	0.2450	0.1290
	32 × 32	57.7587	0.1827	0.0962
Brain image 3	8 × 8	58.8570	0.1438	0.0757
	16 × 16	58.4001	0.2022	0.1065
	32 × 32	54.5226	0.2137	0.1125
Brain image 4	8 × 8	51.9742	0.1812	0.0954
	16 × 16	57.7184	0.2114	0.1113
	32 × 32	57.7508	0.2076	0.1093

TABLE 14.7: Objective quality metrics of the compressed image for various image blocks when the threshold TH = 80.

Image Name	Block Size	PSNR(dB)	CR	bpp
Brain image 1	8 × 8	63.1783	0.086746	0.04567
	16 × 16	62.643	0.13413	0.070618
	32 × 32	61.0072	0.14729	0.077545
Brain image 2	8 × 8	63.5811	0.087036	0.045822
	16 × 16	53.3885	0.21746	0.11449
	32 × 32	60.4346	0.15636	0.082321
Brain image 3	8 × 8	62.4735	0.095064	0.050049
	16 × 16	60.7811	0.16923	0.089096
	32 × 32	55.3104	0.18367	0.096695
Brain image 4	8 × 8	52.2333	0.13483	0.070984
	16 × 16	59.2318	0.17294	0.091049
	32 × 32	59.2318	0.17294	0.091049

In this experiment, the behavior of the compression model is examined for various parameters, including different threshold values, different blocks, and comparisons made over other existing systems.

The proposed compression algorithm is finally tested for natural images. The quality metrics of the proposed technique for those images are depicted in Table 14.8.

From a different analysis, it is found that for the image with block size (8 × 8) and the threshold value 80, the proposed compression technique gives better performance. Figure 14.17 shows a comparison of the PSNR and the

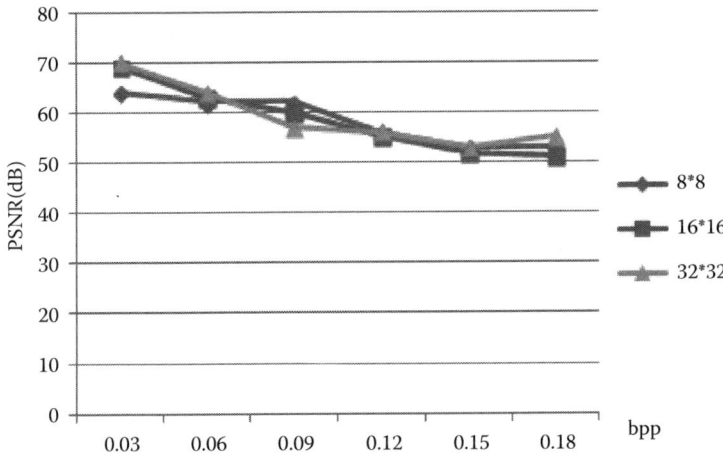

FIGURE 14.15: Comparison of PSNR value and the bpp of the proposed reconstructed image for various image blocks

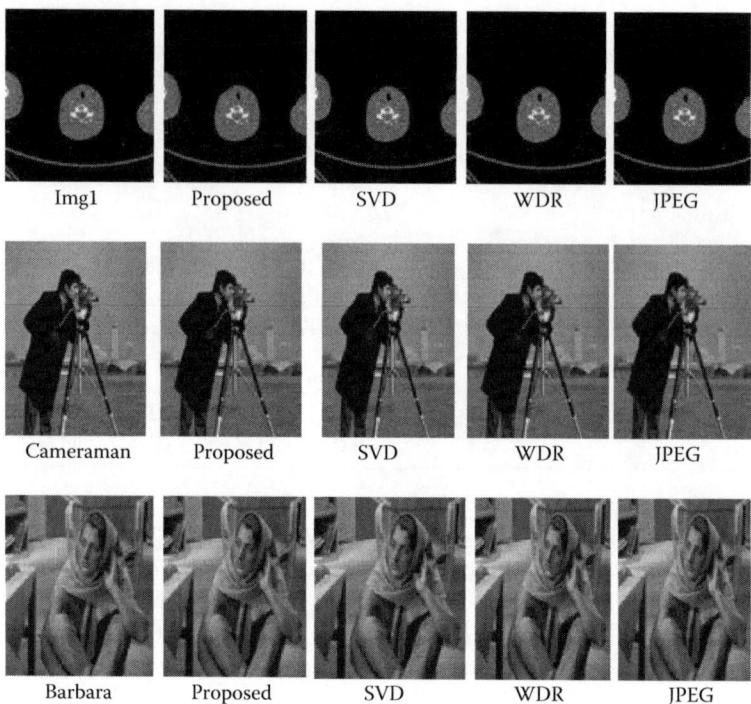

FIGURE 14.16: Results of tested images for the proposed HID method and the existing methods SVD, WDR, and JPEG

CR of the proposed technique over SVD (singular value decomposition), WDR (wavelet difference reduction), and JPEG image compression. The proposed HID method gives a higher PSNR value than the other existing techniques, with the average of 63.1783. The quality of the image is decreasing over an increase in compression ratio.

TABLE 14.8: Quality metrics of different images for the proposed method.

Image Name	PSNR(dB)	CR	bpp
	55.7651	0.1915	0.09
	54.6780	0.1337	0.167
	52.7653	0.1445	0.1543

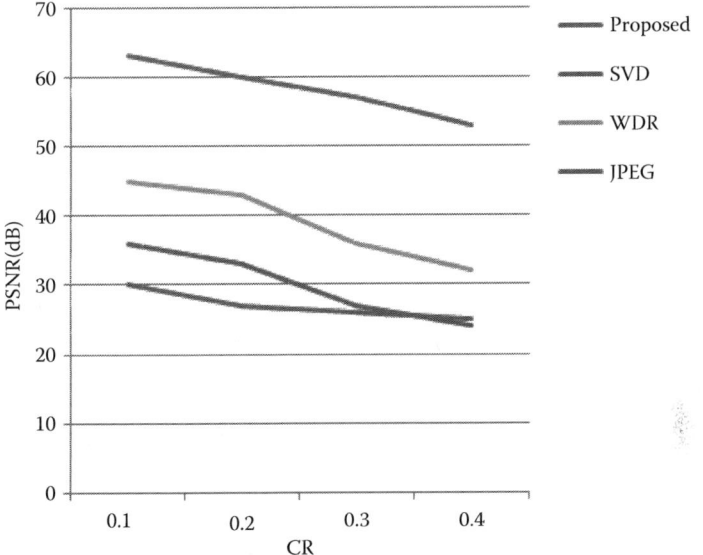

FIGURE 14.17: Comparison of PSNR value and the CR of the proposed reconstructed image over other existing methods

14.5 Conclusion

In this chapter, an efficient medical image compression algorithm based on the integration of the histogram indexed dictionary and the Huffman encoding has been proposed and implemented. The indexed dictionary is created based on the histogram analysis of images, which significantly improves the performance of the Huffman encoding. The idea of sliding neighborhood filtering with DCT helps to retain the reconstructed image with higher quality. The PSNR value is greater than 50 dB with the proposed compression method. The proposed algorithm was also implemented for natural images, which yields higher PSNR. Therefore, the proposed HID algorithm provided better results than the other existing methods.

Bibliography

[1] Francesc Aulí-Llinàs, Michael W. Marcellin, Joan Serra-Sagrista, and Joan Bartrina-Rapesta. Lossy-to-lossless 3d image coding through prior coefficient lookup tables. *Information Sciences*, 239:266–282, 2013.

[2] Tim Bruylants, Adrian Munteanu, and Peter Schelkens. Wavelet based volumetric medical image compression. *Signal Processing: Image Communication*, 31:112–133, 2015.

[3] Giuliano Grossi, Raffaella Lanzarotti, and Jianyi Lin. High-rate compression of ecg signals by an accuracy-driven sparsity model relying on natural basis. *Digital Signal Processing*, 45:96–106, 2015.

[4] Juliet Sujitha , Elijah Blessing Rajsingh, and Ezra Kirubakaran. A novel image compression method for medical images using geometrical regularity of image structure. *Signal, Image and Video Processing*, 9(7):1691–1703, 2015.

[5] Basar Koc, Ziya Arnavut, and Hüseyin Koçak. The pseudo-distance technique for parallel lossless compression of color-mapped images. *Computers & Electrical Engineering*, 46:456–470, 2015.

[6] Jia Li and Liping Chang. A sar image compression algorithm based on mallat tower-type wavelet decomposition. *Optik-International Journal for Light and Electron Optics*, 126(23):3982–3986, 2015.

[7] Jerónimo Mora Pascual, Higinio Mora Mora, Andrés Fuster Guilló, and Jorge Azorín López. Adjustable compression method for still jpeg images. *Signal Processing: Image Communication*, 32:16–32, 2015.

[8] Nibedita Pati, Annapurna Pradhan, Lalit Kumar Kanoje, and Tanmaya Kumar Das. An approach to image compression by using sparse approximation technique. *Procedia Computer Science*, 48:769–775, 2015.

[9] K.M.M. Prabhu, K. Sridhar, Massimo Mischi, and H.N. Bharath. 3-d warped discrete cosine transform for mri image compression. *Biomedical Signal Processing and Control*, 8(1):50–58, 2013.

[10] Milan S. Savić, Zoran H. Perić, and Nikola Simić. Coding algorithm for gray scale images based on linear prediction and dual mode quantization. *Expert Systems with Applications*, 42(21):7285–7291, 2015.

[11] Milan S. Savić, Zoran H. Perić, and Nikola Simić. Coding algorithm for gray scale images based on linear prediction and dual mode quantization. *Expert Systems with Applications*, 42(21):7285–7291, 2015.

[12] G. Uma Vetri Selvi and R. Nadarajan. CT and MRI image compression using wavelet-based contourlet transform and binary array technique. *Journal of Real-Time Image Processing*, pages 1–12, 2014.

[13] Hualei Shen, Dianfu Ma, Yongwang Zhao, Hailong Sun, Sujun Sun, Rongwei Ye, Lei Huang, Bo Lang, and Yan Sun. Miaps: A web-based system for remotely accessing and presenting medical images. *Computer methods and programs in biomedicine*, 113(1):266–283, 2014.

[14] Roman Starosolski. New simple and efficient color space transformations for lossless image compression. *Journal of Visual Communication and Image Representation*, 25(5):1056–1063, 2014.

[15] Jin Wang, Zhensen Wu, Gwanggil Jeon, and Jechang Jeong. An efficient spatial deblocking of images with dct compression. *Digital Signal Processing*, 42:80–88, 2015.

[16] Lei Wang, Licheng Jiao, Jiaji Wu, Guangming Shi, and Yanjun Gong. Lossy-to-lossless image compression based on multiplier-less reversible integer time domain lapped transform. *Signal Processing: Image Communication*, 25(8):622–632, 2010.

[17] Lige Wang, Jin Y Ooi, and Ian Butler. Interpretation of particle breakage under compression using x-ray computed tomography and digital image correlation. *Procedia Engineering*, 102:240–248, 2015.

[18] Xin Zhan, Rong Zhang, Dong Yin, and Chengfu Huo. Sar image compression using multiscale dictionary learning and sparse representation. *IEEE Geoscience and Remote Sensing Letters*, 10(5):1090–1094, 2013.

[19] Jing-Ya Zhu, Zhong-Yuan Wang, Rui Zhong, and Shen-Ming Qu. Dictionary based surveillance image compression. *Journal of Visual Communication and Image Representation*, 31:225–230, 2015.

Index

A
ABC. *See* artificial bee colony (ABC)
adaptive Levy evolutionary
    programming (ALEP), 271
ANN. *See* artificial neural network
    (ANN)
ANOVA, 52–53, 55
ant colony optimization (ACO), 24
    applications, 207
    description, 207–208
    FCM, use with, 211–214, 215,
        219, 220
    fuzzy morphological-based
        fusion (FMF) algorithm,
        use with, 217, 219
    overview, 207
    path choices, 208
    scaling factor, determining, 134
artificial bee colony (ABC), 32
artificial neural network (ANN), 32
average error (AE), 59, 77, 82
average fusion, 5, 7
average gradient, 4, 20

B
BA. *See* bat algorithm (BA)
backprojection, 97
bacterial foraging optimization
    (BFO), 5, 32
    scaling factor, determining, 134
bacterial foraging optimization
    algorithm (BFOA), 4, 5, 6,
    25
    genetic algorithm for, 7
    host image creation, 7
    overview, 24
Bak-Sneppen simulation model, 273

bat algorithm (BA), 32, 34
Bayes' theorem, 187
behavior, animal, 324
BFO. *See* bacterial foraging
    optimization (BFO)
binary array technique (BAT), 373
biometric traits
    fingerprint, 4, 5
    iris, 4, 5–6
    palmprint, 4, 5, 6
    voice, 6
biometrics
    applications, 4
    reliability, 4
    traits (*see* biometric traits)
BOFA. *See* bacterial foraging
    optimization algorithm
    (BFOA)
Boltzman constant, 107
Bresenham algorithm, 61

C
canny edge detection, 6, 8
CASIA database, 6, 11, 28
CCA. *See* cricket chirping algorithm
    (CCA)
Chebychev polynomial problem, 114
chip rate (CR), 135
circular Hough transform, 6
Columbia Object Image Library
    (COIL), 330, 338, 354, 361,
    362
compression ratios, 382
compressive sensing theory, 6
computer tomography (CT), 298
    bone segmentation, 162

image enhancement with Gabor
filters (*see* Gabor filters)
overview, 162–163
smooth filtering, 163
spiral optimization (SO)
strategy (*see* spiral
optimization (SO) strategy)
Congreve, William, 229
content-based image retrieval
(CBIR), 343
Gabor-wavelets transform,
306–308, 318
greedy search selection, 312
magnetic resonance imaging
(MRI) (*see* content-based
magnetic resonance imaging
(CBMRI))
overview, 297–298
content-based magnetic resonance
imaging (CBMRI)
approach to, 302
challenges, 300
color, 302
Euclidean distance measure, 317
forward greedy feature selection
(FGFS), 302, 312–313, 314
K-NN model, 302, 313, 314–315,
317, 319
objectives of, 298, 300, 301
principles, 298–299
progressive retrieval strategy,
316
quantity of images, 299–300
retrieval performance, 300, 301
shape, 301, 308–309, 311
texture, 301, 303–306
visual feature representation,
319
contourlet transform, 138, 156, 247
cooperative particle swarm
optimization (CPSO-H),
271
copy protection technical working
group (CPTWG), 229
cricket chirping algorithm (CCA)

aggression phase, 35, 36
execution time, 43
image segmentation, 36, 43, 45,
48
mating phase, 34–35
MTEMO, *versus*, 52
overview, 31–32
population, 41
pseudocode, 36, 37, 42
cryptography, 230
CS. *See* cuckoo search (CS)
cuckoo search (CS), 32, 34
convergence of the algorithm,
236
cuckoo breeding behaviors,
232–234
DWT-SVD, use with, 242
egg laying, 234–235
egg-laying radius (ELR), 234
elimination of cuckoos in
inappropriate areas, 236
genetic algorithm (GA), 234
immigration, 235–236
optimization, 234
particle swarm optimization
(PSO), 234
profit function, 234
pseudocode, 239
simulation experiments, 239–241
watermarking with, 236–239

D

Darwin, Charles, 100
DE. *See* differential evaluation (DE)
density-based score level fusion, 6
differential evolution (DE), 32, 267,
270–271
applications, 114, 117, 291
crossover, 116
evaluating, 276–277
features, 272
initialization, 115
mutation, 115–116
origins, 114
overview, 113–114, 114–115

Index

performance, 271, 272
digital watermarking
   applications, 5
   copy protection technical
      working group (CPTWG),
      229
   cuckoo search, using (see cuckoo
      search (CS))
   description, 5
   DWT-SVD watermarking (see
      DWT-SVD watermarking)
   formulating, 231–232
   history, 21–22, 228–229
   image intensity, 231
   importance of, 230–231
   invisible watermarks, 231
   least significant bit (LSB)
      manipulation, 247
   overview, 21–22, 227–228,
      228–230, 245–246, 246–247
   security analysis, 136
   spatial domain techniques, 132
   technique description, 132
   techniques, 22
   visible watermarks, 231
   watermark extraction, 132
direct sequence spread spectrum
   (DSSS), 134
discrete cosine transform (DCT),
   370, 376–377
discrete cosine transforms (DCT),
   133
   medical images watermarking,
      use with, 247
   video watermarking, 134
discrete Fourier transform (DFT),
   133
discrete wavelet transforms (DWT),
   4, 5, 7, 16
   -SVD watermarking (see
      DWT-SVD watermarking)
   image decomposition, 133
   overview, 248–249
   performance, 21

DWT fusion. See discrete wavelet
   transforms (DWT)
DWT-SVD watermarking
   embedding process, 251–252
   extraction process, 253
   normalized correlation
      coefficient (NCC), 254, 256
   overview, 247–248
   particle swarm optimization
      (PSO), 250–251
   peak signal-to-noise ratio
      (PSNR), 254, 256
   performance analysis, 254,
      256–259, 260–261
   pseudocode, 252

E

Eberhart, Russell, 108
Eigen vectors, 14
eigenvalue decomposition (EVD),
   328
EIODORS. See Electrical Impedance
   Tomography and Diffuse
   Optical Tomography
   Reconstruction Software
   (EIDORS)
electrical impedance tomography
   (EIT), 94
   advantages, 95
   applications, 95
   computational intelligence
      techniques, 100
   disadvantages, 95
   electrical conductivity vectors,
      103
   fitness function, 99
   GAs of, 100–101, 101–103
   image generation, 97
   mathematical formulation, 96
   objective function, evaluation of,
      104
   overview, 94, 95
   particle swarm optimization
      (PSO) algorithm, 110–112
   pseudocode, 107

qualitative experiments, 104, 105
quantitative experiments, 103–104, 105
reconstructing images, 96–97
simulated annealing, 106–107
Electrical Impedance Tomography and Diffuse Optical Tomography Reconstruction Software (EIDORS), 97, 98–99, 104
electromagnetism optimization (EMO), 33, 34
EMO. See electromagnetism optimization (EMO)
entropy, 4, 21
entropy criterion method, 38–39, 41
evolutionary algorithms. See also specific algorithm types
overview, 270
evolutionary programming (EP), 24, 100
evolutionary strategies (ES), 24

F

FA. See firefly algorithm (FA)
factor-based image watermarking system (I-MSF), 153
firefly algorithm (FA), 33, 34
algorithm example, 138
applications, 136
description, 137
embedding strength, for optimal selection of, 145–146
firefly-based embedding algorithm, 140, 142–143
scaling factor, determining, 134
fish school search (FSS) algorithm
collective instinctive operator, 120
collective volitive operator, 120–121
feeding operator, 119–120
individual movement operator, 119

iterations, 121, 122, 123
overview, 118–119
pseudocode, 121
fitness function, 32
forward greedy feature selection (FGFS), 302, 312–313, 314
FSS. See fish school search (FSS) algorithm
fuzzy C-means (FCM) clustering algorithm, 206
ant colony optimization with, 211–214, 215, 219, 220
cluster centers, 222
development, 210
overview, 210–211
particle swarm optimization, use with, 214, 215, 216, 218
fuzzy hyper sphere neural network, 6
fuzzy morphological-based fusion (FMF) algorithm, 206
ant colony optimization, use with, 217, 219
particle swarm optimization (PSO), use with, 220

G

GA. See genetic algorithm (GA)
Gabor filters, 343, 352, 363
aspect ratio, 182, 184
CT image enhancement, 193–195
filter bank, 185–186, 188
image analysis by, 180–181
parameter selection, 182
real and complex kernel, 182
rotation angle, 185
sinusoid period, 184
usage, 177, 179
Gabor-wavelets transform, 306–308, 318
galaxy-based search algorithm (GSA), 32, 34
Gauss-Newton, 97
Gaussian radial basis function (GRBF) kernel, 355, 356

Index

GBT. *See* generalized balanced ternary (GBT)
generalized balanced ternary (GBT), 62, 63
genetic algorithm (GA), 4, 5, 7, 50, 100–101
   advantages of, 272–273
   algorithm example, 23
   approaches to, 101–103
   convergence, 273
   cuckoo search (CS), use with, 234
   hybrid, 33
   optimization, 32
   performance, 271
   process, 22
   random immigrants, 273
   roulette method, 102
   scaling factor, determining, 134
   selection, 101–102
   tournament method, 102
greedy search selection, 312
GSA. *See* galaxy-based search algorithm (GSA)

H

Haar wavelet, 248–249
Hamming error correction code, 135
harmony search optimization (HS), 32
Hembrooke, Emil, 229
hex-splines. *See* hexagonal grid algorithms
hexagonal discrete cosine transform (HDCT), 64
hexagonal grid algorithms, 59
   advantages, 67
   character generation, 83, 87
   deviance, analysis of, 72–73, 77, 79–81, 81–82
   generalized balanced ternary (GBT), 62
   generation of straight lines, algorithm for, 63–64
   lattices, 62–63
   mapping techniques, 70–72
   pixels, 68
   rasterization algorithms, 73–76, 83
   sampling, 62
   sampling circles, 79–81
   spline model, 65–67
   usage and applications, 60
high-dimensional indexing methods, 344
Hilbert transform, 139–140, 142, 156
hill climbing segmentation, 347, 363
histogram indexed dictionary (HID), 369, 374, 376, 379, 381, 383, 384, 388
histogram of oriented gradients (HOG), 343, 353–354
Holland, John H., 100
Hough transform, 8
HS. *See* harmony search optimization (HS)
Huffman encoding, 369, 370, 374, 379, 380, 381, 383, 384

I

IHS fusion. *See* intensity hue saturation (IHS) fusion technique
image compression, 370
image enhancement factor (IEF), 268
   denoising, 278
   filters, 287
   performance evaluation, 288
image fusion algorithms
   select maximum, 12
   select minimum, 12
   simple average, 12
image processing
   applications and use, 32
   image segmentation (*see* threshold-based segmentation methods)
image quality index (IQI), 268
   denoising, 278, 280
imposters, 4, 5

independent component analysis
(ICA), 326
  neural network, data
    presentation to, 326–327
  recognitions, 337
InfoMax algorithm, 326
Intel Atom, 11
intensity hue saturation (IHS) fusion
  technique, 5, 7
  fast IHS fusion algorithm, 13
  inverse transform, 13
  overview, 12–13
  process, 12–13
international organization for
  standardization (ISO), 229
iris recognition systems, 5–6
  advantages, 8
  applications, 7–8
  canny edge detection, 8
  normalization, 9
  overview, 7–8
  segmentation, 8
  speed of matching, 8

J

JMuzak Corporation, 229

K

k-nearest neighbor search (k-NN),
  302, 313, 314–315, 317, 319,
  344, 365
Kapur's method, 34, 38, 41, 43, 45
  Otsu's, *versus*, 50–51
KD-ORS tree, 344, 345
  cell and block descriptors, 353
  color feature extraction, 348
  feature vector size, 356
  Gabor filters, 352
  Gaussian radial basis function
    (GRBF) kernel, 355, 356
  gradient computation, 353
  histogram of oriented gradients
    (HOG), 343, 353–354
  KD-tree indexing method, 357,
    358, 359

kernel principle component
  analysis (KPCA), 346,
  354–356
LAB histograms, 351
optimal range search algorithm,
  359–360
performance evaluation, 361,
  362–363
saliency calculation, 347–348
texture feature extraction,
  351–352
Kennedy, James, 108
kernel principle component analysis
  (KPCA), 344, 346, 354–356,
  364

L

LAB histograms, 343, 345, 351
Laplacian filters, 193
Laplacian gradient fusion, 5, 7, 11
Laplacian pyramid (LP), 15, 138–139
least significant bit (LSB)
  manipulation, 247
lossy compression, 371

M

magnetic resonance imaging (MRI),
  298, 372. *See also*
  content-based magnetic
  resonance imaging
  (CBMRI)
Markov random fields, 64
maximum entropy-based artificial
  bee colony thresholding
  (MEABCT) method, 33
maximum errors, 59
maximum fusion, 5
Maxwell's equations, 96
McClellan transformation, 268
mean square error (MSE), 59, 77
medical image access and
  presentation system
  (MIAPS), 373
Metropolis algorithm, 106–107
MI, 4

micro-genetic algorithm, 273
minimum fusion, 5
modality extraction level, at, 5
MT. *See* multilevel thresholding (MT)
multi-agent system (MAS) theory, 33
multilevel thresholding (MT), 37–38, 43, 55
multilevel thresholding using electro-magnetism optimization (MTEMO), 48, 50
   CCA, *versus*, 51–52
multilinear ICA (MICA) algorithm, 339
multimedia processing, advances in 132
multimodal biometric watermarking systems
   computational models, 4
   fusion mutual information, 19–20
   fusion schemes (*see specific fusion schemes*)
   insufficient population coverage issues, 5
   modality extraction level, 4
   non-universality issues, 4
   overview, 4
   performance metrics, 17, 18 (*see also specific performance metrics*)
   unimodal systems, *versus*, 4–5
multiplier-less filters
   overview, 268–270

N
NAE. *See* normalized absolute error (NAE)
NCC. *See* normalized cross correlation (NCC)
normalized absolute error (NAE), 4, 5, 26, 28
normalized correlation coefficient (NCC), 254, 256
normalized cross correlation (NCC), 4, 5, 26–27, 28
   digital watermarking, use in, 137
NOSER, 97

O
optimization-based video watermarking technique. *See also* digital watermarking
   attacks on, 147, 149, 151, 152–153
   bit error rate, 149
   efficiency, 155–156
   extraction algorithm, 143–144
   frame dropping, 153
   overview, 131–132
   performance evaluation, 147
   validating, 146–147
optimized range search (ORS) algorithm, 344
Otsu's method, 33, 34, 39–41, 48
   Kapur's, *versus*, 50–51
output primitives, 73–74

P
palmprint recognition systems, 6
   binarization, 10
   convolving, 11
   cropping, 11
   data sets, 11
   overview, 9–10
   preprocessing, 10
partial differential equations (PDE), 32
particle swarm optimization (PSO), 24, 32, 50
   algorithm, 110–112
   applications, 109
   cuckoo search (CS), use with, 234
   description, 109
   diversity, 110
   FCM, use with, 214, 215, 216, 218

fuzzy morphological-based fusion (FMF) algorithm, use with, 220
history of, 108
hybrid, 33
inertia factor, 110
minimum complexity, 271
multi-wavelet digital watermarking, 247
overview, 209–210, 249–250
performance, 271
scaling factor, determining, 134
self-iterative method, 33
steps, 210
vectors, 110
PCA. See principal component analysis (PCA)
PDE. See partial differential equations (PDE)
peak signal-to-noise ratio (PSNR), 4, 5, 32, 43, 45, 48, 147, 149, 156, 254, 256, 381–382, 384, 385, 386–387
analyzing CCA, 50, 555
denoising, 278
performance evaluation, 267–268, 288
use as performance metric, 24–25, 28
watermark embedding, 136, 137
Petrovic metric, 18
picture archiving and communication systems (PACS), 298
Poisson equation, 96
Pondicherry Engineering College, 11
Prewitt edge region, 302, 314, 317, 319
Price, Kenneth, 114
principal component analysis (PCA), 5, 7, 135, 136
advantages, 327
algorithm, 14
feature extraction, 324
overview, 13
projection, 326
pseudocode, 329
recognitions, 337
PSNR. See peak signal-to-noise ratio (PSNR)
pyramid fusion algorithm, 14–15
pyramid gradient fusion, 5, 7, 11

Q
Qabf, 4
quantization index modulation (QIM), 135

R
rasterization algorithms, 73–76
reversible integer-to-integer time domain lapped transform (RTDLT), 373
rotation, scaling and translation (RST), 133, 247

S
segmentation, image. See threshold-based segmentation methods
self-organizing criticality (SOC), 273
self-organizing random immigrants genetic algorithm (SORIGA), 267, 273–274, 275–276
applications, 291
evaluating, 276, 277
filters, 287, 288
SIFT for Watermarking (SIFTW), 247
simple fusion scheme
elements of, 11
simulated annealing (SA), 106–107
single-scaling-factor-based video watermarking system (V-OL1-SSF), 153, 155
single-scaling-factor-based video watermarking system (V-OL2-SSF), 153, 155
singular value decomposition (SVD), 133, 247

advantages, 328
DWT-SVD watermarking (see DWT-SVD watermarking)
EVD, applied to, 328–329
feature extraction, 324
HID, use with, 390
overview, 249
properties, 249
singular value decomposition principal component analysis (SVD-PCA)
    angle sampling, 331–332, 333–334
    database, 330, 338
    dominant eigenvectors, 335, 337
    memory constraints, 330–331
    position estimation, 337–338
    potential, 324
    pseudocode, 329
    recognition rates, 339
sliding neighborhood filtering, 377–378
Smith, William Henry, 229
SO initialized Gaussian threshold (SOIGT)
    bone tissue segmentation, 195–197, 198
    Gabor filter bank, 188 (see also Gabor filters)
    Gaussian modeling, 190–191
    in-filling, 193
    methodology, 186–187
    performance, 196, 198–199
    post-processing of image, 191, 193
    preprocessing, image, 187–188
    process, 190
    seeds, 188–190, 194
sparse representation fusion, 5, 7, 11, 17
spiral optimization (SO) strategy, 162
    diversification, 169
    Gabor filters (see Gabor filters)
    implementation, 167–170

    intensification, 169
    maximum iteration number, 168–169
    modification, 175–177
    overview, 164–167
    pseudocode, 169, 171
    rotation angle, 168
    rotation matrix, 166
    SO initialized Gaussian threshold (SOIGT) (See SO initialized Gaussian threshold (SOIGT))
    spatial number, 168–169
    sphere function, optimization, 170–177
stationary wavelet transforms (SWT) fusion, 5, 11, 16, 17
Storner, Rainer, 114
structural similarity indices (SSIM), 43, 45, 48, 55, 147, 149, 156
    denoising, 278, 281–282
    performance evaluation, 268, 289
swarm intelligence, 223
SWT fusion. See stationary wavelet transforms (SWT) fusion
synthetic aperture radar (SAR) amplitude images, 373

T

three-dimensional object recognition (3DOR), 325
threshold-based segmentation methods, 32
    optimization methods, 32–33
    SAR, 33

U

ultrasound, 298
unimodal systems, 4

V

VIF, 4
visual information fidelity (VIF), 18
VLC code substitution, 134

W
wavelet packet transform (WPT), 135
wavelet-based contourlet transform (WBCT), 373
Wilcoxon's rank test, 50

X
X-ray radiography, 298
xydeas metric, 18